New Frontiers in Energy Engineering

New Frontiers in Energy Engineering

Edited by **Nora Ayling**

SYRAWOOD
PUBLISHING HOUSE
New York

Published by Syrawood Publishing House,
750 Third Avenue, 9th Floor,
New York, NY 10017, USA
www.syrawoodpublishinghouse.com

New Frontiers in Energy Engineering
Edited by Nora Ayling

International Standard Book Number: 978-1-68286-258-2 (Hardback)

Printed in China.

Contents

Preface

Energy engineering is a rapidly emerging field that focuses on outlining regulations and measures for efficient use of energy resources and sustainable consumption of energy. This book unravels the recent studies in this field with respect to topics such as energy and sustainable development, evaluation of renewable energy production, ecosystem research, assessment and management of environmental risk and safety, etc. The extensive content of this book provides the readers with a thorough understanding of the subject.

The researches compiled throughout the book are authentic and of high quality, combining several disciplines and from very diverse regions from around the world. Drawing on the contributions of many researchers from diverse countries, the book's objective is to provide the readers with the latest achievements in the area of research. This book will surely be a source of knowledge to all interested and researching the field.

In the end, I would like to express my deep sense of gratitude to all the authors for meeting the set deadlines in completing and submitting their research chapters. I would also like to thank the publisher for the support offered to us throughout the course of the book. Finally, I extend my sincere thanks to my family for being a constant source of inspiration and encouragement.

Editor

1

Seasonal Energy Experimental Evaluation of a SBFBR for Treatment of the Azo Dye Direct Red 23

Morales-Guzmán F.[1], Romero R. J.[2], Domínguez-Patiño M. L.[2], Rodríguez-Martínez A.[2] & Melgoza-Alemán Rosa María[1]

[1] Facultad de Ciencias Químicas e Ingeniería, Universidad Autónoma del Estado de Morelos, Av. Universidad 1001. Col. Chamilpa C.P. 62209, Cuernavaca, Morelos, México

[2] Centro de Investigaciones en Ingeniería y Ciencias Aplicadas, Universidad Autónoma del Estado de Morelos, Av. Universidad 1001. Col. Chamilpa C.P. 62209, Cuernavaca, Morelos, México

Correspondence: Melgoza-Alemán Rosa María, Facultad de Ciencias Químicas e Ingeniería, Universidad Autónoma del Estado de Morelos, Av. Universidad 1001. Col. Chamilpa C.P. 62209, Cuernavaca, Morelos, México. E-mail: rmelgoza@uaem.mx & rosenberg@uaem.mx

Abstract

This paper shows the experimental energy evaluation for a Sequencing Batch Fluidized Bed Reactor (SBFBR) anaerobic/aerobic. The SBFBR was installed for the treatment of the Direct Red 23 azo dye (DR23 C. I. 29160). The evaluation of energy was made for summer and winter seasons. DR23 was selected for study because is not quickly biodegraded. DR23 is one of the most used dyes in the textile industry and it is used to dye cotton, linen and rayon fibers. DR23 is stable to light, easily absorbed in the water and its degree of fixation in the fiber goes from 70 to 95%. The SBFBR was packed with granular activated carbon (CAG) as support of the biomass and as redox mediator for acceleration reaction rate by the transport of equivalent reducers between the electrodonor and electroacceptor compounds. A mixture 50/50 of activated sludge from two treatment plants was used as inoculums. Synthetic wastewater was composed with DR23 as substrate, 25 mgL^{-1}; acetic acid as co-substrate15 mgL^{-1} and basal medium. The strategy of fixed efficiencies was used for the acclimatization of the biomass to the DR23 and to the changes of environmental anaerobic/aerobic. The reactor was operated by 160 days (253 cycles). The acclimatization was reached in 43 days in the cycle 27. The reaction time was reduced from 72 h to 24 h. The concentrations of azo DR23 were increasing of 25 to 200 mgL^{-1} with global removal efficiency of 95% of DR23. The reactor SBFBR anaerobic/aerobic was evaluated for the treatment of the DR23 azo dye and their metabolites secondary's that are aromatics amines. The thermodynamic evaluation was made for 27 cycles of operation of SBFBR in each operation temporary. Finally the total energy balance of process during summer and winter was evaluated and shows a energy consumption of 20 and 26 MJ respectly added to the SBFBR.

Keywords: energy evaluation, fluidized bed reactor, sequencing batch reactor, DR23 azo dye, anaerobic/aerobic process, textile effluent

1. Introduction

Elimination of colour from textile wastewater is necessary for industrial reuse purposes. The textile industry consumes approximately 60% of the total production of dyes. For textile processing, inefficiencies in dyeing leads to wastewater, which go to the environment. The amount of dye lost is function of the class of dye application, from 2% to 50% loss when reactive azo dyes are used (McMullan et al., 2001). Residual of dyes are delivered in residual waters for treatment systems before it goes to the environment, and that causes contamination of rivers and underground water in the areas with high concentration of textile industries (Chudgar, 1985; Stolz, 2001).

Azo dyes are xenobiotics compounds characterized by the presence of one or more azo groups (-N=N-). More than 50% are azo dyes from production of dyes, and around 2000 of them are used in the textile, leather, plastics, paper, cosmetics and foods industries (Stolz, 2001). The azo dyes into textile effluents are not toxic, but the formed products of their anaerobic biotransformation are amines aromatic, these products are toxics and are identified as carcinogenic and mutagenic compound (Cheng et al., 1997; Pinheiro et al., 2004).

DR23 is used in the dyed of cotton fibbers, linen and rayon; it is stable to the light and it is absorbed easily in the water, its fixation grade in the fibber is from 70 to 95% (O'Neill et al., 1999). In the Figure 1 the chemical structure of the DR23 is shown.

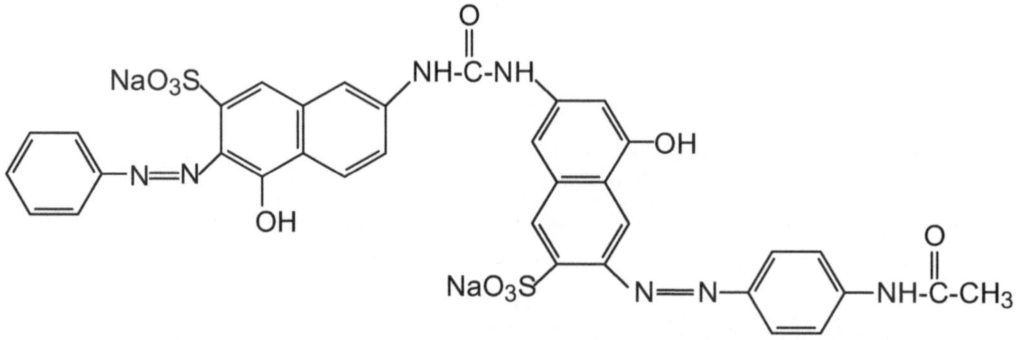

Figure 1. Chemical structure of direct red azo dye 23 (RD23 CI 29160)

An alternative treatment of RD23's effluents is the application of biological processes conventionally combining ambient anaerobic and aerobic integrated in a reactor. In these systems, reductive phase generates aromatic amines are secondary byproducts of the anaerobic reduction of the azo dye due to biotransformation. These byproducts are not generally biodegraded in this stage but in an oxidative phase during aerobic conditions they are mineralized to NO_3, N_2, CO_2 yH_2O and in this way their carcinogenic and mutagenic character is eliminated (Tan et al., 1999; O'Neill et al., 1999; Melgoza et al., 2004). This can be a method for the complete removal of azo dyes of the wastewaters. However, they present two limitations: the long retention time that frequently is applied in the anaerobic phase and the lack of knowledge on the fate of the aromatic amines in the aerobic phase for the biological and chemical transformations that will continue having the remainders of toxic aromatic amines in final effluents. Therefore it is important to assure a grade of mineralization of the aromatic amines (Van der Zee et al., 2005).

2. Methodology

2.1 Experimental System

It was installed a discontinuous sequencing batch fluidized bed reactor (SBFBR) with expanded bed at pilot level. The size of the SBFBR are shown in the Table 1. Three pumps and an aeration pump were connected to a programmable clock, with the purpose of controlling the load, recirculation, aeration and the discharges, all detailed in Table 2.

For the duration of the aerobic phase the air was diffused from the bottom of the reactor through a porous diffuser. The temperature was controlled to 30 ±1 °C, by means of a recirculation and heating of the water system (Poly Science Model 210). Electrodes of pH, potential redox (ORP), nitrates, nitrogen amoniacal and dissolved oxygen were installed to the reactor.

Table 1. SBFBR Dimension

Wall	Acrilic	double
height	58	cm
Internal diameter	14	Cm
Total volume	9	L
Packed	2.0	Kg of actived carbon

Table 2. Instrumentation

Liquid Pump	peristaltic of variable speed	Master Flex Cole Parmer Model 72200-62, 77601-10
Aeration pump	Immersion	Elite 802
Control	Timer	Cole Parmer Programable timer
Temperature control	Recirculation	Poly Science Model 210
pH, ORP, O_2	Electrodes	Cole Parmer

2.2 Substrate Characteristics, Biomass and Operating Conditions

The reactor was inoculated with a mixture of aerobic activated sludge from two wastewater treatment plants, a municipal wastewater and other industrial waste water in 50:50 ratios. The reactor was packed with 2 Kg of granular activated carbon. The biomass concentration was 4.200 mg L^{-1}. Synthetic wastewater was prepared with 25 mgL^{-1} DR23, 15 mgL^{-1} CH_3COOH as co-substrate (used as source of carbon and electrons for complete the reduction reactions) in 1:8 molar ratio (DR23: CH_3COOH) and basal medium with the following composition (mg L^{-1}): K_2HPO_4 (65.25), $Na_2HPO_4{\cdot}2H_2O$ (100.2), KH_2PO_4 (25.5), NH_4Cl (7.5), $MgSO_4{\cdot}7H_2O$ (22.5), $CaCl_2{\cdot}2H_2O$ (27.5), $FeCl_3{\cdot}6H_2O$ (0.25), H_3BO_3 (0.06), $MnSO_4{\cdot}H_2O$ (0.04), $ZnSO_4{\cdot}7H_2O$ (0.04), $(NH_4)6Mo7O_{24}$ (0.03) and EDTA (0.1). For the acclimatization of biomass to DR23 and varying environments anaerobic and aerobic, we used the fixed efficiency strategy, which consisted in allow to acclimate the biomass the time required for reduction DR23 80% during the anaerobic phase and in the aerobic phase to eliminate 80% of the biotransformation of AAT (Melgoza et al., 2000)

2.3 Analytic Methods

The analyzed parameters were pH, ORP (oxide-redox potential), temperature, total alkalinity as $CaCO_3$, CO_2, dissolved oxygen, SST (total suspended solids), SSV (volatile suspended solids), SS (solid sedimentables), sulfurs, sulfates and nitrogen as $N\text{-}NH_4$ and $N\text{-}NO_3$, they were carried out for standardized methods (APHA, 2005).

The determination of the DR23 was carried out in HPLC Agilent 1100 Series to λ= 501 nm. The total aromatic amines were determined by the method of the p-dimethyl aminebenzaldehyde (Oren et al., 1991) spectrophotometer UV-VIS Agilent 8453E Spectroscopy System, measures to a wavelength of 440 nm.

2.4 Energy Analysis

In order to evaluate the energy consumption for the SBFBR operation, 20 sensors in the experimental system were installed, as shown in the Figure 2. The sensors were connected to a data logger Agilent HP34970 with a personal computer. Measures of temperature were taken with the software HP-VEE Pro with programming directed to objects (Object Oriented Design OOD) focus in energy evaluation systems (Romero et al., 2007; Silva-Sotelo et al., 2010, Romero et al., 2013). Each one of the sensors was associated to the thermal balances to determine the thermodynamic behavior of the SBFBR in the cycle's operation.

$$Q = C_P \frac{\partial(mT)}{\partial t} \tag{1}$$

The used energy balance consider a constant of heat capacity (Cp) with the value of 4.18 kJ s^{-1} $°C^{-1}$ and the characteristic flow of each pump, multiplied by the difference of temperatures in each reading.

$$\frac{\partial(mT)}{\partial t} = m\frac{\partial T}{\partial t} \tag{2}$$

Finally, it was evaluated the time that operated each pump to obtain the quantity of energy consumed in each cycle. The determination of time that heater maintains the controlled temperature of the SBFBR, the time of elevation of the registered temperatures was evaluated, obtaining as a result statistical the percentage of time in that it was turned on the heater.

$$\frac{dT}{dt} \leq 0 \tag{3}$$

for a cooling time for both seasons.

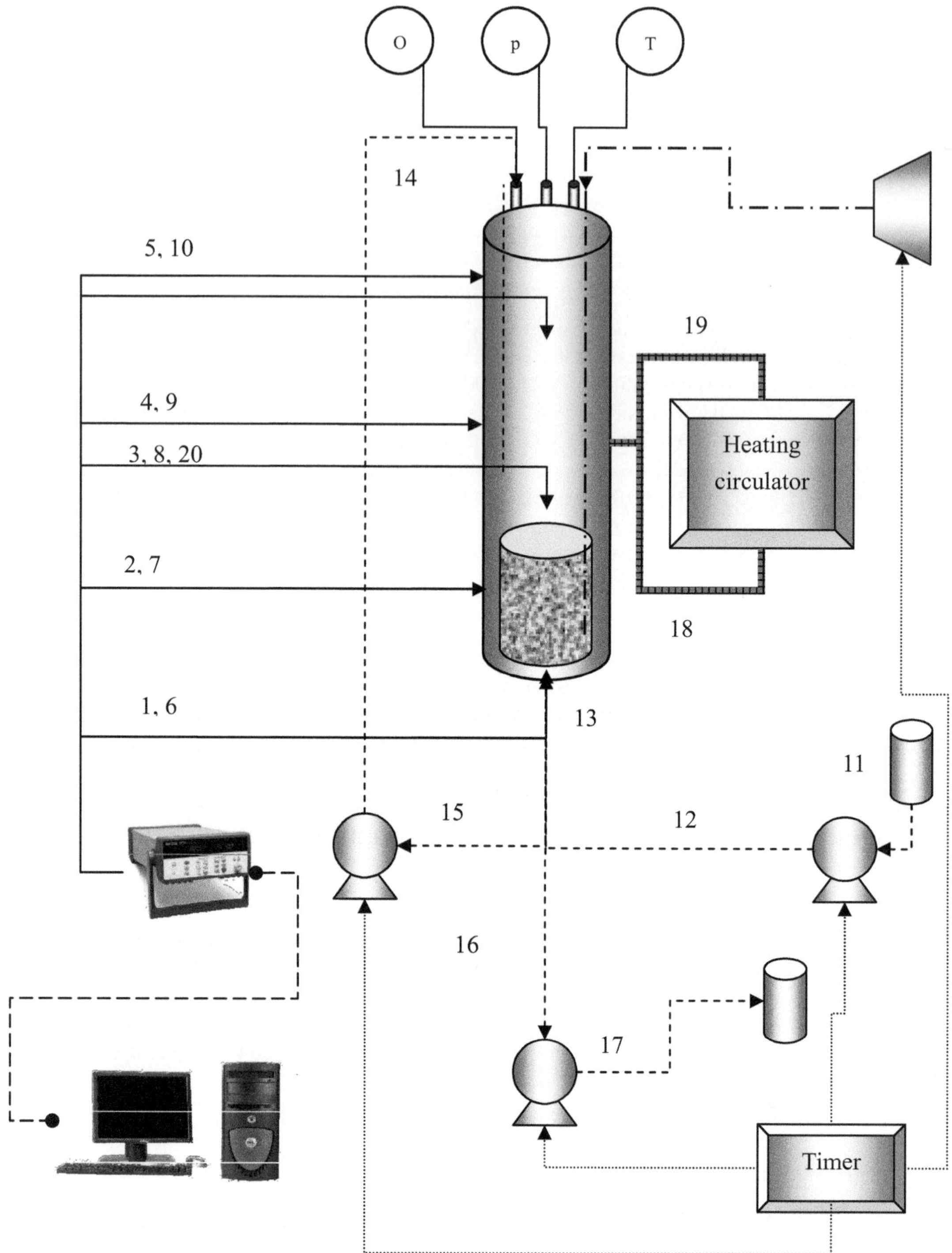

Figure 2. Diagram of the SBFBR anaerobic/aerobic for the treatment of the DR23

1, 6	Bottom reactor (inside and outside)	13	Reactor inlet
3, 8, 20	To 20 cm bottom reactor (inside and outside)	14	Water of recirculation outlet
5, 10	To 50 cm bottom reactor (inside and outside)	15	Recirculator water inlet
2, 7	To 10 cm bottom reactor (inside and outside)	16	Water unloading
4, 9	To 30 cm bottom reactor (inside and outside)	17	Unloading point
11	Feed tank	18	Water to jacket inlet
12	Feed tank outlet	19	Water of the jacket outlet

$$\frac{dT}{dt} > 0 \tag{4}$$

for a heating time for both seasons.

Thermal analysis was conducted in the summer and winter seasons. In summer season for acclimatization of biomass and in the winter when the reactor was stabilized RD23 degradation. In both periods the operating conditions in concentration of DR23 and hydraulic retention times were similar.

Energy calculation was based on the determination of the energy that entered to the system so that they could be carried out the anaerobic and aerobic phases in each cycle, it was not considered useful the energy that left the system since didn't contribute energy to the pumps and the heater-recirculator of water.

3. Results and Discussion

3.1 Performance of the SBFBR Anaerobic/Aerobic

In the integrated biological processes for the treatment of azo dyes the operation strategy is a very important factor because the acclimatization of the bacterial population to the dye and anaerobic/aerobic environmental induces the microorganisms to develop its metabolic activities in presence or oxygen absence. The reactor operated during 253 cycles. The concentration of DR23 fed in the influent was increased as it was adapted to each operation condition. The acclimatization of SBFBR was achieved during the cycle 27 after 43 days. The thermal study was performed during cycles 19-25. Cycle 21 was a representative during acclimation of the reactor in the summer; subsequently thermal study was performed during the winter season in the same operating conditions that summer and the reactor power consumption was evaluated. The concentration initial was of 25 mg DR23L^{-1}, as it is observed in figure 3. The global removal of DR23 was stabilized to 85%. After the acclimatization, the concentration of DR23 was increased of 25 to 200 mgL^{-1}. The global removal efficiency of DR23 was from 91 to 97%, and the removal efficiency of color measured in units Pt-Co went from 92 to 98%. In works carried out by (Tan et al., 1999, 2001), in a reactor of expanded bed for the treatment of the dye mordent yellow 10 (MY10/L), for a concentration from 59 to 65 mgL^{-1} using ethanol as co-substrate, the removal efficiency of the MY10 was 100%.

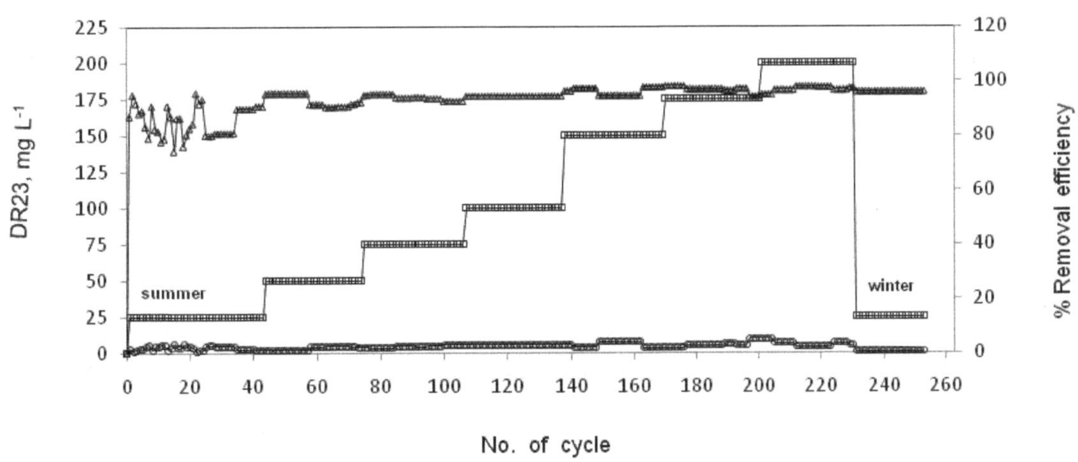

(■) Cycle start (○) Anaerobic phase end (-) Aerobic phase end (Δ) % Removal efficiency

Figure 3. Performance of the SBFBR in the treatment of DR23

A summary of the operation of the SBFBR during the 253 cycles is shown in the Table 3. The largest efficiency of removal of DR23 was reached during the anaerobic phase what suggests that acted as a rough treatment while the aerobic phase acted as a polishing treatment, suggesting that in the treatment of azo dyes in a process anaerobic/aerobic the aerobic phase serves as polish of the anaerobic effluent. In the most of the cases the anaerobic treatment doesn't eliminate the present compounds, but it transforms them in metabolites even more toxic that those of the influent.

Table 3. Performance of the SBFBR in the treatment of the DR23

Operation (days)	Cycles	RHT anaerobic/ aerobic (h)	S_i (mg L^{-1})	S_{ANA} (mg L^{-1})	S_{AER} (mg L^{-1})	Anaerobic removal efficiency (%)	Aerobic removal efficiency (%)	Global removal efficiency (%)
43	1-43	18/6	25	5	4	80	4	84
59	44-74	8/4	50	9	3	82	12	94
75	75-106	8/4	75	12	5	84	9	93
90	107-137	8/4	100	21	6	79	15	94
106	138-169	8/4	150	11	6	93	3	96
122	170-200	8/4	175	7	6	96	1	97
137	201-230	8/4	200	20	6	90	7	97
160	231-253	18/6	25	5	4	80	4	84

For the anaerobic phase the maximum biotransformation of the DR23 by reduction to amines was 3%. After in the aerobic phase the mineralization of amines was 80%. In works carried out by FitzGerald and Bishop (1995), in the treatment of the dye orange acid 10 (AO10) for a concentration of 10 mgL^{-1} using an anaerobic reactor of fluidized bed, the recovery of the amines was <1%. Likewise, in carried out works by Sarsour et al. (2001), in an anaerobic-aerobic coupled system for a concentration of 5000 mgL^{-1} of reactive red 198 (RR198), using starch as co-substrate, the amines recovery were from 2 to 3%. In this work, the time of reaction was optimized of 72 to 12 h (8 h anaerobic phase, 4 h aerobic phase). In general terms the redox potential during the experimental cycles was -299 mV in anaerobic phase and +105 mV in the aerobic phase. The reactor operated under reduction conditions in the anaerobic phase and oxidation conditions in the aerobic phase. These conditions are indispensable in this type of reactors that integrate anaerobic and aerobic environments.

Through entire experiment was not detected production of biogas only presence of CO_2. The average total alkalinity was from 350 to 450 mg $CaCO_3$ L^{-1} which was enough buffer capacity to neutralize the volatile fatty acids during the process to avoid the inhibition of the biomass for acidification. The average pH was 7.0 in the anaerobic phase and 8.0 in the aerobic phase

3.2 Energy Consumption in the Operation of SBFBR

A cycle of operation of the SBFBR begins with the activation of the feed pump (ALI) which stayed on during 7 min to temperature ambient. After of the load of the reactor the pump was switch off. The anaerobic phase and the recirculation of the residual water began by means of pump (REC) during 19.88 h. In the anaerobic phase energy was required to maintain the system to 30 °C (heating recirculator). Once concluded the anaerobic phase the pump (REC) was switch off and the aeration pump was activated (AIR) for one period of 3.88 h (23.76 h of global operation). After the pump (AIR) was switch off and the pump discharge (DAR) was on during 7 minutes. The energy consumed was based on the calculation of the energy that entered for the phases anaerobic and aerobic in each cycle, did not consider the energy useful that loged-off since it did not contribute energy to the pumps and to the water heater-recirculador.

In anaerobic and aerobic processes the temperature of the SBFBR was controlled to 30.0 °C by a recirculator-heater of water with a nominal power of 750 W. It is necessary to mention that the recirculator-heater of water was switch on and it switch off automatically by the controller of temperature shown periods of 14.37% (on the average) of the time switch on and the rest of the time stayed out, therefore the energy consumption for each cycle of 24 h was of 9.32 MJ for 9 L of wastewater for summer season and 15.00 MJ in the winter season. The results of are shown in the Table 4.

Table 4. Energy consumption in each pump during the reactor operation

Energy Consumption (MJ)	Summer season	Winter season
Feed pump	0.02	0.02
Recirculation system	10.74	10.74
Aeration system	0.05	0.05
Discharge system	0.02	0.02
Heating system	9.32	15.00
Total energy	20.15	25.83

The ambient temperature for summer season was 25 °C and in winter was 18 °C. The difference of temperature derives that consumption of energy in the heater for remaining the reactor at 30 °C. In winter was approximately double time heater required than summer, as can be seen in Figures 4 and 5 showing the temperature profiles for each season. The heating zone is larger in winter than in summer.

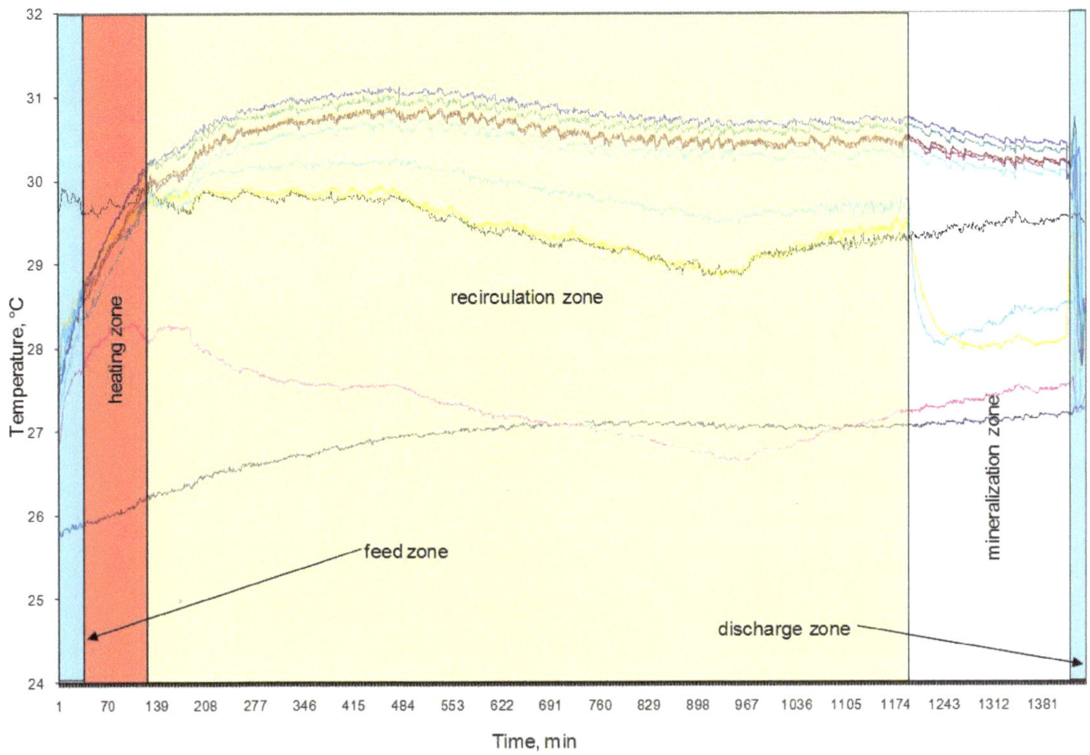

Figure 4. Temperature profile for summer season

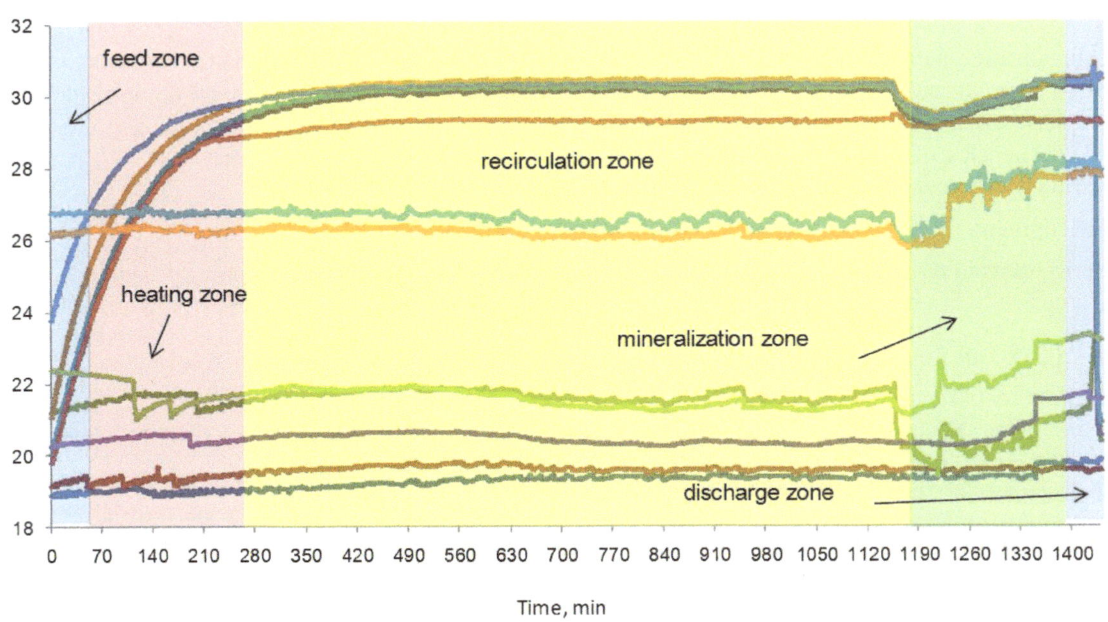

Figure 5. Temperature profile for winter season

The energy consumption at begin of the cycle was of 0.02 MJ by load of the reactor, after were consumed 10.74 MJ by the anaerobic process. The aeration shown a relative low consumption of 0.05 MJ and finally the energy consumption of unload was of 0.02 MJ.

Figure 6 shows the analysis of temperature variation during SBFBR operation in the two seasons. The analysis indicated that in the summer at the entrance to the reactor had a total power consumption of 1.8 MJ and the rest of the energy that entered the reactor is dissipated as heat (Q) to the environment. In winter analysis the temperature variation indicates that the system does not consume energy, it releases energy as heat in accordance with the negative values of Q = -7.51 MJ.

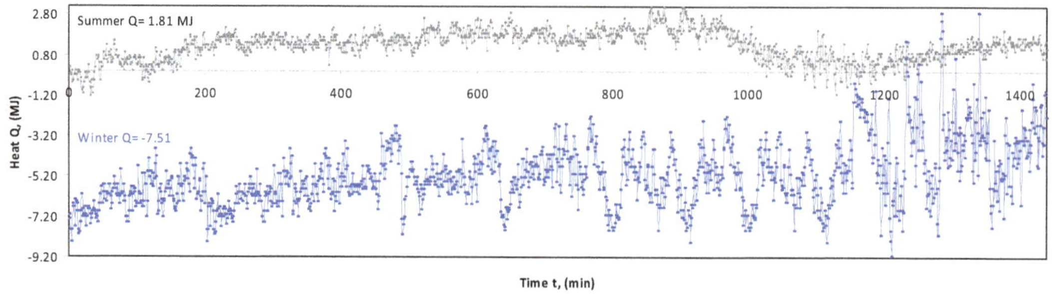

Figure 6. Analysis of variation of temperature during operation of the SBFBR, in the summer and winter seasons respectively

4. Conclusions

Energy consumption evaluation for the mineralization and the DR23 in a sequencing batch reactor of expanded bed was carried out in a pilot level; integrating anaerobic/aerobic environmental, alternating reductive and oxidative phases. In the reductive phase the biotransformation of the DR23 was reached and in the oxidative phases the mineralization of the amines to CO_2, H_2O and NO_3 was also investigated.

The total of energy consumed in this process during one cycle of operation in the summer season was of 20.15 MJ and the winter season of 25.83 MJ.

The analysis of the variation of the temperature indicated that in summer season there is a total energy consumption of 1.8 MJ, while in winter season, the reactor did not consume energy provided the environment as the value obtained was negative which means that the energy leaving the system (-7.51 MJ). In winter season was approximately double time that required in summer season so that the system reaches a temperature of 30 °C.

The acclimatization period was 27 cycles for a concentration of 25 mg/L with a constant temperature of 30 °C (+ / - 1 °C). The concentrations studied were from 25 to 200 mg/L. The reaction times were diminished from 72 h (48 h phase anaerobia, 24 h aerobic phase) to 12 h (8 h anaerobic phase, 4 h aerobic phase). The global removal efficiencies of DR23 were from 84 to 97%, the biotransformation of amines was of 3% and the mineralization of these was of 80%.

Acknowledgment

Dr Romero wants to thank to 167434 CONACyT' Basic Science Project for partial support.

References

APHA. (2005). *Standard Methods for Examination of Water and Wastewater* (21st ed.). Washington D. C. USA: American Public Health Association/American Water Works Association/ Water Environment Federation.

Cheng, J., Suidan, M. T., & Venosa, A. D. (1997). Kinetic of anaerobic cometabolism of 2-4 dinitrotoluene with ethanol as the primary substrate. *Water Science and Technology, 36*(6-7), 271-278. http://dx.doi.org/10.1016/S0273-1223(97)00532-5

Chudgar, R. J. (1985). Azo dyes. In L. I. Kroschwitz (Ed.), *Krk-Othmer Encyclopedia of Chemical Technology* (4th ed., pp. 821-875). New York: Wiley.

Fitz Gerald, S. W., & Bishop, O. L. (1995). Two stage anaerobic/aerobic treatment of sulfonated azo dyes. *J. Environ. Sci. Health Part A-Toxic/Hazard. Subst. Environ. Eng., 30*(6), 1251-1276.

McMullan, G., Meehan, C., Conneely, A., Kirby, N., Robinson, T., Nigam P., …, Smyth, W. F. (2001). Microbial decolourisation and degradation of textile dyes. *Applied Microbiology and Biotechnology, 56*, 81-87. http://dx.doi.org/10.1007/s002530000587

Melgoza, R. M., Chew, M., & Buitrón, G. (2000). Start-up of a sequential anaerobic/aerobic batch reactor for the mineralization of p-nitrophenol. *Water Science and Technology, 42*(5-6), 289-292.

Melgoza, R. M., Cruz, A., & Buitrón, G. (2004). Anaerobic/Aerobic Treatment of Colorants Present in Textile Effluents. *Water Science and Technology, 50*(2), 149-155.

O'Neill, C., Hawkes, F., Esteves, S., Hawkwa, D., & Wilcox, S. (1999). Anaerobic and aerobic of simulated textile effluent. *Journal Chemical Technology and Biotechnology, 74*, 993-999.

Oren, A., Gurevich, P., & Henis, Y. (1991). Reduction of nitrosubstituted aromatic compounds by the halophilic anaerobic eurobacteria Haloanarobium praevalens and Sporohalobacter marismortui. *Applied and Environmental Microbiology, 57*, 3367-3370.

Pinheiro, H., Touraud, E., & Thomas, O. (2004). Aromatic amines from azo dye reduction: status review with emphasis on direct UV spectrophotometric detection in textile industry wastewaters. *Dyes and Pigments., 61*, 121-139. http://dx.doi.org/10.1016/j.dyepig.2003.10.009

Romero, R. J., Silva-Sotelo, S., Rodríguez, M., & Cerezo, R. J. (2013). Energy Saving in Advanced Absorption Heat Pump with Object Oriented Programming. In *Emerging Trends in Computing, Informatics, Systems Sciences, and Engineering, Lecture Notes in Electrical Engineering* (Vol. 151, pp. 1101-1111). http://dx.doi.org/10.1007/978-1-4614-3558-7_94

Romero, R., Martínez, A., & González, E. (2007). Laboratory instrumentation and object oriented design for working fluid control in an "absorption heat pump" using Water / Carrol. In *Innovations and Advanced Techniques in Computer and Information Sciences and Engineering* (pp. 429-432). http://dx.doi.org/10.1007/978-1-4020-6268-1_76

Russ, R., Rau, J., & Stolz, A. (2000). The function of cytoplasmic flavin reductases in the reduction of azo dyes by bacteria. *Appl. Environ. Microbiol., 66*(4), 1429-1434. http://dx.doi.org/10.1128/AEM.66.4.1429-1434.2000

Silva-Sotelo, S., Romero, R. J., & Rodríguez-Martínez, A. (2010). Double Stage Heat Transformer Controlled by Flow Ratio. In *Innovations in Computing Sciences and Software Engineering* (pp. 577-581). http://dx.doi.org/10.1007/978-90-481-9112-3_100

Stolz, A. (2001). Basic and applied aspects in the microbial degradation of azo dyes. *Applied Microbiology Biotechnology, 56*, 69-80. http://dx.doi.org/10.1007/s002530100686

Tan, N. C. G., Prenafeta-Boldú, F. X., Opsteeg, J. L., Lettinga, G., & Field, J. A. (1999). Biodegradation of azo dyes in cocultures of anaerobic granular sludge with aerobic aromatic amines degrading enrichment cultures. *Applied Microbiology Biotechnology., 51*, 865-871. http://dx.doi.org/10.1007/s002530051475

Tan, N. C. G. (2001). Integrated and sequential anaerobic/aerobic biodegradation of azo dyes. *Ph D. thesis*. Wageningen, The Netherlands: Wageningen University.

Van der Zee, F., & Villaverde, S. (2005). Combined anaerobic-aerobic treatment of azo dyes - A short review of bioreactor studies. *Water Research., 39*, 1425-1440. http://dx.doi.org/10.1016/j.watres.2005.03.007

2

The Financial Viability of Solar Photovoltaic Canopies as Urban Climate Change Mitigation: An Analysis of the Potential Utilization of Public Space in Boston, Massachusetts

Clayton Hunter[1], James G. Baldwin[1] & Michael L. Mann[2]

[1] Department of Earth and Environment, Boston University, Boston, MA, USA

[2] Department of Geography, George Washington University, Washington, D.C., USA

Correspondence: James Baldwin, Department of Earth and Environment, Boston University, 675 Commonwealth Ave. Boston, MA 02215, USA. E-mail: jbaldwin@bu.edu

Additional support was provided by the U.S. National Science Foundation ULTRA-ex program (DEB 0948857)

Abstract

We model the electrical power generation potential, calculate the associated avoided emissions and conduct an economic valuation of a scenario of large scale photovoltaic canopy implementation in metro Boston, Massachusetts on publicly managed parking real-estate operated by the Massachusetts Bay Transit Authority. The results of this work demonstrate that 102.4 MWh of electricity may be generated by such a program which is equal to nearly 0.013% of total electricity demand in the state. Annual CO_2, NO and SO_2 emissions avoided in this scenario amounts to 53.7 tonnes of CO_2, 46.5 kg of NO, and 107.1 kg of SO_2. Despite the substantial amount of power production potential we find that such a scenario is not economically viable under current market conditions and using conventional financing mechanisms with a net present value of -1.4 Billion U.S. dollars.

Keywords: solar energy, integrated generation, net present value analysis, climate change mitigation, photovoltaic canopies

1. Introduction

1.1 Background

The United States consumes approximately 1,400 terawatt-hours (TWh) of electricity annually for residential use and 3,856 TWh in total of which the state of Massachusetts is responsible for more than 740,000 megawatt-hours (MWh) of consumption (EIA, 2012). At that level, Massachusetts emits approximately 1,702,000 pounds of Sulfur Dioxide, 740,000 pounds of Nitrogen Oxide, and 854,000,000 pounds of Carbon Dioxide every year (EIA, 2012). At the national level, approximately 40% of all carbon dioxide emissions in the U.S. can be attributed to the use of fossil fuels in the production of electricity (Jun et al., 2012). There is a general scientific consensus that the emission of these and other pollutants are significant driving forces behind climate change. Large disruptions in climate are associated with adverse effects on human health, economic stability, and food security (IPCC, 2007). With this in mind we consider the possible utilization of existing publicly owned property as a location for renewable energy production in a metropolitan area using photovoltaic canopies. In this analysis, we investigate the use of large parking facilities operated by the public transportation system for Boston, Massachusetts. In particular, we consider those parking lots serving the Massachusetts Bay Transit Authority customers. These lots were chosen because of the quality of the data (see section 2) and the fact that this land is already under public management and thus available for the use of photovoltaic canopies as part of an urban distributed generation scheme. We identify the potential for electricity generation at these sites, the economic viability of such developments and quantify the offset in NO, SO_2 and CO_2 that this renewable electricity generation entails. We do not however quantify some other values associated with these installations, such as the benefits of network resiliency and reliability.

1.2 Energy Valuation and Grid Integration

There are numerous approaches that are applied to energy valuation. These techniques may be broken down into five overarching methodological categories: stated preference techniques, revealed preference techniques, financial option theory (or portfolio analysis), emergy analysis, and other economic approaches (Menegaki, 2008). However the most common approach, and the one which provides the most clear cut results for interpretation by policy makers, is net present value (NPV) analysis. This technique permits the lifetime return and lifetime cost of a project to be considered in its entirety, thus providing useful information for those attempting to determine the best use of limited funds to maximize returns to those funds. This analysis is done by simply subtracting the net present costs of a course of action from the net present benefits, which are both calculated in current dollars (2008$) by way of a defined social discount rate. It is this technique that we employ in our valuation of a scenario in which a large scale distributed solar energy project is undertaken. However it is useful to briefly outline some of the other valuation techniques and the insight they provide with respect to renewable energy systems.

Let us first turn to stated and revealed preference techniques which seek to interpret consumer responses and behavior, respectively. Willingness to pay methods rely upon response to changing price structure while revealed prefence methods utilize trends in consumer behavior to guage how individuals value the options available to them. Multiple studies conducted over the last few years have used these techniques to demonstrate the public's positive willingness-to-pay for energy production that is cleaner and more sustainable than current practices (see Bergmann et al., 2008; Li et al., 2009). In addition to people's willingness to pay for cleaner technology,using portfolio analysis it has been estimated that increasing renewable installations by 10% in the U.S. energy portfolio could equate to US$29-53 billion in avoided losses due to the price volatility, market risk, and other disruptions associated with fossil fuels (Menegaki, 2008). This knowledge begs the question, if people are willing to pay for cleaner energy and there is high external cost of pollution, why is there not a greater installed capacity of renewables? Aside from the historic prescedence of large, fossil fuel-based power plants, arguably the most important part of the answer to that question the inherent challenge associated with the integration of renewables with the current grid system (Denholm & Margolis, 2006).

Integration of renewables in the current electrical grid is often argued to be problematic because of intermittency and load following issues. Intermittency is the given variability associated with uncontrollable energy sources, such as solar and wind which cannot be guaranteed to provide a steady flow of power. Load following is the ability of an energy source to produce electricity as it is demanded. In traditional power plants, this is not an issue because electricity production is completely controlled by the plant operators who are able to increase or decrease supply as needed. In their hypothetical simulation of a large-scale PV production plant, Denholm and Margolis (2006) find that, although solar electricity could provide enough electricity to power the entire U.S. and its peak output coincides with times of largest electricity demand, a 16 GW system installed in Texas (which would meet 11% of overall electricity demand) would not be economically viable due to the costs associated with shutting down and powering up traditional, fossil fuel-powered plants to maintain constant supply-demand matching. In a more recent study that counters this finding, Sovacool (2009) surveyed a variety of energy experts with the goal of answering: (1) What do energy experts really think about renewables in the United States? (2) To what degree are conventional baseload units reliable? and (3) Is intermittency a justifiable reason to reject renewable electricity resources? He finds that experts in the energy field are hesitant about the potential of renewables mainly because of the relative ease with which conventional power sources can be controlled. Sovacool concludes that renewable energy systems are no less reliable than conventional power plants and intermittency issues can be overcome with spatially diverse installations. It is this solution of spatially diverse installations through the use of photovoltaic canopies that we evaluate herein.

1.3 Parking Canopies

There have been a number of studies performed that consider the impact of areas covered by parking lots, including associated environmental and economic costs (see McPherson, 2001) and loss of ecosystem services (see Davis et al., 2010). The increase in abundance of large department stores has been cited as a reason for the disproportionate increase in area covered by parking lots (Noguera, 2007). Taking that observation into consideration, and in order to minimize detrimental effects of parking lots, researchers have encouraged city planners to be intelligent about the placement of new lots, and to ensure they are necessary (Knepper, 2007). The bulk of research has focused on the aforementioned costs and benefits associated with new parking lots, but existing lots, especially in urban areas, also provide an opportunity to use existing space without changing land use. This creates an opportunity to offset some of the negative environmental impacts associated with parking facilities. There is existing research that demonstrates parking canopies have the ability to reduce the urban heat island (UHI) effect [increased temperatures in urban areas, as compared to rural areas, attributable to lower urban albedo and greater waste heat generation] and suggests that parking canopies with PV installations have a greater potential to reduce parking lot pavement temperatures than either canopies without PV or shade trees (Golden et al., 2007). Similarly, PV canopies are effective at reducing energy use through the shading of buildings (Mandalaki et al., 2012) providing an added benefit to building integrated parking structures. Numerous PV canopies have already been built, such as the Chevron facility at California State University Fresno, the parking structure at University of California San Diego, the Kyocera headquarters facility in San Diego and the small scale canopy at the U.S. National Renewable Energy Laboratory in Colorado. With the surge in popularity of PV parking canopies it is worth considering the economic viability of large scale implementation in urban areas with solar energy potential.

1.4 PV System Installation and Electricity Production Estimation

The purpose of any solar PV installation is to maximize electricity production while minimizing system costs. To achieve these goals there are choices that need to be made between various solar PV modules, inverters, and other components (such as electricity storage and charge controllers where necessary or desired). Although some new technologies are emerging, the most commonly used PV modules are built from silicon, specifically single crystalline silicon as opposed to amorphous silicon or other polycrystalline silicon structures. There are many different brands and models of PV modules. However, there are fewer choices for inverters and the specific model is selected based on the size of the system it serves. This is also true of any other electrical components the particular system may require (Ayompe et al., 2010; MCEC, 2012). In their study of 16 installed crystalline PV systems in Europe, Ayompe et al. found that the balance of system costs were approximately equal to 54% of the PV module cost. Balance of system (BOS) cost is the cost of all components (other than the PV modules) necessary for a system to function. For example the BOS cost would include the costs of inverters, structural support and wiring as well as the cost of labor to install the system. They also found that the average installation spends approximately one percent of the total cost on annual variable costs associated with maintenance of the system. There are essentially two types of support structures for PV systems, fixed structures and mobile structures with tracking systems which can track the sun's path across the sky. The predominant installation method uses a fixed structure because moving parts in a tracking system increases the risk for structural failures and increases both initial installation costs as well as maintenance costs in addition to consuming 2-3% of the energy from the panel for the drive system (Mousazadeh et al., 2009). In order to maximize the exposure of PV modules in a fixed system installed in the northern hemisphere, the modules should be tilted facing south at an angle equal to that of the location's latitude (Ayompe et al., 2010; El-Sebaii et al., 2010). Once the structural layout is determined, the annual output can be estimated to within approximately 3% using the average monthly solar radiation and average monthly day-time temperature (Huld et al., 2010; Perpinan et al., 2008).

2. Methods

2.1 Lot Area Estimation 1: Parking Lot and Impervious Surface

The step in developing and evaluating the scenario of widespread PV parking canopy development is to ascertain exactly how much land area is available. Our area estimation made use of data freely available from the Massachusetts Office of Geographic Information (MassGIS) and ESRI's ArcGIS Desktop software. The data were the MBTA parking lot in point form (MassGIS, 2006) and the impervious surface (MassGIS, 2007) layers. The MBTA parking lot layer, as provided to MassGIS by the Executive Office of Transportation, includes many attributes such as facility type and number of parking spaces. The impervious surface layer was derived from the 2005 Color Ortho Imagery project by the Sanborn Map Company and includes all forms of cement, brick, asphalt, and packed dirt visible at one meter resolution. MBTA parking lots are assumed to have a compact shape, such as a square, as opposed to a line. As such, the points were buffered with a radius one-half that of the number of spaces for all types of facilities, except that of garages. As garages are multi-story complexes, one-tenth of the number of spaces was used as a radius for the buffer to decrease the inclusion of surrounding structures in the area estimate. The buffered parking lots were then converted to raster and overlaid on the impervious surface layer and the area of all impervious surface cells that lie within these buffered regions were summed in meters. This methodology assumes there is relatively little to no pervious surface (such as grass or tree islands) within the bounds of the parking lot and, simultaneously, that any given parking lot is surrounded by some pervious surface.

MassGIS (2009) freely distributes very high resolution (15-30 cm) areal imagery of Boston and the surrounding area which is used by Google in its Google Earth and Maps products. We use this data to perform an accuracy assessment of our GIS modeled parking areas by exporting the MBTA parking lot data from ESRI's ArcGIS format to Google Earth. First, the points where checked to determine whether they were in the approximate center of each parking lot. Then six individual lots were randomly chosen from the garage lots and nine lots were selected from the remaining non-garage lots. Polygons of all 15 lots were drawn and exported back to ArcGIS format, which calculates the area covered by each polygon. The resulting areas were assumed to be the most accurate estimates and were used to perform an accuracy assessment for both area estimations performed in this study.

2.2 Lot Area Estimation 2: Google Earth and Single Space Averages

According to the Victoria Transportation Policy Institute (2010), the average space (including access lanes) in an off-street parking lot in the US requires between 27.9 and 32.5 square meters. To test whether this held true for some sample lots in Boston, polygons of three parking lots near Boston University's Campus were again drawn in Google Earth and exported to ArcGIS. The number of parking spaces was counted for each test parking lot and individual space requirements averaged accordingly. Polygons of the 15 full parking lots were again used to perform an accuracy assessment of the area estimation.

2.3 PV System Area Requirements

As previously discussed, shading of panels can substantially reduce output. Therefore an important consideration in determining the land area required by any PV installation is limiting or excluding the effect of shadows (Alonso-Garcia et al., 2006; Perpinan et al., 2008). Without knowledge of the texture of the areas surrounding the parking lots, only shadows created by the PV modules could be considered. However, this is consistent with the location of most MBTA parking areas which are generally away from large buildings and thus we believe it does not introduce undue error in our area estimates. Assuming that the PV modules will be installed on a fixed structure and tilted south to a degree equal that of its location's latitude, simple geometry can be used to calculate the area required by any one PV module (as shown in Figure 1). We assume PV panel dimensions that are representative of modules from other such studies (Huld et al., 2010; Perpinan et al., 2008; Ayompe et al., 2010).

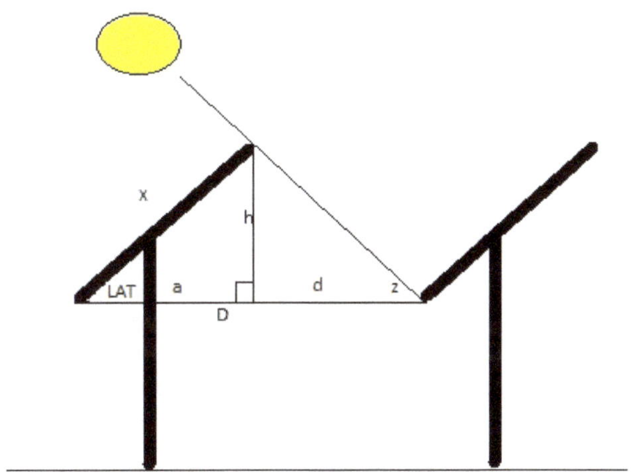

Figure 1. Shadow effects of one array on another

$$h = X * SIN(LAT) \quad (1)$$

$$a = X * COS(LAT) \quad (2)$$

$$d = \frac{h}{TAN(z)} \quad (3)$$

$$D = a + d \quad (4)$$

$$A_{mod} = PV_w * D \quad (5)$$

Equations 1-5 (above) calculate the horizontal area required by a PV module (A_{mod}) where X is the module length, LAT is the latitude, z is the sun's angle (90-zenith), PV_w is the module width, and D is the distance from the south (low) edge of one module to the south edge of the next module either directly north or south (NREL, 1990). Precise latitude was the only parameter that varied and is specific to each parking lot. The zenith angle (65 degrees) was chosen because it approximates the lowest the sun gets in Boston over the course of a year. Using the calculated area required by one PV module, we are then able to estimate the number of modules possible in a given installation.

2.4 Electricity Production and Installation Costs

Electricity production potential is calculated using straightforward mathematical relationships that take as inputs the average global radiation (from NREL, 2010), installed module surface area, rated power output per module, average temperature (from NREL, 2010), inverter and wiring efficiencies, and module degradation rate (from Ayompe, 2010). As for the module dimensions, the rated power output per module was chosen based on reviewed studies (Ayompe, 2010; Huld et al., 2010). The power production equations (Eq. 6-10) were adapted from Huld et al. (2010) and are defined as follows (see Table 1 for variable definitions):

$$P(G',T') = P_{stc} * G' * N_{rel}(G',T') \tag{6}$$

$$N_{rel}(G',T') = 1 + k_1 lnG' + k_2 \ln(G')^2 + T'(k_3 + k_4 lnG' + k_5 \ln(G')^2) + k_6(T')^2 \tag{7}$$

$$G' = \frac{G}{1000} \tag{8}$$

$$T' = T_{mod} - T_{mod\_stc} \tag{9}$$

$$T_{mod} = T_{amb} + cT * G \tag{10}$$

To convert the resulting output from direct current to active current) the inverter and wiring efficiencies were multiplied to approximate the DC to AC conversion efficiency of the system. The present value of electricity was determined by applying the degradation rate to all future years' production over the lifetime of the PV modules, and multiplying by the price of electricity. Installation costs are based on the peak wattage of a given system and are tracked and reported by industry on the Solar Buzz website (Note 1). The remainder of the installation cost is the BOS cost. To calculate the net present cost of an installation two costs remain: the BOS replacement cost and the variable costs. The limiting component of the BOS is the inverter, which typically only has a lifespan of 10 years (Ayompe et al., 2010). The present value of the electricity produced is calculated by multiplying the average annual electricity production by the cost of electricity. The future benefits and costs were discounted to represent the time value of money before the net present value was calculated by subtracting the costs from the revenue. The precise discount rate for policy decision making (and whether a discount rate should be used *at all*) is a matter of some debate (Note 2), however we employ the commonly used rate of 0.05 over the lifetime of the installation (Ayompe et al., 2010). Finally we use the energy output and the EIA figures for emissions from electricity generation in Massachusetts (EIA, 2012) to estimate avoided emissions of carbon dioxide, nitrogen oxide, and sulfur dioxide. Table 1 (below) lists the parameters and variables used in this study (Note 3).

Table 1. Study parameters

Name	abbreviation	value	units
MBTA parking lot layer (MassGIS, 2006)	N/A	N/A	N/A
Impervious surface layer (MassGIS, 2007)	N/A	N/A	N/A
Horizontal area per module	A_{mod}	varies	m^2
Latitude (MassGIS, 2006)	LAT	varies	DD
Zenith angle (NREL, 2010)	zenith	65	degrees
Power output at standard testing conditions	P_{stc}	200	W
Average daily irradiance (NREL, 2010)	G	varies	W/m^2
Average daily temperature (NREL, 2010)	T_{amb}	varies	deg. F
Normalized irradiance	G'	varies	N/A
Normalized temperature	T'	varies	N/A
Module temperature	T_{mod}	varies	deg. F
Module temperature at standard temperature and pressure	$T_{mod\ stc}$	varies	deg. F
Relative efficiency	N_{rel}	varies	N/A
Coefficients (Huld et al., 2010)	cT	0.035	N/A
	k1	-0.01716	N/A
	k2	-0.04029	N/A
	k3	-0.00468	N/A
	k4	0.000148	N/A
	k5	0.000169	N/A
	k6	0.000005	N/A
Inverter efficiency (Ayompe, 2010)	I	0.935	N/A
Wiring efficiency (Ayompe, 2010)	W	0.9	N/A
PV module lifetime (Ayompe, 2010)	L	25	Years
degradation rate (Ayompe, 2010)	D	0.0082	N/A
Module Cost (Solarbuzz)	Cm	4.21	$/Wp
BOS cost factor (Ayompe, 2010)	BOSc	0.54	N/A
BOS replacement factor (Ayompe, 2010)	BOSr	0.7	N/A
Variable cost factor (Ayompe, 2010)	Vc	0.01	N/A
Discount Rate (Ayompe, 2010)	R	0.05	N/A
Price of Electricity (EIA, 2012)	Pe^-	0.1484	$/kWh
Price of Electricity increase rate(Ayompe, 2010)	Pe^-_r	0.045	N/A
CO2 emission intensity (EIA, 2012)	CO_2	1154	lbs/MWh
NO emission intensity (EIA, 2012)	NO	1	lb/MWh
SO2 emission intensity (EIA, 2012)	SO_2	2.3	lbs/MWh

3. Results

3.1 Lot Area Estimations and Accuracy Assessments

The first area estimation (Figure 2), point buffer was found to be inaccurate. For the 15 sample parking lots, the average area was 20,783 m^2 with a standard deviation of 23,002 m^2. As the parking lots are not expected to be of equal size, these raw values simply show that there is a large variation in the area estimates. The zonal sum method, on average, overestimates the parking lot area by 150% of the actual size, if each area estimate is compared to the actual area (as defined by the estimate made using ground truth combined with orthophotos and Google Earth). However, when parking lot type is considered, it becomes apparent that the zonal sum method performed much better. On average the zonal sum method overestimated parking garage areas by only 14%, whereas the same approach overestimated ground level parking lots by 248% or 3.48 times the actual area.

Figure 2. Points are MBTA parking lots, light blue circles seen around some lots are buffered regions, and both are overlayed on the impervious surface layer. The center of the image is the city of Boston

The second area estimation approach, which simply multiplied the number of parking spaces by the average single space area (31.5 m^2), returned an average lot size of 12,765 m^2 with a standard deviation of 12,657 m^2. By again normalizing the values to those measured in Google Earth, the average lot was overestimated by 111% and performed better on the ground level lots than the parking garages. This single space method overestimated average garage areas by 222%, but only overestimated ground lots by an average of 37%. Figure 3 (below) shows the difference in estimated areas by the two different approaches as well as the ground truth as measured using Google Earth.

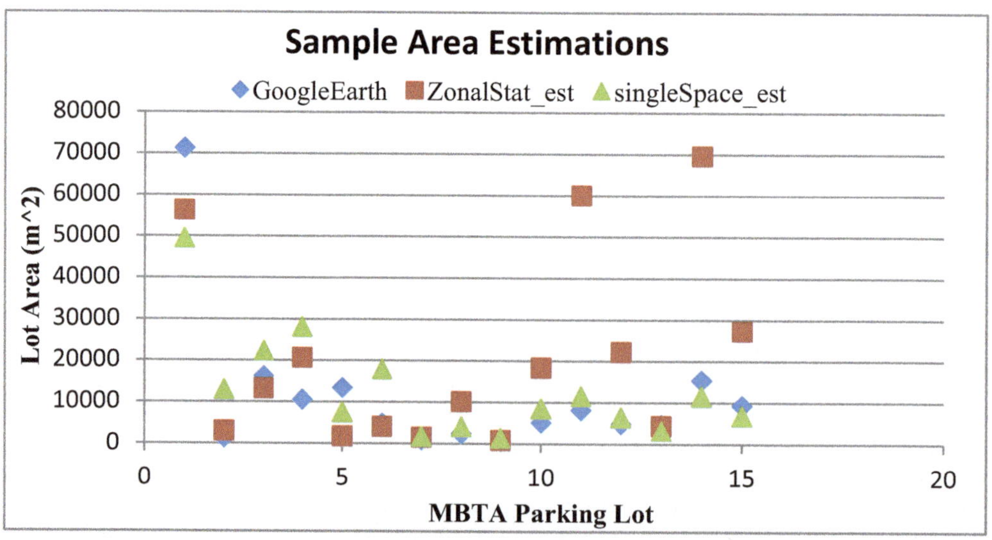

Figure 3. Sample area estimation as measured in Google Earth and as calculated using two approaches

The first six records correspond to parking garages and the remainder to off-street, ground level lots. Especially notable are lots 8-15, where, except for two closely approximated examples, the zonal sum method (depicted by red squares) can be seen to greatly overestimate ground level lots and the same lot areas are strongly correlated with the actual areas. Based upon the observed performance of each technique previously described, the available area for each parking garage was defined by the zonal sum method and for each ground level lot by the single space method.

3.2 PV System Area, Electricity Production, and Net Present Value

The average required area per module was calculated as 1.8996 m^2 with a standard deviation of 0.0027 m^2. The average electricity production (active current) for the first year was found to be 742.4 kWh with a total of 102.4 MWh. At that production level, the electricity produced by PV canopies over the 138 MBTA parking lots would produce approximately 0.013% of electricity demand in Massachusetts. Over 25 years, the total electricity production would equal approximately 2.3 GWh. Interestingly, as seen in table 2 (below), the total monthly power output peaks in March rather than in the middle of summer when there is the highest sun exposure. This is attributable to the decline in performance with higher temperatures that occurs with PV panels.

Table 2. Monthly power production for the first year (kWh)

	January	February	March	April	May	June
Total:	7097.54	8235.51	10069.03	9723.96	9470.70	8577.43
Average:	51.43	59.68	72.96	70.46	68.63	62.16
	July	August	September	October	November	December
Total:	8687.27	9086.46	9169.23	9002.80	6978.89	6349.55
Average:	62.95	65.84	66.44	65.23	50.57	46.01

To calculate the net present value (NPV) of any installation, the net present costs and benefits must be determined. At a peak output of 191.4 MW, the PV modules are estimated to cost 805.9 million USD (M$) and the BOS components just under 435.2 M$, which would put the initial installation costs at 1.2 B$. The future BOS replacement and annual variable costs are estimated at 56 M$ and 174.9 M$, respectively. This means that the net present cost of installing PV canopies at all 138 MBTA parking lots would equal 1.46 billion (2008$).

At the modeled production level of 102.4 MWh in the first year and a price of 0.1484 $/kWh, revenue from the first year would equal $15,203. Accounting for future increases in electricity prices as well as degradation rates of the panels using the values from Table 1, the net present revenue is equal to 15.4 M$. Thus the total NPV is -1.46 B$ and the average NPV for an individual installation is -10.6 M$ in our modeled scenario. The average installation has a net present benefit of 10.56 M$.

Table 3. Summary of net present costs, revenue, and value

	Net Costs (M$)	Net Revenue (M$)	NPV (M$)
Total:	1472	1.60	-1,456
Average:	10.70	0.12	-10.60

These numbers do not provide a strong case for PV canopy development, however they represent only the direct market value of such a project. Knowledge of the external costs and benefits is critical information needed for planning and public decision making. Among the potential positive externalities of renewable energy projects are the offset of critical air pollutants such as sulfur and nitrogen oxides as well as the savings with respect to carbon emissions and offset climate change potential. Additionally, distributed redundancy in energy systems provides resiliency to a critical and vulnerable infrastructure. Identifying a precise dollar value for these benefits is beyond the scope of this project, however we are able to provide a quantitative measure of the emissions abatement that would result from our PV canopy scenario. Using the figures on electricity production related emissions from the EIA we are able to calculate the annual emissions avoidances for CO_2, NO and SO_2 resulting from the offset in fossil fuel combustion. These values are seen in table 4 with both the total emissions avoided and the individual installation average avoided emissions.

Table 4. Emissions avoided in first year

	CO_2 (tonnes)	NO (kg)	SO_2 (kg)
Total:	53.73	46.57	107.10
Average Site:	0.38	0.34	0.78

In a location that is classified as a "non-attainment area" by the U.S. EPA with respect to the National Ambient Air-Quality Standards (Note 4), these avoided emissions are an important and valuable benefit of renewable energy development. If we assume that the excess net present cost of the project (anything less than a breakeven NPV of zero) is to be offset by the benefit of avoided emissions we can estimate the per pound cost of the offsets over the 25 year life of the installations. The degradation of the panels in that time equates to an average output for the panels of approximately 90% of the first year output. We use this output level to determine the lifetime present cost per pound of abated emissions which is equal to the negative net present value divided by the sum of average lifetime emissions (Table 5). It is quite apparent that under current conditions in NPV terms this is an expensive emissions abatement scheme.

Table 5. Present value cost of avoided emissions over the 25 year lifetime of the project

	CO_2	NO	SO_2
Total Emissions Avoided(kg):	1,209,123.45	1,047.82	2,409.75
US$/lbs	1,204	1.39 million	0.60 million

4. Conclusions

Increased attention and action aimed at addressing climate change through the reduction of greenhouse gas emissions has brought ever more focus on renewable energy technologies. This work developed a scenario of large scale distributed photovoltaic energy production making use of existing publicly managed land in a large metropolitan area where Massachusetts Bay Transit Authority parking areas. We evaluated the economic viability of such a scenario using net present value analysis and quantify the associated emissions offset that this scenario would entail. We find a generation potential equal to 102.4 MWh or 0.013% of electricity consumption in Massachusetts and annual emissions offsets equal to 53.7 tonnes of CO_2, 46.5 kg of NO, and 107.1 kg of SO_2. However the market conditions for such a scenario are not favorable with a calculated NPV of -1.4 billion U.S. dollars for the project.

Our results demonstrate two key points. Urban environments have great potential for generating renewable energy supplies while offsetting the emissions associated with conventional energy production. Secondly that such a scenario is not economically viable under current market conditions in urban areas with low solar potential. For the scenario to be a financially viable solution for energy production in Massachusetts, several fundamental conditions must be changed. Through the evolution of the market or through policy intervention, the cost of PV systems decrease or the value of avoided pollution must increase. Examples of such conditions include creative financing that pushes payment into the discounted future (for example see Singh and Singh, 2010), energy pricing that favors low carbon production, a decline in panel costs or advances in technology, or the use of more generous social discount rates. Both the costs and efficiency of PV panels have been steadily improving in recent years and the other options are readily manipulated via policy makers. It is our hope that this paper highlights both the potential for creative integration of distributed renewable energy production, and the challenges that remain for their implementation.

References

Alonso-Garcia, M. C., Ruiz, J. M., & Chenlo, F. (2006). Experimental study of mismatch and shading effects in the I-V characteristic of a photovoltaic module. *Solar Energy Materials and Solar Cells, 90*(3), 329-340. http://dx.doi.org/10.1016/j.solmat.2005.04.022

Ayompe, L. M., Duffy, A., McCormack, S. J., & Conlon, M. (2010). Projected costs of a grid-connected domestic PV system under different scenarios in Ireland, using measured data from a trial installation. *Energy Policy, 38*(7), 3731-3743. http://dx.doi.org/10.1016/j.enpol.2010.02.051

Bergmann, A., Colombo, S., & Hanley, N. (2008). Rural versus urban preferences for renewable energy developments. *Ecological Economics, 65*(3), 616-625. http://dx.doi.org/10.1016/j.ecolecon.2007.08.011

Caplin, A., & Leahy, J. (2004). The Social Discount Rate. *Journal of Political Economy, 112*(6), 1257-1268. http://www.jstor.org/stable/10.1086/424740

Davis, A. Y., Pijanowski, B. C., Robinson, K., & Engel, B. (2010). The environmental and economic costs of sprawling parking lots in the United States. *Land Use Policy, 27*(2), 255-261. http://dx.doi.org/10.1016/j.landusepol.2009.03.002

Denholm, P., & Margolis, R. (2006). Very Large-Scale Deployment of Grid-Connected Solar Photovoltaics in the United States: Challenges and Opportunities. *NREL/CP-620-39683*. Retrieved from http://www.nrel.gov/docs/fy06osti/39683.pdf

El-Sebaii, A. A., Al-Hazmi, F. S., Al-Ghamdi, A. A., & Yaghmour, S. J. (2010). Global, direct and diffuse solar radiation on horizontal and tilted surfaces in Jeddah, Saudi Arabia. *Applied Energy, 87*(2), 568-576. http://dx.doi.org/10.1016/j.apenergy.2009.06.032

Energy Information Administration (EIA). (2012) *Official Energy Statistics from the U.S. Government*. Retrieved from http://www.eia.doe.gov/

Golden, J. S., Carlson, J., Kaloush, K. E., & Phelan, P. (2007). A comparative study of the thermal and radiative impacts of photovoltaic canopies on pavement surface temperatures. *Solar Energy, 81*(7), 872-883. http://dx.doi.org/10.1016/j.solener.2006.11.007

Huld, T., Gottshalg, R., Beyer, H. G., & Topic, M. (2010). Mapping the performance of PV modules, effects of module type and data averaging. *Solar Energy, 84*(2), 324-338. http://dx.doi.org/10.1016/j.solener.2009.12.002

Intergovernmental Panel on Climate Change (IPCC). (2007). *IPCC Fourth Assessment Report: Climate Change 2007*. Retrieved from http://www.ipcc.ch/publications_and_data/publications_and_data_reports.htm#1

Jannson, P. M., & Michelfelder, R. A. (2008). Integrating Renewables into the U.S. Grid: Is It Sustainable? *The Electricity Journal, 21*(6), 9-21. http://dx.doi.org/10.1016/j.tej.2008.07.005

Jun, E., Kim, W. J., Jeong, Y. H., & Chang, S. H. (2010). Measuring the social value of nuclear energy using contingent valuation methodology. *Energy Policy, 38*(3), 1470-1476. http://dx.doi.org/10.1016/j.enpol.2009.11.028

Knepper, V. (2007). *Developing parking policies to support smart growth in local jurisdictions: best practices. Metropolitan Transportation Commission for San Francisco Bay Area, CA*. Retrieved from http://www.mtc.ca.gov/planning/smart_growth/parking_seminar/BestPractices.pdf

Li, H., Jenkins-Smith, H. C., Silva, C. L., Berrens, R. P., & Herron, K. G. (2009). Public support for reducing US reliance on fossil fuels: Investigating household willingness-to-pay for energy research and development. *Ecological Economics, 68*(3), 731-742. http://dx.doi.org/10.1016/j.ecolecon.2008.06.005

Mandalaki, M., Zervas, K., Tsoutsos, T., & Vazakas, A. (2012). Assessment of fixed shading devices with integrated PV for efficient energy use. *Solar Energy, 86*(9), 2561-2575. http://dx.doi.org/10.1016/j.solener.2012.05.026

Massachusetts Clean Energy Center (MCEC). (2012). *Massachusetts Residential Guide to Solar Power*. Retrieved from http://www.masscec.com/index.cfm/pk/download/id/13301/pid/11163

Massachusetts Office of Geographic Information (MassGIS). (2006). *MBTA Parking Lots*. Retrieved from http://www.mass.gov/mgis/eot_layers.htm

Massachusetts Office of Geographic Information (MassGIS). (2007). *Impervious Surface*. Retrieved from http://www.mass.gov/mgis/impervious_surface.htm

Massachusetts Office of Geographic Information (MassGIS). (2009). *USGS Color Ortho Imagery 2008/2009*. Retrieved from http://www.mass.gov/mgis/colororthos2008.htm

McPherson, E. G. (2009). Sacramento's parking lot shading ordinance: environmental and economic costs of compliance. *Landscape and Urban Planning, 57*(2), 105-123. http://dx.doi.org/10.1016/S0169-2046(01)00196-7

Menegaki, A. (2008). Valuation for renewable energy: A comparative review. *Renewable and Sustainable Energy Reviews, 12*(9), 2422-2437. http://dx.doi.org/10.1016/j.rser.2007.06.003

Moore, M. A., Boardman, A. E., Vining, A. R., Weimer, D. L., & Greenberg, D. H. (2004). "Just give me a number!" Practical values for the social discount rate. *Journal of Policy Analysis and Management, 23*(4), 789-812.

Mousazadeh, H., Keyhani, A., Arzhang, J., Mobli, H. Abrinia, K., & Sharifi, A. (2009). A review of principle and sun-tracking methods for maximizing solar systems output. *Renewable and Sustainable Energy Reviews, 13*(8), 1800-1818. http://dx.doi.org/10.1016/j.rser.2009.01.022

Noguera, J. (2005), Parking lots, store chains and spatial agglomeration. *Papers in Regional Science, 84*(2), 145-158. http://dx.doi.org/10.1111/j.1435-5957.2005.00010.x

Perpinan, O., Lorenzo, E., Castro, M. A., & Eyras, R. (2008). On the complexity of radiation models for PV energy production calculation. *Solar Energy, 82*(2), 125-131. http://dx.doi.org/10.1016/j.solener.2007.06.007

Rambaud, S. C., & Torrecillas, M. J. M. (2005). Some considerations on the social discount rate. *Environmental Science & Policy, 8*(4), 343-355. http://dx.doi.org/10.1016/j.envsci.2005.04.003

Singh, P. P., & Singh, S. (2010). Realistic generation cost of solar photovoltaic electricity. *Renewable Energy, 35*(3), 563-569. http://dx.doi.org/10.1016/j.renene.2009.07.020

Sovacool, B. (2009). The intermittency of wind, solar, and renewable electricity generators: Technical barrier or rhetorical excuse? *Utilities Policy, 17*(3-4), 288-296. http://dx.doi.org/10.1016/j.jup.2008.07.001

U.S. National Renewable Energy Lab (NREL). (1990). *National Solar Radiation Data Base User's Manual (1961-1990)*. Retrieved from http://rredc.nrel.gov/solar/old_data/nsrdb/

Victoria Transportation Policy Institute. (2010). 5.4 Parking. *Transportation Cost and Benefit Analysis II – Parking Costs*. Retrieved from http://www.vtpi.org/tca/tca0504.pdf

Notes

Note 1. www.Solarbuzz.com is published by the NPD Group, Port Washington, NY, USA, data available from http://www.solarbuzz.com/Moduleprices.htm

Note 2. See variously Moore et al., 2004; Caplin and Leahy, 2004 and Rambaud and Torrecillas among many other works on the topic.

Note 3. For consistency all values are for 2008.

Note 4. Current attainment and non-attainment areas as well as the criteria pollution levels are available via the U.S. Environmental Protection Agency website, www.epa.gov

3

Characterization of Biomass Bottom Ash from an Industrial Scale Fixed-Bed Boiler by Fractionation

Adrian K. James[1], Ronald W. Thring[1], P. Michael Rutherford[1] & Steve S. Helle[1]

[1] Environmental Science & Engineering, University of Northern British Columbia, 3333 University Way, Prince George, Canada

Correspondence: Ronald W. Thring, Environmental Science & Engineering, University of Northern British Columbia, 3333 University Way, Prince George, BC V2N 4Z9, Canada. E-mail: thring@unbc.ca

Abstract

It is expected that increasing amounts of energy will be generated from the direct combustion of biomass residues. However, biomass combustion processes are known to produce large amounts of bottom ash, resulting in ash storage and disposal problems. The presence of unburned carbon in some bottom ash suggests its potential for beneficial uses, such as an energy source. This comparative study characterizes two bottom ash samples obtained from an industrial scale fixed-bed boiler. The physical and chemical properties of each bottom ash, as well as their respective particle fractions obtained by sieving, are analyzed and discussed. Analyses included proximate and ultimate analysis, Brunauer–Emmett–Teller (BET) surface area, thermogravimetric analysis (TGA) and bulk density. The percent fixed carbon in the samples was 30% and 50%. The higher heating value (HHV) ranged from 5 - 25 MJ/kg for the ash samples when characterized within fractions. The boiler ash showed that 68 % or more of the energy could be recovered in fractions $\geq$ 425 μm for high carbon ash. Low carbon fractions of ash have four times the bulk density compared to the high carbon fractions. By reburning the larger fractions, ash volumes can be decreased by over 50%.

Keywords: bottom ash, fixed carbon, heating value, bulk density, ash, energy content

1. Introduction

Increased use of bioenergy has significantly increased ash generation in many countries. Large-scale utilisation of wood and wood residues for district heating plants, process heating plants and combined heat and power plants has resulted in large volumes of ash production. It is projected that by 2020, the thermal application of biomass in European countries will result in an annual ash production of approximately 15.5 million tons per year (Obernberger & Supancic, 2009).

Ash formed during biomass combustion can be divided into bottom ash and fly ash (Picco, 2010). Bottom ash is the fraction produced on the grate in the primary combustion chamber. While many studies have investigated fly ash, bottom ash usually accounts for 60 to 90 % of the total ash generated (Obernberger & Supancic, 2009).

A number of barriers to the utilization of wood ash have been identified, including heavy metal concentrations and the presence of organic pollutants (Obernberger & Supancic, 2009). High amounts of organic carbon in the ash indicate incomplete combustion of biomass, which suggests inefficient fuel use (Picco, 2010; Demirbas, 2005). The presence of carbon in ash limits its application in forest soils since carbon dilutes the liming and fertilization effects and also restricts the hardening of the ash (Demirbas, 2005; Sarenbo, 2009). Ash hardening is used to improve handling, reduce dust and to decrease the ash dissolution rate. Unburned matter, as indicated by high carbon content, increases ash volumes and results in higher handling, transportation, disposal and spreading costs (Demirbas, 2005; Gomez-Barea, Vilches, Campoy, & Fernandez-Pereira, 2009). For many environmental applications, it has been shown that the carbon content should be below 5 wt% (dry basis) (Van Loo & Kopejan, 2008). In addition, the high combustibility of char, restricts the application of high carbon ash to some lands due to increased forest fire risk (Dimitrakopoulous, 2001).

According to Emilsson, a high content of uncombusted material in ash leads to difficulty in ash hardening for use in construction materials and suggests that high charcoal content in the ash can be recombusted (Emilsson, 2006). High content of uncombusted material also poses problems when used in cement compositions.

The relationship between bulk density and the presence of unburnt carbon in ash is important when considering ash transport and storage. High bulk densities may reduce ash volumes, thus decreasing the need for additional storage space and reducing transportation costs. Bulk density decreases with the fineness of the ash fraction (Van Loo & Kopejan, 2008). The authors reported that ash from straw and cereal combustion showed low density due to its specific chemical matrix, containing more salts and less minerals when compared to wood ash. The bulk density of bottom ash from bark combustion and woodchip combustion, both from a moving grate underfeed stoker, had densities of 950 kg/m^3 (Pels, de Nie, & Kiel, n.d). On the other hand, the bulk density of bottom ash from sawdust combustion (from an underfeed stroker) was 650 kg/m^3. Thus, fuel type may influence the bulk density of the produced ash.

While improving combustion system efficiency is integral to reducing the carbon content of bottom ash, identifying effective recycling methods are also essential in dealing with high carbon ash. The unburned carbon content in bottom ash varies amongst combustion systems. Short residence time (Duan, Liu, Chen, & Zhao, 2012) and high ash content are considered as major factors for the incomplete burnout of char. High ash content may result in an ash layer covering the char and preventing the free movement of gases to the core of the char during combustion. State-of-the-art fluidized bed combustion systems have combustion efficiencies of over 95% irrespective of ash content (Saidur, Abdelaziz, Demirbas, Hossain, & Mekhilef, 2011). In these systems, the only ash fraction that may contain a high carbon content is the coarse fly ash fraction (as precipitated in cyclones) which is reintroduced into the combustion chamber.

A possible use of ash, depending on the carbon content, is as a fuel (Gomez-Barea et al., 2009). Preliminary research on this idea has been carried out predominantly on fly ash (Duan et al., 2012). Gasification of high-carbon fly ash has been investigated and the resulting ash shows positive results in the production of C-fix blocks, a concrete-like material that uses a heavy petroleum residue as a binder (Pels, de Nie, & Kiel, n.d). The high carbon content in some forms of ash could also make it useful as a fuel after briquetting or if pelletized (Batra, Urbonaite, & Svensson, 2008; Cetin, Moghtaderi, Gupta, & Well, 2004). Combustion of the high carbon ash presents a number of operational problems. Related problems include corrosion (Khan, Jong, Jansens, & Spliethoff, 2009) and scouring. Based on the design of the boilers, the ash may be carried with the flue gas through the boiler tubes (Saidur et al., 2011). The variability in particle sizes as the feedstock becomes more burnt-out also causes inherent problems in fixed bed systems. Fixed bed systems usually require a feedstock that is as uniformed as possible so as to avoid channeling (Ryu, Yang, Khor, Yates, Sharifi, & Swithenbank, 2006; Warnecke, 2000).

This study investigates the high carbon bottom ash from fixed bed boilers generated by a large pulp and paper producer in Canada. Large volumes of high carbon ash are produced and are typically sent to landfill due to the low potential economic and environmental benefit for other applications. The costs associated with changing the system to a more recent boiler design are considerably high. If the high carbon fraction can be separated from the low carbon fraction of the ash other applications may become viable. Separation methods for isolating carbon from ash include floatation in a continuous mode and fractionation by sieving (Gomez-Barea et al., 2009). Industrially, carbon separation can be carried out using industrial scale sieve shakers for removal of the coarse carbon rich fraction or by scaling up floatation methods.

The objective of this research is to characterize bottom ash from an industrial scale fixed bed boiler and to identify the ash fractions with potential energy importance. This work examines the physical and chemical properties of the ash and ash fractions. Fractions with high carbon contents are identified and an analysis of the combustion behavior is carried out. The data obtained enhances knowledge about the utilization and storage of woody biomass bottom ash.

2. Experimental Section

Wood ash from a Canfor Pulp Limited Partnership pulp mill fixed-bed boiler in Northern British Columbia was used in the study. Boiler feed is hog fuel, a mixture of bark and sawdust primarily derived from pine wood with variable particle sizes. Based on an ultimate analysis, the hog fuel is 49% C and 6% H on a dry basis. Hog fuel moisture content varies depending on sawmill feedstock, ranging from 25 to 50%; ash content is approximately 2.5%. Typically fixed bed boilers operate within 850 to 1400 °C (Quaak, Knoef, & Stassen, 1999). This fixed bed boiler produces approximately 27 MW of power and has a lower grate temperature of 255 °C with an outlet flue gas temperature of 170 °C. Ash samples, denoted B1 and B2 were collected from the system on November 17, 2010 – B1 and on April 27, 2012 – B2. The boiler has a dry ash removal system, and the ash was sampled from the ash bins shortly after removal from the boiler. The samples were separated into different particle fractions as described in section 2.1.

2.1 Particle Size Distribution

Approximately 100 g of "as-received" (original sample obtained from boiler) ash was separated into different size fractions using a stack of five sieves (2000 μm, 850 μm, 425 μm, 250 μm, 150 μm) arranged in decreasing diameter openings (Herrman & Baker, 2002). The ash was poured on the top sieve (largest opening), which was then covered; the sieve stack was placed on an automatic shaker for 15 minutes after which the stack was removed. Each sieve with the retained material was gently tapped on the sides before being removed from the stack and weighed. Retained ash was removed and stored, while the sieve trays were thoroughly cleaned and reweighed to obtain the mass of ash retained on each tray. The analyses were repeated twice and an average value calculated and reported.

The following retained fractions were used for subsequent analyses:

AR = as-received wood-ash,

≥ 2000 μm,

≥ 850 μm and < 2000 μm,

≥ 425 μm and <850 μm,

≥ 250 μm and < 425 μm,

≥ 150 μm and less than 250 μm

<150 μm.

For convenience, the above fractions will simply be denoted as 2000 μm, 850 μm, 425 μm, 250 μm, 150 μm and <150 μm.

Each of the retained fractions were weighed and then stored for further analysis.

2.2 Proximate and Ultimate Analysis of Wood Ash

The proximate analysis of each sample fraction was carried out according to American Society for Testing and Materials (ASTM) method D1762-84. The fixed carbon (FC), volatile matter (VM) and ash content were determined on a dry basis. Samples were dried for 3 h at 105 °C to remove all moisture. VM analyses was carried out at 950 °C for 7 min in capped crucibles and then ash content was determined at 550 °C for 6 h in open crucibles. The FC was determined by difference as shown in Equation (1). The ash content is defined as the remaining inorganic contents after the complete removal of fixed carbon, volatile matter and moisture. All analyses were performed on three replicates and the averages presented.

$$FC = 100 - VM - Ash \tag{1}$$

where FC, VM and Ash content are wt.% expressed on a dry basis.

Ultimate analysis of carbon (C), hydrogen (H), nitrogen (N), and sulfur (S) was conducted using a Costech™ Elemental Analyzer, ECS 4010 Elemental Combustion System. Reactor conditions were: 1000 °C, helium carrier gas at 105 mL/min, Gas Chromatography (GC) Column #051080, SS 5 mm x 2 m, at 100 °C, 450 L x 18 mm reaction tube, and HayeSep Q 60/80GC column packing. A Thermal Conductivity Detector was used. Results are presented in wt.% and on a dry basis.

2.3 Thermal Analysis

Thermal characteristics were determined using a TA Instrument TGA-50/50H Shimadzu thermogravimetric analyzer. The analyses were conducted using air at a flow rate of 50 mL/min, under atmospheric conditions. For each test, approximately 10 mg of the sample was heated from room temperature to 800 °C at a heating rate of 10 °C/min. The TGA analysis was done only on B1 sample for the AR portion, ≥ 425 μm fraction and < 425 μm fraction. The thermogravimetric (TG) plots describe the weight loss of the sample with increasing temperatures while the differential thermogravimetric (DTG) plot illustrates the derivative of sample mass with time at a specific temperature.

2.4 Surface Area

Brunauer–Emmett–Teller (BET) surface area measurements were carried out with N_2 at -196 °C using a single–point Micromeritics FlowSorb 11 2300 surface area analyzer. The surface area was determined only on char particles that were ≥ 2000 μm for both B1 and B2 samples. Particles in this fraction were more homogeneous due to the absence of finer ash particles and more char-like particles.

2.5 Heating Value

The gross calorific value was determined for each fraction using a Parr Oxygen Bomb with approximately 1 g of boiler ash. Results were determined in accordance with ASTM D-5865/E711 and were calculated relative to sample weight on an oven dried basis.

2.6 Bulk Density

The bulk density for each fraction was determined as follows. Approximately 50 ml of sample was placed into a 100 ml graduated measuring cylinder (1ml divisions). The cylinder was then tapped 6 times on a rubber surface from a height of approximately 1.5 in. The mass and volume of the sample were then obtained and used to calculate bulk density. This procedure was repeated three times and an average value obtained for each sample. All bulk density determinations were expressed on a dry basis.

3. Results and Discussion

3.1 Particle Size Distribution of Ash

Distribution of the particles in the boiler samples are shown in Figure 1. Fractions ≥ 250 μm showed a similar distribution for both ash samples. The B2 sample was approximately 10% higher in mass than B1 for the fractions 250 μm, 425 μm and 850 μm. As the particle sizes became smaller, that is for fractions 150 and < 150 μm, the mass of B1 was higher than B2. For B1, fraction < 150 μm accounted for the highest weight with 25% of the total mass passing through the 150 μm sieve. The same fraction for B2 was approximately 20% lower than B1. The difference in the particle size distributions of the samples is likely associated with the variation in the char contents of the AR samples. As char becomes more combusted, the sample particle sizes will decrease due to an increase in ash formation. Therefore, B2 is expected to have greater carbon content in the AR sample.

3.2 Results of Proximate Analysis and Ultimate Analysis

The results of the proximate analysis and ultimate analysis of the wood ash samples are shown in Table 1. Percent volatile matter ranged from 0 to 20% for all samples. The volatile matter for the samples appeared to slightly increase as particle size decreased. This may be due to an increase in inorganic content such as carbonates and oxides that are usually present in wood ash (Liao, Wu, & Yan, 2007). Ash content increased with decreasing particle size. Fractions 425 μm and higher for B1 and B2 ranged from ~ 17 to 34% ash content, while fractions 250 μm and below ranged from ~ 60 to 70%. The ash content in the 425 μm fractions of B1 and B2, respectively, were 32% and 23% lower when compared to their 250 μm fractions. In general, the FC content increased with increasing particle size. B2 showed a gradual increase from 10 to 70% in FC as particle size increased. For both B1 and B2, fractions 425 μm and larger contained 50% or more FC while fractions 250 μm and smaller had ~ 26% or less FC. The high FC seen in the larger fractions are a result of incomplete combustion of the sample and the presence of char-like particles.

Table 1. Proximate analysis for fixed bed boiler bottom ash samples (B1 & B2)

Particle size (μm)	B1			B2		
	FC (wt.%)	**Vm** (wt.%)	**Ash** (wt.%)	**FC** (wt.%)	**Vm** (wt.%)	**Ash** (wt.%)
<150	13.8	19.1	67.1	10.0	19.6	70.4
150	9.5	17.5	73.0	10.6	17.1	72.3
250	16.0	17.4	66.6	25.3	17.4	57.4
425	51.1	14.9	34.0	51.9	13.8	34.3
850	63.6	16.7	19.7	67.8	13.5	18.7
2000	59.0	14.8	26.2	67.4	14.9	17.7
AR	29.3	15.9	54.8	54.0	15.8	30.2

All results are on a dry weight basis.

AR = as received (i.e. not separated into particle size fractions).

Table 2 displays the results of the ultimate analysis, and shows the total C, H, and N for each particle fraction. In general, the total % C increased with increasing particle size fraction. The weight percent distribution of C for most samples was greater than those from the proximate analysis (i.e. C greater than FC). This is anticipated since the total carbon includes the organic and inorganic C. All samples had H concentrations of 1% or less, and N concentrations of less than 0.3%. S was below detection limits. These low concentrations of H are expected since the H present in the wood is converted to H_2O during the combustion process. Also, sulphur is usually not present or is only present in small amounts in woody biomass samples unless the soil is very high in sulphur content.

Table 2. Ultimate analysis of fixed-bed boiler bottom ash samples (B1&B2)

Particle size (μm)	B1			B2		
	C wt.%	H wt.%	N wt.%	C wt.%	H wt.%	N wt.%
<150	22.70	0.41	0.06	17.17	0.65	0.04
150	15.99	0.27	0.04	13.85	0.56	0.03
250	13.13	0.21	0.01	21.56	0.59	0.01
425	55.32	0.81	0.70	52.76	0.79	0.06
850	63.92	0.98	0.06	66.41	0.71	0.18
2000	68.15	0.76	0.12	70.73	0.53	0.25
AR	38.17	0.42	0.11	51.79	0.68	0.08

All results are on a dry weight basis.

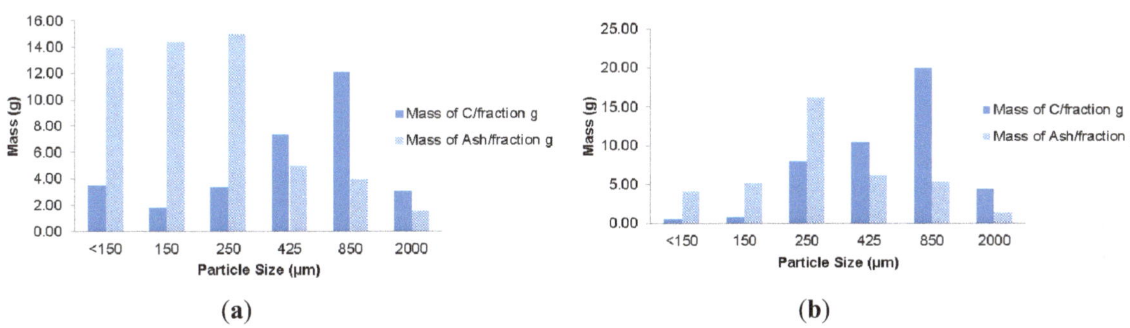

Figure 2. Distribution of fixed carbon (g) and ash (g) in 100 g (dry basis) of (**a**) B1 sample; (**b**) B2 sample based on particle size fraction distribution

Figure 2a and 2b shows the distribution of fixed carbon (FC) in 100 grams of boiler ash samples B1 and B2. The 850 μm fraction for both B1 and B2 were the highest in FC content, 12 g and 20 g respectively. Fraction 425 μm had the second highest FC contents of 7 and 10 g respectively. The two boiler ash samples showed a variation in FC distribution. This could be due to a number of factors such as variations in retention time in the reactor, incomplete combustion, temperature variations and fluctuations, moisture content of fuel etc. Also, the moisture content and composition of the biomass feedstock may vary depending on when and where it was sourced. Higher moisture content will reduce the rate of combustion of the fuel. Also, because of the variability in energy demand of the plant, ash may be removed from the grate before complete combustion. This is in addition to the other ash related challenges that occur in the boiler as previously mentioned in the introduction. The 250 μm fractions for B1and B2 had the highest ash contents of approximately 14 and 26 g. As previously seen in the proximate analysis, fractions 425 μm and higher in both boiler samples contain more than 50 % FC. Combined, fractions 425, 850 and 2000 μm contained ~ 72% of the total FC in sample B1. These same fractions accounted for only ~ 36% of the total sample mass. For B2, the 425 μm and higher fractions contained ~ 82% of the total FC and accounted for 59% of the total mass. The high FC present in boiler ash provides an opportunity for re-burning or reusing its energy.

3.3 Thermogravimetric Analysis

Figure 3 shows the thermal behaviour of the as-received, combined fractions ≥ 425 μm and combined fractions < 425 μm for boiler ash B1 when heated under air at atmospheric conditions. The results are only shown for B1, however, the results for B2 as-received show similar thermal distributions (not shown). The TGA analyses showed that the mass of the boiler sample decreased as temperature increased. The weight loss occurred in phases which is linked to the removal of specific components of the sample such as water, carbonates, aluminates, silicates, oxides, organic carbon and other components (Zhang, Zhao, & Jingyu, 2010). All three samples displayed a similar pattern in thermal distribution. The first weight loss was seen between ~ 25 – 110 °C and accounted for the removal of moisture from the sample. A further mass loss was seen up to ~ 350 °C. This change is generally attributed to the removal of moisture, carbon monoxide, and volatiles such as tars and other organic carbons that may have condensed on the sample (Khan et al., 2009). The most noticeable change in mass of the AR, ≥ 425 μm and < 425 μm fractions was seen between the temperatures 350 to 440 °C and accounted for ~ 32, 43 and 14% mass loss, respectively. This is assumed to be organic carbon (char). Further mass losses occurred between 440 to 610 °C and between 610 to 670 °C. The decomposition of mineral components such as carbonates ($CaCO_3$) and some of the other metal carbonates and oxides present in the sample may occur during this phase. For the range of temperatures studied, the AR, ≥ 425 and < 425 μm fractions lost ~ 44, 56 and 26% of initial mass by 800 °C, respectively. Therefore, fractions ≥ 425 μm had the highest combustible content. The sample burnout time corresponding to the burnout temperatures, that is, the temperature that shows no further mass loss on the TGA (Idris, Rahman, & Ismail, 2012; Miranda, Esteban, Rojas, Montero, & Ruiz, 2008) are shown in Table 3. The fraction ≥ 425 μm had a slighter shorter burnout time but was not significantly different from the other two samples under study.

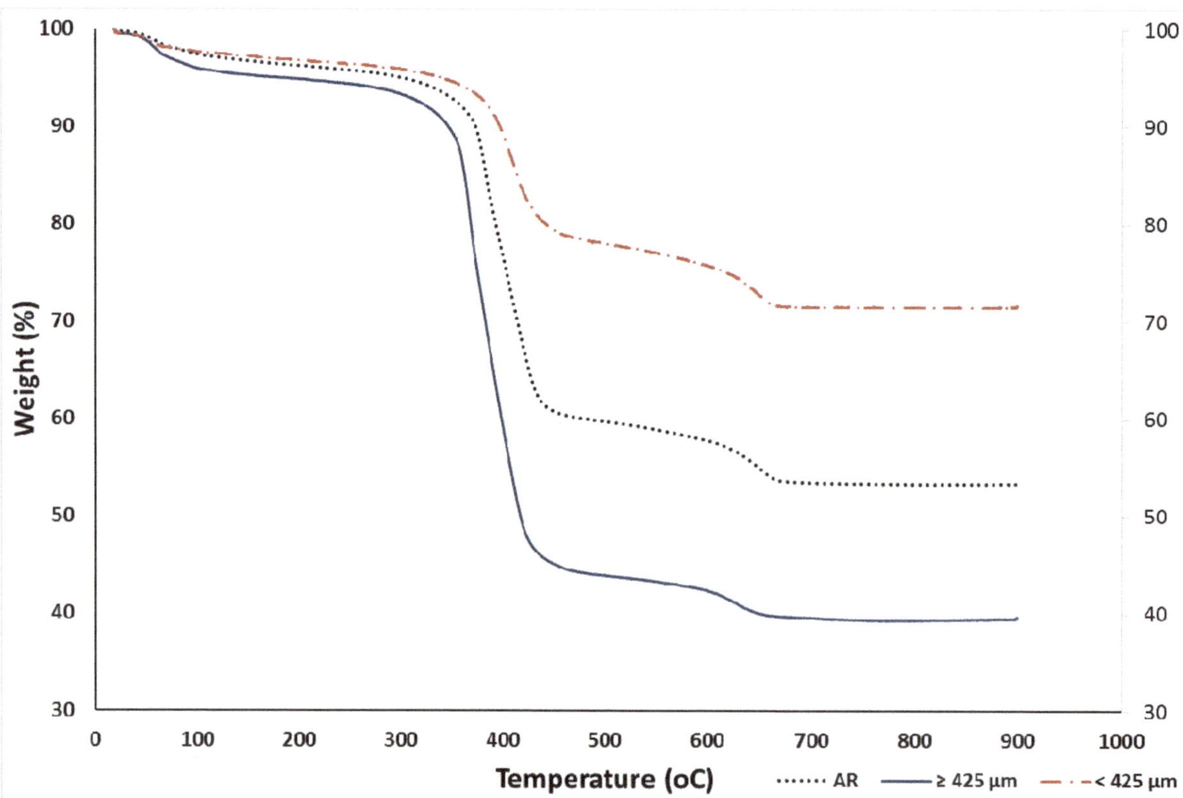

Figure 3. TGA graph of weight change as a function of temperature for B1 boiler ash sample from a fixed bed boiler. Lines displayed represents AR sample, fractions ≥ 425 μm and < 425 μm

Table 3. Combustion characteristic of fixed bed boiler bottom ash (B1) samples

Sample	Temperature range (°C)	Weight Loss (wt. %)	Peak Temperature Tp (°C)	DTGmax (mg/min)	Sample burning time (min)
AR	110-350	4.28			35.33
	350-440	31.98	383	1.2082	
	440-610	3.54			
	610-670	3.79	649	0.2023	
	670-800	0.36			
≥ 425 μm	110-350	6.49			34.23
	350-440	43.59	362	1.3843	
	440-610	3.71			
	610-670	2.25	648	0.1357	
	670-800	0.4			
< 425 μm	110-350	2.94			34.47
	350-440	14.44	405	0.6642	
	440-610	4.81			
	610-670	3.83	646	0.1815	
	670-800	0.15			

Results are on a dry basis.

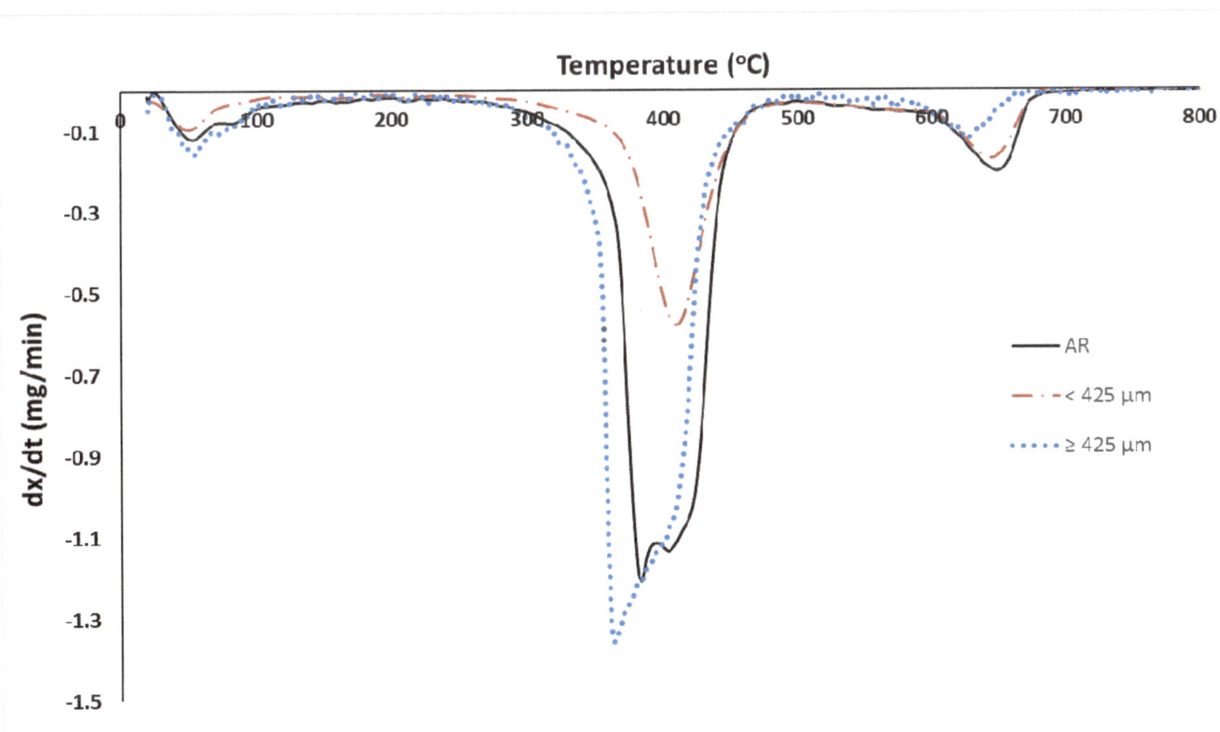

Figure 4. DTGA graph displaying rate of mass loss as a function of temperature for B1 boiler ash sample from a fixed bed boiler. Lines displayed represents AR sample, fractions ≥ 425 μm and < 425 μm

The DTGA analysis in Figure 4 shows two distinct peaks. These peaks are associated with the thermal oxidation of different components in the particles such as carbonized material, heavy organic molecules, condensed tars and inorganic compounds (Miguel, Dominguez, Hemandez, & Sanz-Perez, 2012). As seen in Table 3, the AR sample, fractions ≥ 425 and < 425 μm had the highest rate of mass losses at ~ 1.2082 mg/min at 383 °C, 1.3843 mg/min at 362 °C, and 0.6642 mg/min at 405 °C, respectively. Lignocellulosic samples heated in air typically have well resolved peaks at ~ 400 °C (Elder, Kush, & Hermann, 2011) (Di Blasi, 2009). The fractions containing higher carbon contents were observed to have lower peak temperatures. Peak temperatures determines how easily a fuel is ignited and is defined as the maximum rate of weight loss due to volatilization accompanied by the formation of carbonaceous residue, on the DTGA curve (Idris et al., 2012). Low peak temperatures mean that the fuel is easier to ignite. Since the ≥ 425 μm fraction was found to have the highest rate of mass loss at a lower temperature, faster reaction times and lower ignition temperatures could be observed if re-burned, when compared to the other two samples. The samples showed clearly defined thermal phases when heated and oxidized and contained substantial amounts of combustible material, particularly char for the AR sample and particle fractions ≥ 425 μm.

3.4 Surface Area

The N_2-BET surface area absorption B1 and B2 (≥ 2000 μm fraction) were 850 and 770 m^2/g, respectively. The high surface areas obtained are expected for wood ash chars (Link, et al., 2010; Hanaoka, Sakanishi, & Okumura, 2012; Demirbas, 2005). This may influence the reactivity rate during combustion or gasification processes. A larger available surface area provides more sites for oxygen to bond with carbon (Moulijn & Kapteijn, 1995) increasing oxidation. It has been reported that wood chars that are produced at higher temperatures have a higher surface area (James, Sabatini, Chiou, Rutherford, Scott, & Karapanagioti, 2005). Lower pressures also increase the surface area of charge (Hanaoka et al., 2012).

3.5 Higher Heating Value (HHV)

Table 4 displays the higher heating values for the various size fractions of the B1 and B2 ash samples. Heating values within each size class were relatively similar between the two ash samples. A range of approximately 5 - 25 MJ/kg was obtained for both samples. As the particle size fraction increased, the heating value of the fraction also increased. A slight decrease in the heating value was observed when moving from 850 μm to 2000 μm for both B1 and B2. The obtained HHVs suggest that the waste being disposed of has good energy potential and is comparable to some forms of biomass used in energy production processes (Liao, Wu, Yanyongjie, & Huang, 2004).

Table 4. Higher heating value of boiler bottom ash samples separated within particle fractions

Particle size (μm)	B1 MJ/kg	B2 MJ/kg
<150	7.02	8.89
150	5.58	6.86
250	7.40	10.40
425	15.77	19.33
850	22.96	24.56
2000	20.83	23.54
AR	10.77	14.42

All results are on a dry weight basis.

3.6 Bulk Density

The bulk density was calculated for each fraction and is represented as a function of particle size in Figure 5. B2 has lower bulk densities in all fractions. The as-received samples of B1 and B2 were 244 and 172 kg/m^3 respectively. For both B1 and B2, fractions 425 μm and higher had significantly lower bulk densities than fractions 250 μm and below. B1 fractions 425 μm and higher had a bulk density range of ~ 100 - 150 kg/m^3. Bulk density was ~ 50% less for fractions 250 μm and lower. For B2, bulk density was also 50% less for fractions 250 μm and lower compared to fractions 425 μm and higher.

The bulk densities were lower with samples of higher FC contents. Figure 6 shows the correlation of bulk density with percent FC within the various particle fractions. A linear relationship is established and shown in Equation 2. While the value of R^2 is 0.8692, the graph shows a significant reduction in bulk densities of fractions with higher FC contents. If the carbon is reduced or eliminated in ash, the bulk density increases. This may be important in the transportation and storage of ash. With increasing bulk densities, greater amounts of ash could be transported from combustion plants to landfills.

$$\rho_B = -4.9579\ FC + 406.73 \qquad R^2\ 0.8692 \tag{2}$$

ρ_B represents Bulk density.

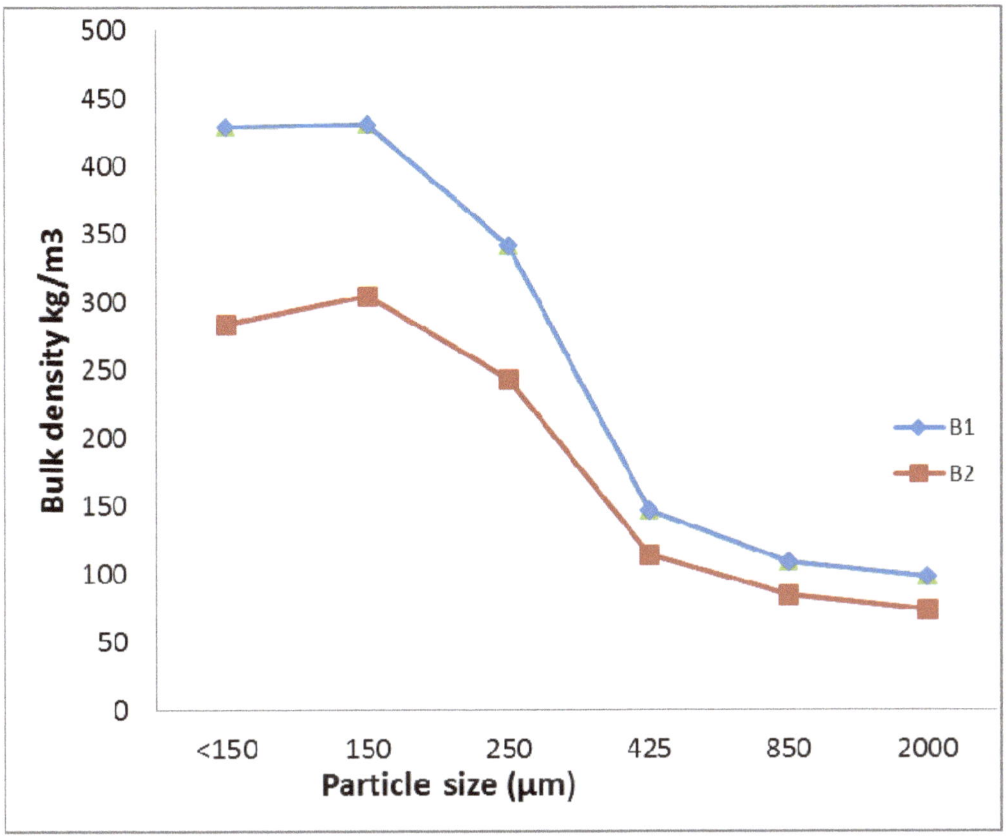

Figure 5. Bulk density variation with particle size (µm) for boiler ash

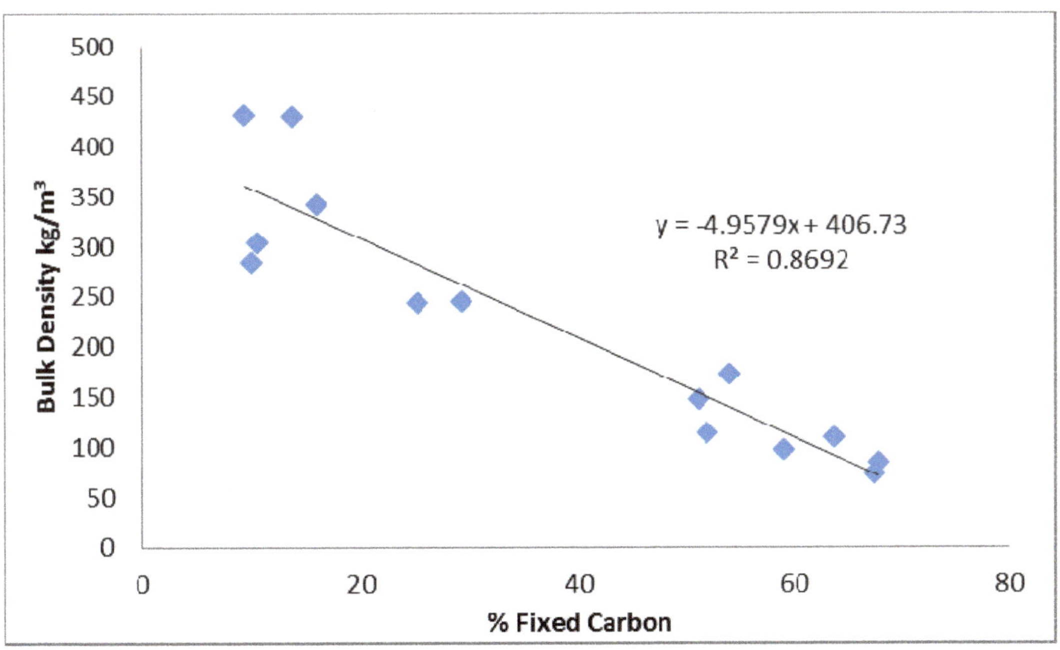

Figure 6. Bulk density versus percent fixed carbon for boiler ash

4. Summary of Results

Table 5 summarizes some of the main results by reporting data obtained for the as-received fractions, combined fractions ≥ 425 μm and combined fractions < 425 μm for B1 and B2. When samples are collected from fractions ≥ 425 μm, the energy density can be increased by approximately 50 % and 64% for B1 and B2 respectively when compared to the AR for both samples. In addition, the majority of the energy content is in the larger fraction (≥ 425), 64% for B1 and 78% for B2. On a mass basis, less than half of the ash is in the larger fractions (≥ 425). The bulk density analysis shows that the low carbon fractions of ash have four times the bulk density, compared to the high carbon fractions. This suggests a good opportunity for volume reduction. By reburning the larger size fraction, ash volumes can be decreased by over a half, while recovering more than two-thirds of the energy present in the as-received sample.

Table 5. Properties of boiler bottom ash of major ash fractions, based on 100 kg sample

Parameter	B1 AR	B1 ≥425 μm	B1 <425 μm	B2 AR	B2 ≥425 μm	B2<425 μm
Mass (kg)	100.0	37.2	62.8	100.0	60.0	40.0
Energy density (MJ/kg)	10.8	20.0	6.7	14.4	22.4	9.7
Energy content (GJ)	1.08	0.74	0.42	1.44	1.34	0.39
FC (%)	29.3	58.3	13.2	54.0	61.6	21.5
Ash (kg)	54.8	9.7	42.8	30.2	14.8	24.4
Bulk density (kg/m^3)	244.0	106.6	448.3	172.0	81.5	385.3
Volume (m^3)	0.41	0.35	0.14	0.58	0.74	0.10

Bulk density calculated from Equation 2.

5. Conclusions

The boiler ash studied shows variability in carbon contents that may be a result of variations in retention time in the reactor, incomplete combustion, temperature variations and fluctuations and variations in moisture content. However, unique to both ash samples is a high carbon content in the larger size fractions, specifically for the fractions ≥ 425 μm. Separating and re-using particle sizes 425 μm and higher would recover over 50 % of the unburnt carbon that is present in this bottom ash. Sieve fractionation may allow for easy accessing of the carbon and may present itself as a cost effective pre-treatment method for reburning high carbon ash, in order to obtain the highest energy components. The reintroduction of ash in the combustion system may present challenges due to problems related to corrosion, scouring and fouling as experienced with the fixed bed boiler that produces this high carbon ash. Thermogravimetric analysis shows that this high carbon ash may be combusted and displays thermal phases similar to the combustion of lignocellulosic biomass. The highest rate of mass loss was 1.3843 mg/min for fractions ≥ 425 μm. This occurs at a lower peak temperature when compared to the other samples, resulting in lower ignition temperatures. There was no significant difference in the burn-out time for all 3 samples. The high surface area of the char particles could increase combustion or gasification reactivity rates, since a large surface area is available for oxidation. A linear correlation was identified between the FC and bulk density, where samples with higher concentrations of fixed carbon have lower bulk densities.

In an effort to ensure environmental sustainability and sustainable use of our biomass resources, efficient use must be realized. The use of woody biomass in bioenergy process must be such that all possible energy is extracted. It is preferred that bioenergy processes are able to so. However, should that be not possible due to system limitations or fuel type etc, other options must be employed in conjunction with existing technologies to ensure maximum use of the resource. Finally, the need for an in-depth analysis of the inorganic distribution within fractions of bottom ash should also be investigated for possible environmental and technical utilization. For example, there are likely to be similarities in the properties of ash and biochar. Biochar may have beneficial soil applications, for example biochar has been suggested to enhance seedling growth (Robertson, Rutherford, Lopez-Gutierrez, & Massicotte, 2012; Knapp & Insam, 2011). Perhaps high carbon ash could also have similar results for soil applications.

Acknowledgments

We would like to thank the following organizations for their support in conducting this research Canfor Pulp Mill and the University of Northern British Columbia Central Equipment Laboratory. Also, we acknowledge funding support from Natural Science and Engineering Research Council (NSERC) and Canfor Pulp Limited Partnership (CPLP).

References

Batra, V., Urbonaite, S., & Svensson, G. (2008). Characterization of unburned carbon in bagasse fly ash. *Fuel, 87*, 2972-2976. http://dx.doi.org/10.1016/j.fuel.2008.04.010

Cetin, E., Moghtaderi, B., Gupta, R., & Well, T. (2004). Influence of pyrolysis conditions on the structure and gasification reactivity of biomass chars. *Fuel, 2004*, 2139-2150. http://dx.doi.org/10.1016/j.fuel.2004.05.008

Demirbas, A. (2005). Potential applications of renewable energy sources, biomass combustion problems in boiler power systems and combustion related environmental issues. *Progress in energy combustion science, 31*, 171-192. http://dx.doi.org/10.1016/j.pecs.2005.02.002

Di Blasi, C. (2009). Combustion and gasification rates of lignocellulosic chars. *Energy Combustion Science, 35*, 121-140. http://dx.doi.org/10.1016/j.pecs.2008.08.001

Dimitrakopoulous, A. (2001). Thermogravimetric analysis of Mediterranean plant species. *Journal of Analytical and Applied Science, 60*, 123-130.

Duan, L., Liu, D., Chen, X., & Zhao, C. (2012). Fly ash recirculation by bottom feeding on aa circulating fluidized bed boiler co-burning coal sludge and coal. *Applied Energy, 95*, 295-299. http://dx.doi.org/10.1016/j.apenergy.2012.02.063

Elder, T., Kush, J., & Hermann, S. (2011). Thermogravimetric analysis of forest understory grasses. *Thermochimica Acta, 512*, 170-1771. http://dx.doi.org/10.1016/j.tca.2010.10.001

Emilsson, S. (2006). *International handbook. From extraction of forest fuels to ash recycling.* Jonkoping: Swedish Foresty Agency.

Gomez-Barea, A., Vilches, L., Campoy, M., & Fernandez-Pereira, C. (2009). Plant optimization and ash recycling in fluidised waste gasification. *Chemical Engineering Journal, 146*, 227-236. http://dx.doi.org/10.1016/j.cej.2008.05.039

Hanaoka, T., Sakanishi, K., & Okumura, Y. (2012). The effect of N2/CO2/O2 content and pressure on charcateristics and CO2 gasification behaviour of biomass-derived char. *Fuel Processing Technology, 104*, 287-294. http://dx.doi.org/10.1016/j.fuproc.2012.05.024

Herrman, T., & Baker, S. (2002). *Evaluating particle size.* Kansas: Kansas State University Agricultural Experiment Station and Cooperative Extension Service.

Idris, S., Rahman, N., & Ismail, K. (2012). Combustion characteristics of Malaysian oil palm biomass, sub-bituminous coal and their respective blends via thermogravimetric analysis (TGA). *Bioresource Technology, 123*, 581-591. http://dx.doi.org/10.1016/j.biortech.2012.07.065

James, G., Sabatini, D., Chiou, C., Rutherford, D., Scott, A., & Karapanagioti, H. (2005). Evaluating phenanthrene sorption on various wood chars. *Water Research, 39*(4), 549-558. http://dx.doi.org/10.1016/j.watres.2004.10.015

Khan, A., Jong, W., Jansens, P., & Spliethoff, H. (2009). Biomass combustion in fluidized bed boilers: potential problems and remidies. *Fuel processing technology, 90*, 21-50. http://dx.doi.org/10.1016/j.fuproc.2008.07.012

Knapp, B., & Insam, H. (2011). Recycling of biomass ashes: current technologies and future research needs. In I. H. Knapp B, *Recycling of biomass ashes.* Heidelberg: Springer. http://dx.doi.org/10.1007/978-3-642-19354-5_1

Liao, C., Wu, C., & Yan, Y. (2007). The characteristics of inorganic elements in ashes from a 1 MW CFB bimass gasification power generation plant. *Fuel Processing Technology, 88*, 149-156. http://dx.doi.org/10.1016/j.fuproc.2005.06.008

Liao, C., Wu, C., Yanyongjie, & Huang, H. (2004). Chemical elemental characteristics of biomass fuels in China. *Biomass and Bioenergy, 27*, 119-130. http://dx.doi.org/10.1016/j.biombioe.2004.01.002

Link, S., Arvelakis, S., Hupa, M., Yrjas, P., Kulaots, I., & Paist, A. (2010). Reactivities of the biomass chars originating from reed, douglas fir and pine. *Energy and Fuels, 24*, 6533-6539. http://dx.doi.org/10.1021/ef100926v

Miguel, G., Dominguez, M., Hemandez, M., & Sanz-Perez, F. (2012). Characterization and potential applications of solid particles produced at a biomass gasification plant. *Biomass and Bioenergy, 47*, 134-144. http://dx.doi.org/10.1016/j.biombioe.2012.09.049

Miranda, T., Esteban, A., Rojas, S., Montero, I., & Ruiz, A. (2008). Combustion analysis of ifferent olive residues. *International Journal of Molecular Science, 9*, 512-525. http://dx.doi.org/10.3390/ijms9040512

Moulijn, J., & Kapteijn, F. (1995). Towards a unified theory of reactions of carbon with oxygen-containing molecules. *Carbon, 8*, 1155-1165. http://dx.doi.org/10.1016/0008-6223(95)00070-T

Obernberger, I., & Supancic, K. (2009). *Possibillities of ash utilisation from biomass combustion plants. Proceedings of the 17th European Biomass conference & Exhibition.* Hamburg: ETA Renewable Energies.

Pels, J., de Nie, D., & Kiel, J. (n.d.). *Improvement of the economics of biomass/waste gasification by higher carbon conversion and advanced ash management.* Retrieved October 20, 2012, from http://www.ecn.nl/docs/library/report/2006/c06038.pdf

Picco, D. (2010). *Technical assistance for the development and improvement of technologies, methodologies and tools for enhanced use of agricultural biomass residues.* Gorizia.

Quaak, P., Knoef, H., & Stassen, H. (1999). *Energy from biomass. A review of combustion and gasification technologies.* Washington, D.C.: World Bank Technical Paper; 422. Energy series.

Robertson, S., Rutherford, M., Lopez-Gutierrez, J., & Massicotte, H. (2012). Biochar enhances seedling growth and alters root symbioses and properties of sub-boreal forest soils. *Canadian Journal of Soil Science, 92*, 329-340. http://dx.doi.org/10.4141/cjss2011-066

Ryu, C., Yang, Y., Khor, A., Yates, N., Sharifi, V., & Swithenbank, J. (2006). Effect of fuel properties on biomass combbustion: Part 1. Experiments - fuel type, equivalence ration and particle size. *Fuel, 85*, 1039-1046. http://dx.doi.org/10.1016/j.fuel.2005.09.019

Saidur, R., Abdelaziz, E., Demirbas, A., Hossain, M., & Mekhilef, S. (2011). A review on biomass as a fuel for boilers. *Renewable and Sustainable Energy Reviews, 15*, 2262-2289. http://dx.doi.org/10.1016/j.rser.2011.02.015

Sarenbo, S. (2009). Wood ash dilemma-reduced quality due to poor combustrion performance. *Biomass and Bioenergy, 33*, 1212-1220. http://dx.doi.org/10.1016/j.biombioe.2009.05.007

Van Loo, S., & Kopejan, J. (2008). *The handbook of biomass combustion and co-firing.* Earthscan publications.

Warnecke, R. (2000). Gasification of biomass: comparison of fixed bed and fluidized bed gasifier. *Biomass and Bioenergy, 18*, 489-497. http://dx.doi.org/10.1016/S0961-9534(00)00009-X

Zhang, H., Zhao, Y., & Jingyu, Q. (2010). Termal characterization of fly ash from municipal solid waste incinerator (MSWI) in Shanghai. *Process Safety and Environmental Protection, 88*, 269-275. http://dx.doi.org/10.1016/j.psep.2010.03.004

Efficiency Evaluation of RDF Plasma Gasification Process

Spyridon Achinas[1] & Evangelos Kapetanios[2]

[1] School of Chemical Engineering, National Technical University of Athens, Athens, Greece

[2] Association of Communities and Municipalities of Attica Region, Athens, Greece

Correspondence: Spyridon Achinas, National Technical University of Athens, Heroon Polytechneiou 9st., PC. 15773, Zographou Campus, Athens, Greece. E-mail: sp.achinas@gmail.com

Abstract

The waste management in Greece is a very important issue by virtue of the increasing of waste material. This paper is concerned with the efficiency evaluation of a plasma gasification process using RDF (Residual Derived Fuel) to produce Syngas. Plasma gasification is an advanced waste management and thermal treatment technology with exceptional environmental and energy performance. This process converts materials such as coal, biomass or other organic waste into carbon monoxide and hydrogen using very high temperatures with an amount of oxygen and/or steam (Leal-Quiros, 2004).

Regarding to the previous studies and using a previous thermodynamic chemical equilibrium model, we devised a new to be appropriate for the RDF to study the efficiency of the plasma gasification process. The total amount of waste materials was supposed 300 tn/day and the produced synthesis gas was composed of CO, H_2, CH_4, H_2O, CO_2, N_2, S and H_2S. Then, at constant temperature and humidity parameters and oxygen, using the results of thermodynamic analysis, is calculated the efficiency of the process according to the thermodynamics laws.

Conclusively, the thermodynamic modeling of plasma gasification process for the RDF treatment resulted in positive energy values. These energy results, along with the environmental characteristics of the process, demonstrate that the plasma gasification process is very useful technology for waste treatment and electricity production.

Keywords: plasma, gasification, exergy, modelling

1. Introduction

Nowadays, many countries face the problem of the excessive amounts of waste material. The developed world is presently facing major environmental problems with garbage as landfills fill up, forcing governments to transport their garbage farther away at greater expense and environmental impact. Recent studies have focused on an innovative technology, the plasma gasification that has been demonstrated as one of the most effective and environmentally friendly methods for solid waste treatment and energy utilization (Pfender, 1999).

Plasma gasification works by treating garbage and any carbon-based materials, such as plant matter or fossil fuels, with high heat from plasma torches in a controlled-air environment. Rather than burn, the materials vaporize into their basic molecular elements as gasses. The gasses are cooled and cleaned and can then be converted into electricity, or a variety of valuable fuels such as ethanol, hydrogen or natural gas. The inorganic materials are melted by the plasma torches and pour out. The molten materials then cool and harden into a vitrified glass called slag. The high heat from the gasses is recycled back into the system as steam. At the most basic level, a plasma waste converter is a plasma torch applied to garbage. A plasma torch uses gas and powerful electrodes to create plasma, sometimes called the fourth state of matter (Higman & Van der Burgt, 2007; Mountouris, 2007).

The principal product of plasma gasification is a low to medium calorific value synthesis gas composed of carbon monoxide and hydrogen. This gas can be burned to produce heat and steam, chemically scrubbed and filtered to remove impurities before conversion to various liquid fuels or industrial chemicals/polymers, or used once cleaned as a turbine or engine fuel to produce electricity. Heat energy can also be recovered via water tube heat exchangers as the hot syngas is cooled from about 2,200 to 400 °F (Rezaiyan & Cheremisinoff, 2005).

2. Process Design

2.1 Materials

The physical and chemical features of RDF are directly dependent from the respective features of his components. Moreover, both the qualitative and quantitative features of gas and solids waste of gasification depend on the composition of RDF. Finally, the creation of RDF depends on the heat capacity. The features of RDF waste from the mechanical and composting plant of Ano Liossion (Attica Region) is:

- Produced quantity RDF: 300 tn/day.
- Moisture of RDF: 27.1%.
- Elemental analysis of RDF: The elemental analysis is shown in table below.

Table 1. Ultimate analysis for materials

Materials	RDF%	C%	H%	O%	S%	N%	Cl%	W%	A%*
Paper	78,45	27,44	3,76	25,60	0,16	0,16	0,27	36,80	5,81
Plastic	11,30	63,57	8,78	9,02	0,34	0,90	3,38	4,20	9,81
Miscellaneous	10,25	24,92	3,14	15,06	0,18	0,73	0,41	33,40	22,16
RDF composition	100	31,26	4,26	22,71	0,37	0,43	1,05	32,75	7,94

*A%: The percentage of aggregate materials in each component of the RDF.

Kapetanios reported lower calorific capacity 14,200 kJ/kg for RDF, initial moisture 27.1%, which is produced in the mechanical and biological pre-treatment plant (MBT) of the Attica Region, while the European Union of 15 members reported that the average lower calorific capacity for RDF from mixed MSW, 24.4% initial moisture is 13,300 kJ/kg. The lower calorific capacity of dried RDF, with humidity 13%, is Hu = 17,918,422 kJ/kg. The lower calorific capacity of RDF is expected to be increased in the near future for the reasons cited above for the expected reduction of the moisture content and was calculated by the equation:

$$Hu = 38834c + 93868h + 101325s - 5945n - 10802o - 2449w$$

with c, h, s, n, o, w the quantities of carbon (C), hydrogen (H_2), sulphur (S), nitrogen (N_2), oxygen (O_2) and moisture respectively. The RDF flow before his drying, with initial moisture of 27.1%, is 12500 kg/h. The RDF flow after drying at 13% humidity is 1,041,375 kg/h.

The analysis of energy quantities is a complicated process for the plasma gasification since the energy outputs should also be reproducible in real scale. The lack of real data makes the problem more difficult because the information such as the adequacy of energy recovery systems of the gas, heating capacity or the combustion efficiency for gaseous mixtures are limited without having proved even in practice. It is decided to study the case of electricity generation from a combined cycle system. The gas engines are considered to have 40% electrical efficiency and thermal efficiency of 50% (Achinas & Kapetanios, 2012).

2.2 Plasma Gasification System Description

The plasma gasification is a modern process and is based in the pure gasification. In the Figure 1 (below) is given the general process flow diagram of an intergarted plasma gasification combined cycle system for energy recovery (Achinas & Kapetanios, 2012). The plasma furnace is the central component of the system where gasification is taking place. Two graphite electrodes, as a part of two transferred arc torches, extend into the plasma furnace. An electric current is passed through the electrodes, and am electric arc is generated between the tip of the electrodes and the conducting receiver, i.e. the slag in the furnace bottom. The gas introduced between the electrode and the slag that becomes plasma can be oxygen, helium or some other, but the use of air is very common due to its low cost.

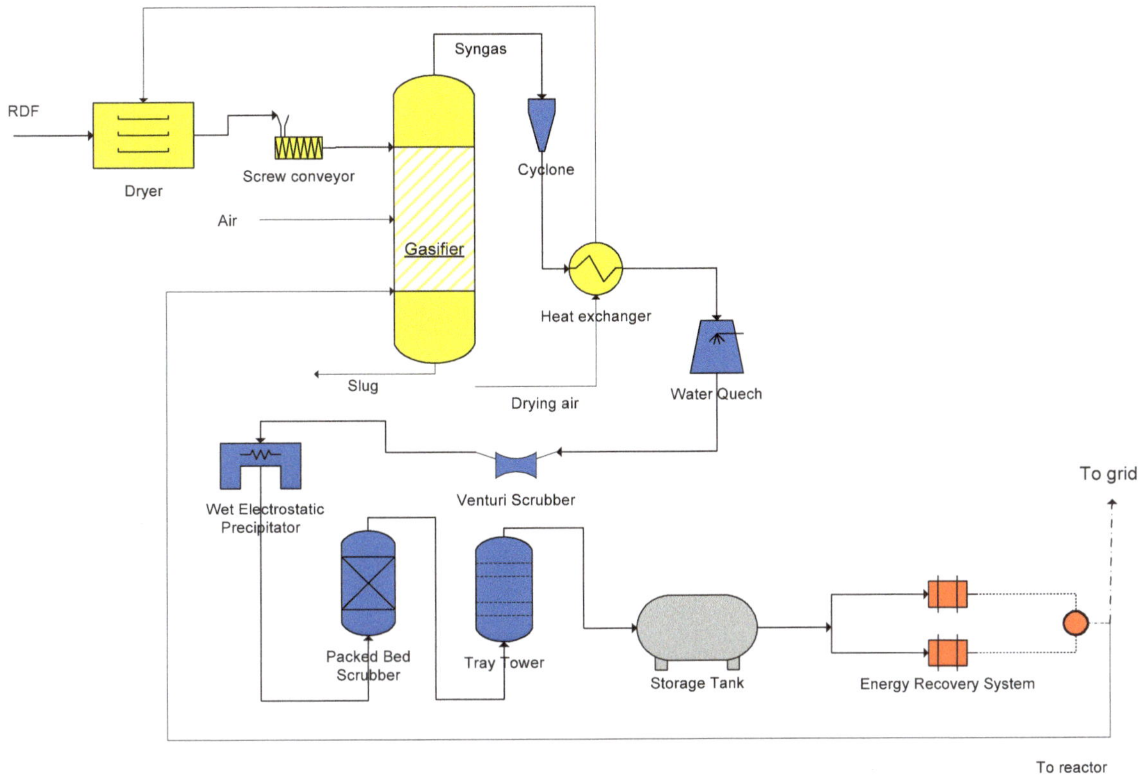

Figure 1. Process flow diagram of RDF plasma gasification for energy recovery

The RDF, after drying, will be led in gasification unit, which will produce a gas mixture and a small percentage of sludge and metal (oils and tar). In the process of drying the RDF, before cracking, will include a multi-cyclone for the retention of suspended particles RDF, which will be then driven gasification reactor. The fluidized bed will be supplied with lime (CaO) from silo to capture mainly emitted sulfur dioxide (SO_2) during gasification, and urea from the tank, to capture the various oxides of nitrogen (NOx).

The following part includes the gas cleaning step in order to reduce up to zero the quantity of acids, metals and other substrates from the syngas before to introduce it in the energy recovery installation (Laurence & Ashenafi, 2012). The syngas from the PG reactor must be cooled down and cleaned before it can be used as fuel in a gas turbine combined cycle. Immediately after leaving the unit, the gas will pass through deducting unit consisting of multi-cyclone and remove the bulk of the produced dust an ash. Dry scrubbing is followed by semi-dry cleaning, through semi-dry scrubber.

The energy recovery system can be based on a combined cycle system, gas turbine and steam turbine (Imris et al., 2005). The sludge is collected from the bottom of the reactor while the gas is led into a combined cycle gas turbine unit, which will produce superheated steam (Global Environmental Solutions Inc.). The steam turbine will be driven at two levels (high and low pressure). The gasification installation of RDF in Athens will include all the necessary anti-pollution systems, in order to comply with the relevant national legislation and the relevant directives of the European Union.

3. System Modelling

3.1 Development of Gasification Process

Gasification is closely related to combustion and pyrolysis, but there are important distinctions. Gasification is like starved-air burning because oxygen is strictly controlled and limited so that as heat is applied the feedstock is not allowed to actually burn (Littlewood, 1977). Instead of combusting, the raw materials break down and go through the process of pyrolysis that produces char and tar. At its simplest form, pyrolysis is commonly used to produce charcoal from wood. As the process continues and the heat is taken higher, the char and tar completely break down into gasses. Depending on the process used and the precise chemistry, the resulting gas may come in a few different forms: synthesis gas, producer gas, town gas, wood gas or others (Bonizzoni & Vassallo, 2002).

The plasma torch, i.e. the device which provides energy in the form of electric current in the reactor is deemed to have 85% yield. The reason it did not be supposed a price equal to 100% is the flares cooling requirements for safe operation. Companies give for torch performance a range from 70 to 90% (Alter NRG & Westinghouse Plasma Corporation, 2008; Minutillo et al., 2009). As we know from the bibliography it is difficult to simulate all the chemical reactions which take place in the gasifier. The chemical reactions below depict the basic gasification process (Mountouris et al., 2008):

$C(s) + H_2O = CO + H_2$ (Heterogeneous water gas shift reaction – endothermic)

$C(s) + CO_2 = 2CO$ (Boudouard equilibrium – endothermic)

$C(s) + 2H_2 = CH_4$ (Hydrogenating gasification – exothermic)

$CH_4 + H_2O = CO + 3H_2$ (Methane decomposition – endothermic)

$CO + H_2O = CO_2 + H_2$ (Water gas shift reaction – exothermic)

The principal reaction for the modeling for RDF gasification process is written as follows:

$$C_xH_yO_zS_kN_l + wH_2O + mO_2 + 3,76mN_2 \Leftrightarrow n_1CO + n_2H_2 + n_3CH_4 + n_4H_20 + n_5CO_2 + n_6N_2 + n_7S + n_8H_2S \quad (1)$$

The three main independent reactions that are selected for the equilibrium calculation are shown below:

$CH_4 + H_2O \Leftrightarrow CO + 3H_2$ (Methane decomposition-endothermic)

$CO + H_2O \Leftrightarrow CO_2 + H_2$ (Water gas shift-exothermic)

$H_2 + S(g) \Leftrightarrow H_2S$ (Sulfur reaction)

The equilibrium constants are:

i) for the methane decomposition

$$K_1=[CO]*[H_2]^3/[CH_4]*[H_2O]=(n_{CO}*n_{H2}^3)/(n_{CH4}*n_{H2O})=(n_1*n_2^3)/(n_3*n_4*n_{total}^2) \quad (2)$$

ii) for the water gas shift reaction

$$K_2=[CO_2]*[H_2]/[CO]*[H_2O]=(n_{CO2}*n_{H2}) / (n_{CO}*n_{H2O})=(n_5*n_2)/(n_1*n_4) \quad (3)$$

iii) for the sulphur reaction

$$K_3=[H_2S]/[H_2]*[S(g)]=n_{H2S}/(n_{H2}*n_{S(g)})=(n_8*n_{total})/(n_2*n_7) \quad (4)$$

3.2 Model Analysis

As we see above, we have eight variables and ase we describe below we need seven equations which represent the seven unknown stoichiometric coefficients of the products and the oxygen content for the reaction (Nikolaou, 2010; Mountouris et al., 2006). Therefore, seven equations are required, which are formulated based on the following:

$$\text{Carbon: } x=n_1+n_3+n_5 \quad (5)$$

$$\text{Hydrogen: } y+2w=2n_2+4n_3+2n_4+2n_8 \quad (6)$$

$$\text{Oxygen: } z+w+2m=n_1+n_4+2n_5 \quad (7)$$

$$\text{Sulfur: } k=n_7+n_8 \quad (8)$$

$$\text{Nitrogen: } l+7,52m=n_6 \quad (9)$$

$$\text{Ολικά kmol: } n_{tot}=n_1+n_2+n_3+n_4+n_5+n_6+n_7+n_8 \quad (10)$$

$$\boldsymbol{H^0_{f,waste}} + \boldsymbol{wH^0_{f,H20(l)}} + \boldsymbol{mH^0_{f,O2}} + \boldsymbol{3,76mH^0_{f,N2}} + Electricity = n_1\boldsymbol{H^0_{f,CO}} + n_2\boldsymbol{H^0_{f,H2}} + n_3\boldsymbol{H^0_{f,CH4}} + n_4\boldsymbol{H^0_{fH20(g)}} + n_5\boldsymbol{H^0_{fCO2}} + n_6\boldsymbol{H^0_{f,N2}} + n_7\boldsymbol{H^0_{f,S}} + n_8\boldsymbol{H^0_{f,H2S}} + Parameter \quad (11)$$

$$Parameter = n_1c_{p,H2}(T)+n_2c_{p,CO}(T)+n_3c_{p,CH4}(T)+n_4c_{p,H20}(T)+n_5c_{p,CO2}(T)+n_6c_{p,N2}(T)+n_7c_{p,S}(T)+n_8c_{p,H2S}(T)dT \quad (12)$$

The formation enthalpies H_i are reported in bibliography.

4. Plasma Gasification Process Performance - Results

4.1 Energy and Exergy Analysis

In energy transformation processes such as combustion, gasi$cation and reforming of fossil and renewable fuels, the conservation of energy (first law of thermodynamics) as well as the qualityof energy (second law of thermodynamics) is important. This study focuses on the conversion of RDF into gaseous components and char (Prins et al., 2003).

Exergy analysis is based upon the second law of thermodynamics, which stipulates that all macroscopic processes are irreversible. Every such irreversible process entails a non-recoverable loss of exergy, expressed as the product of the ambient temperature and the entropy generated. The elementary irreversible phenomena that generate entropy are: mechanical or hydraulic friction, heat transfer with a finite temperature gradient, diffusion with a finite gradient of concentration, and the mixing of substances with different parameters and chemical composition. Combustion is also a typical irreversible phenomenon; however, it is a complex process comprising all the above mentioned elementary irreversibilities (Szargut, 2005).

Some of the components of entropy generation can be negative, but the sum is always positive.Mountouris et al. support applied the exergy analysis for the main section of the process. An exergy analysis is a fundamental tool for evaluating processes and systems. Exergy analysis is a combination of the First and Second Laws of Thermodynamics and accounts for irreversibilities (Mountouris et al., 2006).

The formula of chemical exergy is:

$$\varepsilon_{0,m} = \sum x_i e_{0,i} + RT_0 \sum x_i ln x_i \quad (13)$$

Table 2. Chemical exergy of different substances at atmospheric conditions

Substance	ε (kJ/kmol)
N_2	720
CO_2	19,870
H_2O	9,500
H_2	236,100
CO	275,100
CH_4	831,650

4.2 Efficiency Evaluation with Thermodynamic Approach

Efficiency can be estimated calculated in using the two thermodynamics laws which means that the gas efficiency is calculated by the following formula (Qiu & Hayden, 2004), (Mountouris et al., 2006):

$$E = HV_{Synthesis\ gas} / LHV_{Waste} + Electricity \quad (14)$$

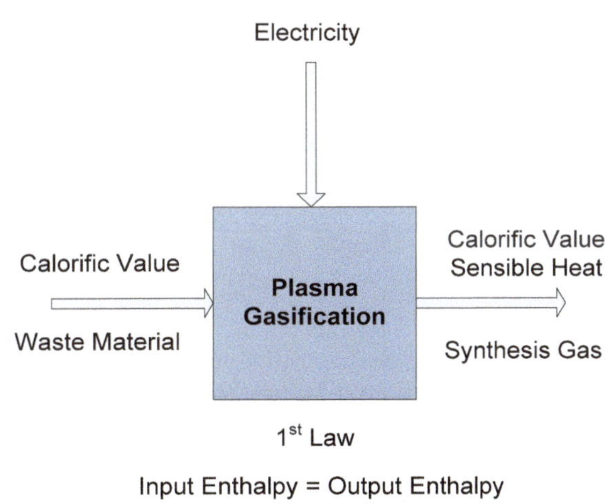

Figure 2. Flow diagram of plasma gasification according to 1nd law

On the other hand, second law is a generic term for a group of concepts that define the maximum work potential of a system, a stream of matter or a heat interaction; the state of the (conceptual) environment being used as the datum state (Reddy et al., 2010).

The second law efficiency is calculated with the same way using the formula (Mountouris et al., 2006).

$$E = Chemical\ exergy_{Synthesis\ gas} / Chemical\ exergy_{Waste} + Electricity \quad (15)$$

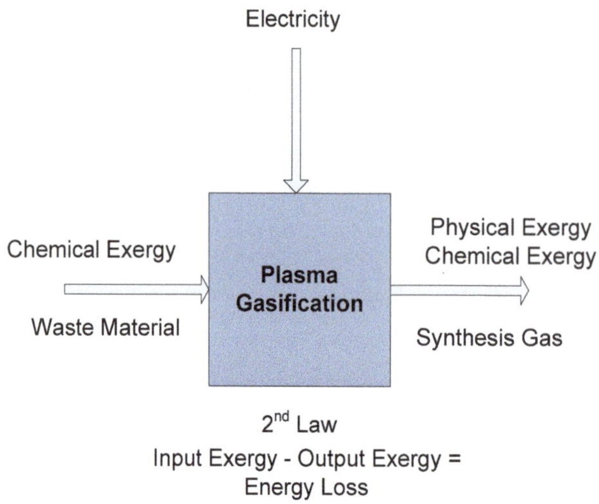

Figure 3. Flow diagram of plasma gasification according to 2nd law

The method which was described in the previous paragraphs gives the following results (Table 3) based on calculations in MATLAB.

Table 3. Results for the efficiency for RDF plasma-gasification

Efficiency according to the 1st law	Efficiency according to the 2nd law
$HV_{Synthesis\ Gas}$ = 140091,36	Chemical Exergy $_{Synthesis\ Gas}$ = 1598401
LHV_{Waste} = 17918,422	Chemical Exergy $_{Waste}$ = 5498778
Electricity = 394115	Electricity = 394115
ε = 0,34	**ε = 0,27**

5. Conclusion

The application of plasma gasification in RDF with fluidized bed technology for power generation is interesting, especially the case studied. The international experience from the implementation of plasma gasification with fluidized bed technology is very small which increases the investment risk. Also for the plasma gasification fluidized bed technology is beyond technical implementation and the limitation of only small and medium sized facilities. But the above should not discourage investments for further development of the existing plasma gasification technology, so it can be increased the expertise and be resolved remaining technical problems because plasma gasification technology can be clearly demonstrated one state-of-the-art, environmental and techno-economic method and in the near future will be fully prepared to give very beneficial solutions to large-scale thermal recycling of solid waste (Malkow, 2004).

The plasma gasification process is a future method for treating solid waste. The option of oxygen and the humidity rates is crucial issue and depends on the exploitation mode of synthesis gas. The process is an effective and environmentally friendly option for treatment and energy recovery of RDF. Based on that energy evaluation from our study, the RDF can be used for energy recovery using a process that includes the milestones of plasma gasification, drying and electricity production. The effective implementation of the proposed process shows that plasma gasification process is not only energy independent but can lead to a net production of electricity.

The first conclusion drawn from the analysis is that the price of electricity and the waste disposal fee are what could make the plasma gasification concept more financially viable. The electricity can be subsidized if the energy production by exploiting waste is considered as a renewable form of energy. The disposal waste fee is important to integrate the unit revenue plasma gasification. From the side of a state disposal fee can replace council tax paid currently in landfills. A good combination of grant, refuse disposal fee and sales price of electricity could make the plasma gasification very attractive, but should not be neglected the high cost of electricity production. This value is very high compared with both the more expensive renewable electricity (photovoltaic - wind) and conventional energy sources (oil-gas).

The impacts study of major economic parameters (price electricity production, waste disposal fee subsidy) can be showed how much the concept of a plasma gasification system can be advantageous with appropriate policies promoted by the state. The various toxic substances are destroyed by high temperatures and this plasma gasification is particularly suitable for the treatment of hazardous waste such as medical waste and some industrial. The treatment of hazardous waste can justify the choice of the high waste disposal fee per tonne which as mentioned above it could contribute positively to the profitability of the investment (Diaz & Leal-Quiros, 2009).

In the process, harmful pollutants can be removed from the syngas before they reach the gas turbine; thus, back-end exhaust gas clean up is not necessary. The SOx, NOx, mercury, metals, and particle emissions from the plant are fractions of those of a conventional pulverized coal boiler power plant. Consequently, IPGCC plants require significantly less effort and time to meet air emissions regulations and to obtain local and state governmental environmental permits. The process is approximately 5% more efficient than other coal power technologies, thus, CO_2 emissions per kW are also 5% lower.

Summarizing what we have been said above, we reach the undeniable fact that the gasification of RDF can produce numerous environmental advantages at a time when such energy technologies are a powerful investment for man and nature.

Acknowledgements

We thank the Association of Communities and Municipalities in Attica Region for providing data about the raw material RDF and for the gasification process technology.

References

Achinas, S., & Kapetanios, E. (2012). Basic design of an integrated plasma gasification combined cycle system for electricity generation from RDF. *International Journal of Engineering Research and Technology, 1*(10), 1-8.

Alter NRG & Westinghouse Plasma Corporation. (2008). The Plasma Gasification Process, Technical report.

Bonizzoni, G., & Vassallo, E. (2002). Plasma Physics and Technology: Industrial Applications. *Vacuum, 64*, 327-336. http://dx.doi.org/10.1016/S0042-207X(01)00341-4

Diaz, G., & Leal-Quiros, E. (2009). Sustainable waste management using plasma gasification at the campus of the University of California – Merced, 29th ICPIG, July 12-17, Cancún, México.

Global Environmental Solutions Inc (GESI). (ND). The utilization of integrated plasma gasification combined cycle system (IPGCC) (white paper).

Higman, C., & Van der Burgt, M. (2007). *Gasification*. Elsevier Science & Technology.

Imris, I., Klenovcanova, A., & Molcan, P. (2005). Energy recovery from waste by the plasma gasification process. *Archives of thermodynamics, 26*, 3-16.

Kapetanios, E. (2007). *Waste Treatment: Process Techniques and Technologies*. Athens: National Technical University of Athens.

Laurence, L. C., & Ashenafi, D. (2012). Syngas Treatment Unit for Small Scale Gasification -Application to IC Engine Gas Quality Requirement. *Journal of Applied Fluid Mechanics, 5*(1), 95-103.

Leal-Quiros, E., (2004). Plasma Processing of Municipal Solid Waste. *Brazilian Journal of Physics, 34*(4B). http://dx.doi.org/10.1590/S0103-97332004000800015

Littlewood, K. (1977). Gasification: Theory and application. *Progress Energy Combustion Science, 3*, 35-71. http://dx.doi.org/10.1016/0360-1285(77)90008-9

Malkow, Th. (2004). Novel and innovative pyrolysis and gasification technologies for energy efficient and environmentally sound MSW disposal. *Waste Management, 24*, 53-79. http://dx.doi.org/10.1016/S0956-053X(03)00038-2

Minutillo, M., Perna, A., & Di Bona, D. (2009). Modelling and performance analysis of an integrated plasma gasification combined cycle (IPGCC) power plant. *Energy Conversion and Management, 50*, 2837-2842. http://dx.doi.org/10.1016/j.enconman.2009.07.002

Mountouris, A. (2007). *Thermodynamic Simulation of Plasma Gasification for Waste Management*. Doctorate thesis, National Technical Univeristy of Athens.

Mountouris, A., Voutsas, E., & Tassios D. (2006). Solid waste plasma gasification: Equilibrium model development and exergy analysis. *Energy Conversion and Management, 47*, 1723-1737. http://dx.doi.org/10.1016/j.enconman.2005.10.015

Mountouris, A., Voutsas, E., & Tassios, D. (2008). Plasma gasification of sewage sludge: Process development and energy optimization. *Energy Conversion and Management, 49*, 2264-2271. http://dx.doi.org/10.1016/j.enconman.2008.01.025

Nikolaou, A. (2010). *Thermodynamic Simulation and Techno-economical study of Plasma Gasification for Solid Waste Treatment* (Diploma thesis). School of Chemical Engineering, National Technical University of Athens.

Pfender, E. (1999). Thermal plasma technology: Where do we stand and where are we going? *Plasma chemistry and plasma processing, 19*, 1-31. http://dx.doi.org/10.1023/A:1021899731587

Prins, M. J., Ptasinski, K. J., & Janssen, F. J. J. G. (2003). Thermodynamics of gas-char reactions: First and second law analysis. *Chemical Engineering Science, 58*, 1003-1011. http://dx.doi.org/10.1016/S0009-2509(02)00641-3

Qiu, K., & Hayden, S. (2004). High-Efficiency Integrated Solid Waste-to-Energy System, CANMET Energy Technology Centre, Ottawa, Canada.

Reddy, V. S., Kaushik, S. C., Tyag, S. K., & Panwar, N. L. (2010). An approach to analyse energy and exergy abalysis of thermal power plants: a review. *Smart Grid and Renewable Energy, 1*, 143-152. http://dx.doi.org/10.4236/sgre.2008.13019

Rezaiyan, J., & Cheremisinoff, N. P. (2005). *Gasification Technologies: a Primer for Engineers and Scientists*. Taylor & Francis Group (CRC press).

Szargut. (2005). Exergy Analysis, Research in Progress Thermodynamics. *Academia, 3*(7).

Yaws, C. L. (1999). *Chemical Properties Handbook* (p. 779). McGraw-Hill.

5

Strategy for Promoting Effective Management of Environmental Effects from Brewery Industries at The Nineth (9TH) Mile Corner Industrial District of Enugu State

Arthur Efeoghene Essaghah[1] & Alabi Michael Oloyede[2]

[1] Department of Urban and Regional Planning, Enugu State University of Science & Technology, Agbani, Nigeria

[2] Department of Geography and Planning, Kogi State University, Anyigba, Nigeria

Correspondence: Alabi Michael Oloyede, Department of Geography and Planning, Kogi State University, Anyigba, Nigeria. E-mail: Alabimo06@Yahoo.Com

Abstract

Related studies on the environmental effects of poor brewery effluent management at the 9th Mile Corner industrial area have indicated that the Nigeria Bottling Company (NBC Plc) is not effectively monitoring liquid wastes from its production operations. There is need to investigate and explore these issues and take effective remedial action before the negative impact of its activities become widespread and elicit adverse reactions from the community. It is the intention of this paper to explore related issues and articulate a framework for effective monitoring of industrial effluents and management of Brewery wastes at the 9th Mile Corner using the Nigeria Bottling Company plant as case study. The research design combined desktop and social survey. Past studies on the environmental effects of the Nigerian Breweries activities were reviewed and analyzed. The company premises and surroundings were surveyed for evidence of industrial effluents flowing from the plant into residential areas. It was found that industrial effluents treatment facilities of the company are old and dilapidating causing effluents to escape outside the plant waste management system. Against this background the paper outlines the type of monitoring and auditing, as well as, relevant organization/agencies that should participate in the implementation of the proposed revised environmental management plan(s) for the factory. Adoption of this strategy for the management of environmental effects of the company effluent would undoubtedly enthrone sound best practices in the management of brewery plants operations in 9th Mile Corner and other industrial districts in Nigeria.

Keywords: post impact assessment, effluent management, environmental management

1. Introduction

Undesirable industrial effluent characteristics from breweries and distilleries like soluble organics, suspended solids, heavy metals, as well as, nitrogen, phosphorous, colour and turbidity may be completely removed through primary and secondary treatment processes (Azard, 1976; Eckenfelder, 2000; Wong & Pfarrer, 1995; Patterson, 1985; Lankford & Eeckenfelder, 1990) before discharge into acceptable media. However, if such treatment processes are poorly handled and or not performed by operator(s) before discharge into open surface water bodies, it contaminates or pollutes the receiving media. When such happens, periodic sampling, monitoring and subsequent treatment of the receiving media is necessary to guarantee that applicable critical physico-chemical parameters do not rise above acceptable thresholds (Bausmith & Neufeld, 1996; Ford, 1998).

A number of studies have shown that Brewery industries at the 9th Mile Corner have not been successful in safely disposing off their industrial effluents and wastes without environmental and social hazards (Planscape Associates, 2005; Onwe, 2004). These studies have all reported that industrial effluents and or wastes from brewery plants at the 9th Mile Corner flow freely from company's premises into residential compound(s) and street(s) in the town in flagrant disregard for standard procedures for disposal of industrial emissions, effluents and wastes as specified by the World Bank Environmental Guidelines (1988), Federal Ministry of Environment (Former Federal Environmental Protection Agency (FEPA S.1.8, S. 1.9 and S. 1.15) regulations on effluent limitations, pollution abatement in industries and management of wastes (FEPA) 1991a & b). These clearly indicate failure of existing company's environmental management procedures and practices.

It is the intention of this paper to explore related issues and articulate a framework for effective monitoring of industrial effluents and management of Brewery wastes at the 9th Mile Corner using the Nigeria Bottling Company plant as case study. The rest of the paper is structured into four parts. Section 2.0 examines the study area and methodology while the direct and indirect effects of the poor disposal of the company's effluents are discussed in section 3.0. Section 4.0 determines and outlines the structure and essential features of the management strategy for enthroning effectiveness in the monitoring of environmental effects of the NB Plc plant at the 9th Mile Corner and the study is concluded in section 5.0.

2. Area of Study and Research Methodology

The Nineth Mile Corner is one of the fastest growing settlements in Enugu State. The 2006 population census put its population at 25,000 people. From its beginning as a transit camp for travelers between the eastern and the northern regions of Nigeria in the early 1930, it has developed into a sprawling industrial settlement locating important industries like the Nigerian Brewery Plc Plant, AMA Brewery Plant, Seven-Up Bottling Company Production Plant, Nigeria Bottling (NB) Company Plant and other associated concerns. Increasingly a sizeable number of medium and high income earners have relocated from Enugu urban to the area. It is today the industrial hob of Enugu State with bustling environmental and social activities and a revenue spinner for Udi Local Government Council, Enugu State and Federal Governments in Nigeria. Unfortunately, its environmental resources like Ajali River (the Ajalli River is the only source of water for domestic and agricultural purposes), vegetation and soils are being degraded by industrial effluents from the various Brewery plants in the area. There is need to protect these vital ecological resources from degradation by industrial effluents and wastes in order to ensure sustainable socio-economic development and ecological integrity of biological environments in the area.

The research design combined desktop and social survey. Past studies on the environmental effects of the NB Plc plant including a post-impact assessment study of the company's effluents management by Planscape and Associates (2005), Essaghah and Ugwu (2010) and AMA Brewery Plant Environmental Impact Assessment (EIA) report were examined to determine the company's environmental commitments and undertakings vis-à-vis environmental regulations and standards regarding management of effluents and industrial wastes. Thereafter the company premises and surroundings were surveyed for evidence of industrial effluents flowing from the plant into residential compound(s), streets and surface water bodies. Finally the company's staff and residents of the area were interviewed to determine why NB Plc plant's effluents are allowed to flow freely through dense residential neighbourhood(s).

3. Direct and Indirect Effects of Poor Disposal of Industrial Effluents from the Nigerian Bottling Company Plant at the 9th Mile Corner

Laboratory analysis of physico-chemical and micro-biological parameters of collected water and soil samples by Planscape and Associates (2005) revealed that values of critical portable water parameters were found to be above maximum thresholds. Critical parameters like Conductivity, Total Solids (TSS), Sulphates (SO_4^{2-}) Manganese (Mn), Coliform were found not to be within World Health Organization (WHO) permissible limits. It was revealed that numerical values of metallic cations in the effluent samples were high for calcium, sodium, magnesium, potassium, copper and zinc ranging between 200–250 mg/l and 897.64 mg/L – 906.60 mg/L for substances like sulphates. Total microbial plant counts were found to range between 29mg/L – 49.0 mg/L. A similar study by Essaghah and Ugwu (2008) indicated that no significant differences between values of physico-chemical and micro-biological parameters of the 2005 study by Planscape & Associates. In particular conductivity and sulphates values were found to significantly higher ranging between 1.2–2.3 x 10^3 μ/m^3 and 901.2–1107 mg/L respectively. Also values of manganese, nitrates and total solids were found to be extremely high ranging from 150–380 mg/L. Effluent samples were found to be highly pungent with intense colouration.

Considering the deleterious and destructive effects of sulphate compounds on ordinary concrete and sand-crete blocks as used for the construction of residential and industrial buildings in the area, there are indications that the strength of some building foundations may have been affected. Similarly considering the effect of heavy metallic cations on the leaching of soil nutrients and related soil fertility, declining yield of farmed crops in the area was implicated. For a water body that was found to be highly conductive (that is, with high conductivity), contamination of Ajalli River by Brewery effluents and other industrial activities in the area cannot be underestimated.

It was not surprising therefore that a number of direct and indirect effects or impacts were identified (Table 1).

Table 1. Some direct and indirect effects of the NBC plc effluents

S/N	Direct Effects	Indirect Effects
i.	Contamination of the air in the building and outdoor environment(s)	Weakening of building foundation(s) and cement screed floor(s) of nearby buildings.
ii.	Progressive development of earth gullies across compound(s)	Weakening of structural stability of buildings
iii.	Contamination of soil and surface water bodies (the Ajalli River is implicated)	Increased building maintenance cost and incurred medical bills directly related to effects of the effluent; and
iv.		Enhanced financial, emotional and psychological problems

Source: Planscape & Associates (2005), Essaghah & Ugwu (2008).

Table 1 shows that the direct effects of the NB Plc effluents are not limited and localized but widespread affecting air, surface and underground water bodies negatively while the indirect effects represent social, economic, emotional and health costs to individuals and the entire section of the residents living near the plant. A participatory rapid appraisal (RAP) study by Essaghah and Ugwu in 2008 also showed that residents living near the production plant complained of the effluents from the brewery destroying their building foundations, polluting their sources of domestic water and accelerated corrosion of building walls, roofing sheets, utility pipelines and structures. It was also observed that most of these complaints were localized since residents living far away (more than 200 m) from the plant showed indifference. On the aggregate level 65% of residents interviewed corroborated these assertion(s), while majority of company staff surveyed refuted these allegations. It was argued by residents that the company industrial effluents treatment facilities are too old and dilapidating thereby allowing poorly treated waste water escape out of the plant production waste management system.

Management staff of the company maintained that all effluents of the factory are treated before disposal and as such, the said effects could not have resulted from its industrial waste operations. All these indicate that the management of brewery plants at the 9th Mile Corner appears not to be monitoring the effluents from its industrial operations to mitigate the adverse impacts of their production activities. There is a clear need to start doing so before the negative impacts from plant production activities become widespread and elicit adverse public reactions from host community. This issue is discussed in the next section.

4. Organisational Framework for Effective Monitoring and Management of Environmental Effects of Industrial Effluents

A starting point for effective monitoring of industrial effluents and effective management of the environment effects of the NBC Plc plant production operations is to re-examine, review, revise and update the existing Environmental Management Plans (EMP) drawn up by the Environmental Impact Assessment (EIA) study preparatory to the construction and operational use of the existing plant. All EIA studies recognize that projects' EMP contains all undertakings and commitments for ensuring compliance with regulatory requirements, means for continuous self assessment of the predictive accuracy of impacts, as well as, strategies and procedures for implementing mitigation measures prescribed for identified impacts (FEPA, 1993; World Bank, 1988).

The review, revision and updating of the brewery plants' EMP should be *data driven*. The Nigerian Brewery Plc is a big player in the Nigerian economy in general and Enugu State and Udi Local Government Council Area in particular. It should not shy away from its responsibilities to the host community, the nation at large and the international community by ensuring that its activities conform to best international practices in engineering production operations. The NB Plc should subject the 9th Mile Corner Brewery Plant to International Standards Organization ISO 14000 audit scheme based on the Environmental Management Systems EMS 14001 criteria to identify operational inadequacies, deficiencies and lapses in the management of the plant industrial effluents and wastes. Result of this audit should form the basis for drawing up a new EMP for the company's operations and activities.

Monitoring and auditing of EMP address the pertinent question of "how do we assure that the EMP is properly implemented as laid and as agreed"? Tables 2 and 3 summarize the list of agencies/organizations that should be involved in the implementation of the revised EMP and the types of monitoring required for its effective implementation respectively. An Environmental Management Plan (EMP) Team should be set up with the sole responsibility of implementing the new waste and environmental management plans. Membership of the EMP management should be drawn from staff of the company with 2 members each selected from EIA consultants and waste management experts to be appointed by Federal Ministry of Environment (FMNEV).

Table 2. Proposed list of agencies to be responsible for implementing of the new or revised EMP

S/N	Agencies/Organization	Status
1.	Federal Ministry of Environment, Abuja	Regulator
2.	Enugu State Waste Management Board	Regulator
3.	Enugu State Ministry of Environment	Regulator
4.	Enugu State Ministry of Lands, Housing and Urban Development	Regulator
5.	Udi Town Planning Authority	Regulator
6.	Nigerian Brewery Management Staff at the 9th Mile Corner Plant	Operator
7.	Community Based Organization(s), Non Governmental Organization (NGOs)	Observer
8.	An EMP Management Team	Consultant

Table 2 above reveals that multi-sectoral and multilateral approaches may be needed to make the implementation of the revised EMP more transparent, accountable and democratic. Involvement of regulatory agencies, and related department NB staff, company consultants and experts, an independent EMP management team, as well as, Non Governmental Organisations (NGOs), Community Based Organisations (CBOs) and development stakeholders/partners will ensure that every aspect and related issues are covered and in the process mainstream sound international best practice.

Table 3. Types of monitoring for the proposed EMP

S/N	Types of monitoring	Description
a.	Effects monitoring	Repetitive, systematic measurement of environmental parameters in effluents, wastes and emissions
b.	Compliance monitoring	Periodic sampling or continuous measurement of environmental parameters to ensure that regulatory requirements are observed and standard met.

Source: Essaghah & Ugwu (2008).

The only way to prevent the company's effluents being disposed off into streets and residential compounds is to ensure that both compliance and effects monitoring are regularly conducted on all operations, tasks activities of the factory (Table 3). Periodic sampling and continuous measurement of environmental parameters will achieve compliance with regulatory requirements as well as, ensure that immediate corrective measures are devised when necessary. Similarly effects monitoring will provide feedbacks on impact predictions, accuracy and effectiveness of mitigation measures including environmental benefits targets contained in the new or revised EMP. In order for these monitoring activities to be effective, the following roles and responsibilities should be assigned to the various stakeholders (Table 4).

Table 4. Proposed roles and responsibilities for implementing revised EMP

S/N	Entity	Role and Responsibilities
i	NB Plc Factory EIA Audit team, EMS 14001 Audit consultants, Environmental Management Team (EMP Team), EIA Consultants, Waste Management Experts and Professionals	Co-ordinate preparation of revised EMP components and endorsement
ii	Federal Ministry of Environment (FMNEV), Enugu State Ministries of Environment and Lands, Housing and Urban Development	Approval of new EMP
iii	Proposed EMP Management Team	Implementation of the new EMP
iv	Federal Ministry of Environment (FMNEV) Enugu State Ministries of Environment and Lands, Housing and Urban Development including Enugu State Waste Management Board and Udi Town Planning Authority.	Internal monitoring and auditing of EMP operations and activities.
v	Environmental Management Experts, NGOs, CBOs, (Community Based Organisation), Civil Society, Interested Parties and Groups	External monitoring and audit of EMP operational activities

Table 4 above reveals that the specific role and responsibility that each stakeholder concerned with sustainable management of environmental effects of industrial activities and operations should perform. The proposed NB Plc factory EIA Audit team, the new EMP Management Team and EIA consultants that prepared the existing factory EMP including waste management experts and professionals should jointly collaborate to prepare the new EMP. Regulatory and relevant government agencies should approve the revised EMP after due consultations, panel reviews and considerations. Once new EMP is approved, constituted EMP management team and contractors should implement the environmental management plan. While implementation of the EMP is on-going, the various regulatory and or supervisory agencies of government should monitor and audit relevant tasks and assignments of EMP management team to ensure that results are delivered as scheduled. The role of Non-Governmental Organizations (NGOs), Community Based Organisations (CBOs) and Civil Society cannot be over emphasized in this regard. They should take responsibility for external monitoring and audit of EMP implementation activities and operations. Participation of these stakeholders in project implementation procedures and management of environment effects will promote checks and balances, transparency and accountability, and mainstream best practices in the various waste management operations and activities between the actors involved. Although participation of these stakeholders in the process will undoubtedly cause extra financial burden (which should be borne by the company), no account of money is too much to pay for prudent management of natural ecological capital and resources, as well as, promotion of human welfare in a developing economy like Nigeria.

5. Conclusions

This paper has examined some environmental, socio-economic and health related issues and problems that has resulted from the poor handling of industrial effluents from Nigeria Bottling Company production plant at the 9th Mile Corner. It has also highlighted relevant direct and indirect effects resulting from the poor management of industrial effluent from the plant with a view to mainstreaming a more democratic, multi-sectoral and multi-lateral approach to the management of factory production effluents and ensuing environmental effects at the 9th Mile Corner that is fast assuming an industrial district status in Enugu State. To this end, the paper proposed a management strategy for promoting sound and effective monitoring and auditing of the factory's production operations and activities.

Finally, the paper outlines the type of monitoring and auditing, as well as, relevant organization/agencies that should participate in the implementation of the proposed revised environmental management plan(s) for the factory. Adoption of this strategy for the management of environmental effects of the company effluent would undoubtedly enthrone sound best practices in the management of brewery plants operations in 9th Mile Corner and other industrial districts in Nigeria.

References

Azard, H. S. (1976). *Industrial Wastewater Managements Handbook.* New York: McGraw-Hill.

Bausmith, D. S., & Neufeild R. D. (1996). *Proceedings WEF, Vol. 3.* Donaldson E. C. Dallas.

Eckenfelder, W. W. (2000). *Industrial Water Pollution Control.* Boston: McGraw-Hill Higher Education.

Enville Environmental Consultants Limited. (2000). *Environmental Impact Assessment of the AMA Brewery Project at Amaeke Ngwo, 9th Mile Corner.* Draft Report.

Essaghah, A. A. E., & Ugwu, L. N. (Mrs) (2008a). Management Strategy for Monitoring And Auditing of Environmental Effects of Brewery Industries in Nigeria: The Case for the Nigerian Bottling Company Plant at the 9th Mile Corner, in Enugu State, Nigeria. *Journal of Ecology, Politics and Environment, 1*(1).

Federal Environmental Protection Agency (FEPA). (1958). *Health Laws Further Amended in 1963.* Federal Government Press Lagos, Nigeria.

Federal Environmental Protection Agency (FEPA). (1988). *Harmful Wastes (Special Criminal Provisions) Act. Cap 165 LFN 1990.* Federal Government Press Lagos.

Federal Environmental Protection Agency (FEPA). (1991a). *National Interim Guidelines and Standard for Environmental Pollution Control in Nigeria.* Federal Government Press Lagos.

Federal Environmental Protection Agency (FEPA). (1991b). *Federal Environmental Protection Agency (Amended) Act.* No. 59 of 1993 FEBA, Abuja.

Federal Environmental Protection Agency (FEPA). (1991b). *National Environmental Protection Regulations.* The Federal Government Press, Lagos Nigeria.

Federal Environmental Protection Agency (FEPA). (1994). *Draft Procedural Guidelines for Environmental Impact Assessment in Nigeria.* FEPA, Abuja.

Federal Environmental Protection Agency (FEPA). (1995a). *Draft Procedural Guidelines for Environmental Impact Assessment in Nigeria.* FEPA, Abuja.

Federal Republic of Nigeria. (1988b). *Federal Environmental Protection Agency Act. 58 of 1988.* Federal Government Press Lagos.

Federal Republic of Nigeria. (1992). *Environmental Impact Assessment Act. No. 86 of 1993.* Official Gazette, No. 72, Vol. 79, Lagos.

Ford, D. L. (1998). *Toxicity Reduction-Evaluation and Control* (2nd Ed.). Lancaster: Technomic Publishing Company.

International Standard Organisation (ISO). (14000). International Standard Organisation Management Audit System for Organization, Geneva.

Lankford, P. W., & Eckenfelder W. W. (1990). *Toxicity Reduction in Industrial Effluents.* New York: Van Nostrand Reinhold.

Onwe, F. O. (2004). The Environmental Effects of Brewery Industries at the 9th Mile Corner of Enugu State (Unpublished Dissertation thesis). Department of Geography, Enugu State University of Science and Technology (ESUT).

Patterson, J. W. (1985). *Industrial Wastewater Treatment Technology* (2nd Ed.). Boston: Butterworth.

Planscape & Associates. (2005). *Environmental Impacts Assessment of Effluents on Mr. Ugwu's compound and environs at Ngwo 9th Mile Corner, Udi L.G.A. of Enugu State.*

Wong. G., & Pfarrer, R. (1995). *Proceedings 50th Purdue Industrial Waste Conference.* Ann Arbor Press, Ann Arbor Michigan.

World Bank. (1988). *Environmental Guidelines.* Environment Department. Washington D. C.: The World Bank.

6

Economic Value Approach to Intermittent Power Generation in the Nordic Power Markets

Maria Kopsakangas-Savolainen[1,2,3] & Rauli Svento[3]

[1] Finnish Environment Institute, Helsinki, Finland

[2] Thule Institute, University of Oulu, Oulu, Finland

[3] Department of Economics and Martti Ahtisaari, Institute of Global Business and Economics, University of Oulu, Oulu, Finland

Correspondence: Maria Kopsakangas-Savolainen, Finnish Environment Institute, Helsinki, Finland. E-mail: maria.kopsakangas@oulu.fi

Abstract

The key elements in solving the increasing problems of power generation are to develop a sustainable and efficient production system and to utilize consumer management and demand response in a clearly stronger way than have been recently done. A significant part of the renewable energy (wind and solar) differs from most conventional (fossil fuel and nuclear) generating technologies such that they produce electricity intermittently and are not dispatchable. If traditional levelized cost approach is used it gives inappropriate and misleading results. In this paper we utilize a long-run oriented Real-Time Price based model to analyze the economic value of intermittent electricity generation in the Nordic power markets. Using this type of model it is possible to simultaneously take into account the increasing role of demand response and the specific characteristics of intermittent technology. We study the impacts of a change in production profile and increase in the share of intermittent generation to the market prices, equilibrium of other capacities, production costs, profits and CO_2 emissions. Our approach which uses hourly output profiles and associated market value of electricity, gives plausible economic values for electricity.

Keywords: intermittent generation, efficient production, economic value, emission reductions

1. Introduction

Energy and electricity have become key elements and drivers in the modern world. In our networked economies with digital products and processes and device driven consumption, stabile, riskless and justly priced electricity supply is the current day basement upon which everything else more or less builds on. At the same time the use of many primary energy sources can be connected with greenhouse gas emissions and global warming. This basement has also gone through heavy structural changes and even bigger changes can be seen in the future. The key elements in solving the increasing problems of power generation are to develop a sustainable and efficient production system and to utilize consumer management and demand response in a clearly stronger way than have been recently done (Kopsakangas-Savolainen & Svento, 2012a, 2012b; Joskow, 2012).

In developing a sustainable power production structure the role of renewable energy is increasing. The European Union (EU) has adopted specific targets for both CO_2 reductions and electricity production from renewable sources, by the year 2020. The share of the renewable energy is going to be increased up to 20% of the total energy consumed (EC, 2007). National targets differ clearly and member states can decide what kind operations and support mechanism they are using in order to reach these national targets.

A significant part of the renewable energy (wind and solar) differs from most conventional (fossil fuel and nuclear) generating technologies such that they are not dispatchable. Typically we have been used to situation where the system operator has scheduled the production to meet demand by dispatching the generators with the lowest marginal generation cost first and then moving up the dispatch curve. Generally speaking this means that generators are dispatched when the wholesale market price for power exceeds short-run marginal cost of generation (Joskow, 2011). The non-dispatchable, so called "intermittent technologies" (wind and solar), produce electricity intermittently and are driven more by weather conditions (such as wind speed and direction, cloud cover etc.) than by economic factors.

In this paper we utilize a long-run oriented Real-Time Price (RTP) based model (Borentstein, 2005; Borenstein & Holland, 2005; Kopsakangas-Savolainen & Svento, 2012b, 2013) to analyze the economic value of intermittent electricity generation in the Nordic power market framework. Even though the model is fit to the Nordic power market framework the results can be generalized also to other markets with similar structures (Note 1). By using simulation techniques we study the impacts of the increase in the share of intermittent generation to the market prices, equilibrium of other capacities, production costs, profits and CO_2 emissions. We show that as traditional levelized cost comparisons fail to take into account the fact that the value of electricity supplied varies widely over the year, our approach which uses hourly output profiles and associated market value of electricity, gives plausible economic values for electricity. We also point out that it is especially important to recognize that intermittent generating technologies can have very different hourly production profiles (and thus economic value profiles) depending on the location and wind/solar characteristics of the area analyzed. Consequently economic value of intermittent power can be very different in different countries or areas and this should be recognized when comparing the investment costs (and other consequences) of different technologies.

The paper is organized as follows. In section 2 we give background information on earlier literature related to intermittent generation. Section 3 presents the main characteristics of the Nordic power markets and gives summarization of the data. In section 4 the simulation model and iteration procedure are described. In section 5 we present and discuss the simulation results. Section 6 gives conclusions.

2. Earlier Literature

Literature on the role of intermittent technologies in the energy system has grown fast in recent years. Vast majority of this literature is still technology oriented and relates to questions of stability of the system. Traditional technologies are dispatchable in the sense that they can be integrated into the system whenever demand it necessitates. This is not the case with intermittent technologies which can be integrated into the dispatch curve when the wind blows or sun shines. The basic technical issues related to costs and reliability standards with large volumes of intermittent energy capacity have been summarized in many reports by e.g. The North-American Electric Reliability Corporation (NERC), US Department of Energy (USDOE), New York ISO (NYISO) and Mount et al. (2011). NERC (2007) uses two concepts to evaluate the reliability of the electric supply system:

(1). Adequacy. The ability of the electric system to supply the aggregate electrical demand and energy requirements of customers at all times, taking into account scheduled and reasonably expected unscheduled outages of system elements.

(2). Operating reliability. The ability of electric system to withstand sudden disturbances such as electric short circuits or unanticipated failure of system elements.

Adequacy is in most cases identified with simple measures like Loss of Load Expectation (LOLE). Using this measure adequacy can be said to be guaranteed if the past investments in the capacity and networks of the system have been such that LOLE is to to be met only in one day in ten years. Optimizing techniques where the objective function minimizes the total expected costs for a base case and a specified set of credible contingencies have also been developed in order to endogenize the optimal dispatch and nodal prices (Chen et al., 2005).

Game theoretic methods have also been widely used for analyzing the role of intermittent technologies in the energy system (Saad et al., 2012; Saad et al., 2011; Li et al., 2011). In these papers co-operational game theory is used in order to model intermittent and distributed energy sources as core parts of the electricity system. The approach is still technical in the sense that the objective of the game is defined somehow related to operating reliability or technical efficiency of the system. Either transfer losses are minimized or system supply reliability maximized.

Chalkiadakis et al. (2011) propose that small scale producers coalesce together to form a co-operative virtual power plant (CVPP) that can profitably be integrated into the grid. They incorporate ideas from mechanism design and co-operative game theory and put forward an energy pricing mechanism to be employed by the grid. The mechanism can be seen as an efficient alternative to feed-in tariffs, and so promotes the incorporation of distributed small scale producers in the grid. It has also other desirable properties. It guarantees that CVPPs have the incentive to truthfully report to the grid accurate estimates of their electricity production, and it promotes CVPP efficiency and reliability encouraging larger rather than smaller CVPPs. Further, individual producers are rewarded for increased production, while the grid maintains the ability to decide the flexibility of the mechanism and its degree of independence from market fluctuations. Chalkiadakis et al. (2011) also propose a payment scheme to allocate payments within the CVPP. This scheme guarantees payments to its members such that no subset of them has a financial incentive to break away from the CVPP.

Another line of research parts from the technical approaches by emphasizing economic questions. It has for instance been argued (Mount et al., 2011) that the adequacy measure is too narrow when large amounts of intermittent technologies are integrated into the system. Mount et al. argue that a new criterion which they call "Financial Adequacy" should be treated as a standard measure to evaluate the desired changes to system capacity. Financial Adequacy relates to missing money coming from non-profitableness of those technologies that cannot be used with high peak hours. These are the intermittent technologies. This approach opens the question of the role of the system structure in defining the economic value of intermittent capacities and many economists have already grabbed this challenge.

Joskow (2011) shows how structural effects are a key driver when calculating the economic value of intermittent technologies. The traditional way to compare investment costs among different technologies is based on determining the "levelized cost" per MWh supplied. This means that we measure the real total life-cycle costs per MWh by supplier using a specific technology. Joskow points out that this method is inappropriate and misleading when we compare intermittent generating technologies with dispatchable generating technologies. Comparisons which use levelized costs are misleading since they fail to take into account the differences in production profiles among technologies. Differences in production profiles cause large variations in market value of the produced electricity. Differences in production profiles and the associated variations in the market value of electricity at the times it is supplied must be integrated with the expected levelized life-cycle costs to provide meaningful economic comparisons between them.

Lamont (2008) develops a theoretical framework to find the expression for the marginal value of intermittent technology as a function of the average system marginal cost, the capacity factor of the generator, and the covariance between the generator's hourly production and the hourly system marginal cost. Using this developed expression it is shown how large-scale intermittent generation tends to reduce the optimal capacity and production of baseload generators and increase the capacity and production of intermittent generators, although the result depends heavily on the pattern of production from the intermittent generators. These results are along the same lines than those that Joskow (2011) reaches with his extended value calculations.

Green and Vasilakos (2010) evaluate the impacts of intermittent wind generation on hourly equilibrium prices and output in the UK electricity market. They use data on expected wind generation capacities in different on-shore and off-shore locations for 2020. Their basic finding is that the volatility of prices increases, and significant year-to-year variation in generator's profits shall emerge. High wind speeds lead to lower prices, but annual revenues for British wind generators are almost as great as for baseload generators.

3. Description of Nordic Power Market and the Data

3.1 Nordic Power Market

The Nordic power market was the first international power market and it consists of four Nordic countries: Sweden, Finland, Norway and Denmark (Note 2). These countries run a common power exchange, the Nord Pool, where market participants can buy and sell electricity. The common power market was mainly motivated by efficiency reasons. The mix of production technologies in the Nordic Power market is quite large and it has been argued that it would improve the efficiency of production if market participants could trade between countries. The main part of the production in Finland and Denmark is based on thermal generation, whereas Norway is a very hydro-intensive country. In Sweden the electricity production is mainly based on hydro and nuclear power.

In the Nordic power market the wholesale trade of electricity is organized through Nord Pool power exchange. It is owned by national transmission system operators. The power exchange operates such that market participants make their quantity-price bids one day in advance to the hourly market. The resulting market clearing price for each hour is called the system price. Even though Nord Pool is a voluntary market more than 70% of the total consumption of electricity was traded through it in 2011. The aggregate demand for electricity in Nordic countries has been quite stable from year to year and the increase has been mainly due to economic growth. Some yearly variations happen along with variation in temperature. Price elasticity of demand has typically been very low because the price that final customers face is typically fixed for some period of time and prices do not follow the pattern of wholesale prices in the short run.

The transmission grid of the Nordic power market is operated by the national transmission operators. If the transmission across borders is not congested there is only one system price. In our simulations below we assume that the Nordic power market always forms only one price area (see e.g. Nordpool webpages for the discussion of bottlenecks). One price area is also actual target in Nordic Power markets in where the idea is to produce electricity always where it is most cost-efficient. Consequently, whenever there is only one price area, the location of production does not impact on the outcome of the market (e.g. on the system price).

Total net electricity production in the Nordic market was 378,6 TWh in 2011 (Eurelectric, 2012). Of that total, 80, 3 TWh was based on nuclear power, 200,2 TWh on hydropower, 55,2 TWh by using conventional thermal power plants and 35,7 TWh was based on other renewables. The installed wind power capacities in 2011 were the following: Sweden 2899 MW, Norway 515 MW, Denmark 3949 MW and Finland 226 MW (Eurelectric, 2012). The amount produced by hydropower can vary somewhat from year to year depending on precipitation. There are five nuclear power plants currently operating in Nordic countries. Three of them (10 reactors) are located in Sweden whereas two of them (4 reactors) are located in Finland. The peak technology includes oil fired condensing power plants as well as gas turbines. In our simulations below we divide our technologies into five representative technology groups based on the main characteristics of the Nordic Power market.

In Nordic power markets is operating under European Commission's internal emission trading. The emissions trading scheme is meant to operate so that the emissions of the companies under the scheme keep the pre-defined total emissions quantity within the limits. Power producers can buy permits from the emission permit markets. This increases the costs of technologies under emission trade.

3.2 Data

For the simulations we need data for the load profile and cost information of the different production technologies. The production specific cost data is presented in Table 1. The variable costs of midmerit and peaker units are determined with emission trade. Assumed emission price is 13€/tCO_2 which is close to the real average emission price of years 2010–2011. Of course the actual mixture of production technologies is more diversified but for the purpose of this study somewhat less diversified production structure is sufficient.

Table 1. Capacity and generation costs

Generation type	Specific investment cost (€/kW)	Economic lifetime (a)	Annual Capital Costs €/MW	Variable costs €/MWh
Wind power	1300	25	92238	11
Hydro power	2000	75	102643	4
Nuclear power (baseload)	3750	40	218543	23
Midmerit power	1370	25	99333	51
Peaker power	700	25	49667	87

As noted above a major part of electricity production in the Nordic power market is based on hydro power. By nature the hydro power differs from most other technologies because of very low variable costs and its dependence on the precipitation of the year. It cannot be clearly identified as belonging to any of the groups: baseload, midmerit or peaker capacity. Part of it can be used as baseload capacity and part as balancing power. The cost of hydro power production of course depends on the type of the power plant and on other environmental factors. The annual capital cost of each technology is calculated by assuming 5% interest rate and by using a standard annuity formula. It is important to use annual capital cost since in the simulation model the demand is determined by the annual hourly load duration curve. Thus, we also need costs determined on a yearly basis. For hydro power we assume that investment costs are 2000 € per installed kW and that economic lifetime of that capacity is 75 years. Using an interest rate of 5% and the standard annuity formula we end up with the 102643 € capital cost per year. Variable costs include fuel costs, operation and maintenance costs, and the impact of emission permits' price on the variable costs. In the case of hydro power variable costs are basically equal to maintenance costs because fuel costs are equal to zero. It is notable that in the simulations we are assuming that the average utilization rate of the hydro power capacity is slightly below 50%.

In simulations nuclear power production is assumed to be baseload production. As a specific investment cost of nuclear power we have used 3750 €/kW. This is clearly higher figure than suggested by Tarjanne and Kivistö (2008) and Vainio (2011) and it is based on recent evaluations of the final investment costs of the new nuclear power plant Olkiluoto 3 (Finland) which is under construction. If we assume a 40-year economic lifetime we end up with the annual capital costs of 218,543 €/MW.

The representative midmerit technology in our simulations is assumed to be coal and peat power plants. We have based our costs on the results of Tarjanne and Kivistö (2008) and Vainio (2011). The costs of the representative midmerit technology are partly (about 35%) based on peat fired conventional thermal production and partly (about 65%) on coal fired production. Specific investment costs of the midmerit technology are assumed to be about 1370 €/kW which results in annual capital costs of 99333/MW over a 25-year economic lifetime.

The capital and variable costs of peaker capacity vary depending on the technology used. Part of the peaker demand can be satisfied by oil fired condensing power plants, part by older power plants kept as reserve capacity, and the rest by the use of gas turbines. We assume that a representative peaker capacity investment cost is 700 € per installed kW which results annual capital costs of 49667 €/MW.

4. Model and Simulation Algorithm

The simulation model is based on the model developed by Borenstein and Holland (2005) and Borenstein (2005, 2007). Our application differs from their specification in the following features. Firstly, we apply the simulation model in the real Nordic Power market context and consequently the relevant features of the market have to be properly modeled. Secondly, we have capacity constrained technologies in the data base. These technologies are hydro and nuclear power (Kopsakangas-Savolainen & Svento, 2012b, 2013). Thirdly, we include a volatile, non-dispatchable energy source (wind power) into our model.

Next we present the general logic behind the simulation model. Following Borenstein and Holland (2005) and Borenstein (2005) we assume a constant elasticity demand function. Since we do not estimate this demand function we have to calibrate it to correspond to the realized demand profiles. We have used the information on the load duration curve to do this. Our hourly load profile is based on hourly consumption data from NordPoolSpot 2011 and it is presented in Figure 1.

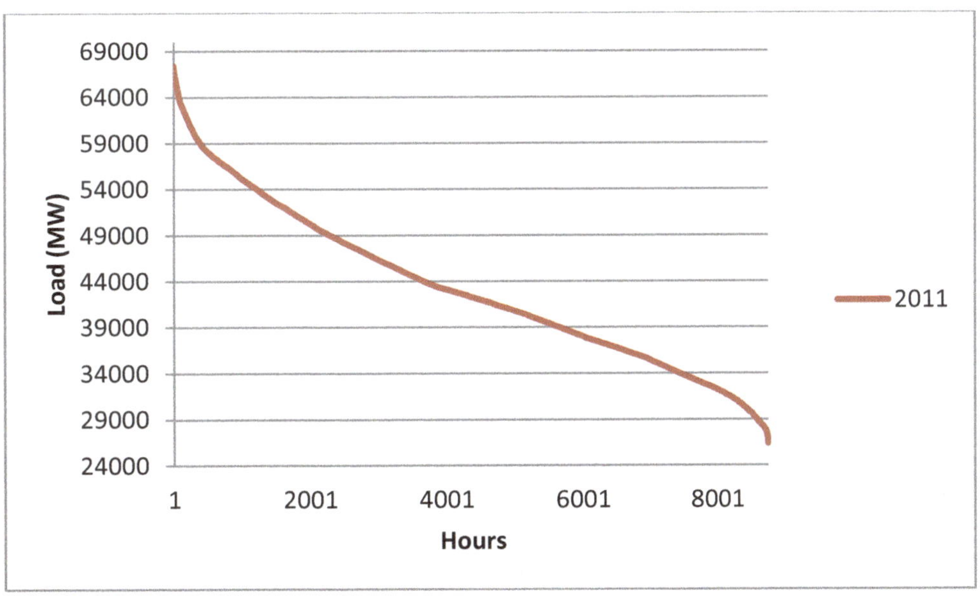

Figure 1. Load duration curve in Nordic Power markets 2011

Because intermittent generators cannot be dispatched, the chronological coincidence between generation and load is essential for analyzing the impact and value of intermittent generation. Intermittent generation can be included in a load duration curve analysis by treating intermittent generation as a "negative" load (Lamont, 2008). In practice this means that intermittent generation is subtracted hour by hour from load duration curve to obtain residual load for each hour. The resulting load duration curve is then constructed and analyzed to optimize the production and capacity of dispatchable generation. The procedure for this analysis is described next.

After identifying the residual load duration curve, the specification of the demand function has been done by calculating an "anchor point" A_h for each hour of the year. In order to specify the anchor point we need to assume some constant price. We have used a price that would allow producers just to break even if they were charging that constant price from all customers. See Borenstein (2005) for a more detailed discussion of the role of this constant price. The anchor point can be determined as follows: $A_h = D_h/p^{\varepsilon}$, where D is demand, h is hour, p is the constant price and ε is the price elasticity for homogenous customers. Notable is that it is more important that we have accurate information of the shape of the hourly demand distribution than very specific information of the constant price. These anchor points are used to scale the demands to the load duration curve. Now let α, $0<\alpha\leq 1$, be the share of customers on Real-Time Pricing, p_r the retail price of electricity for the RTP customers and p_f the flat rate price for non-RTP customers. The demand for electricity for hour h can then be written:

$$D_h(p_r, p_f) = \left[\alpha p_r^{\varepsilon} + (1-\alpha)p_f^{\varepsilon}\right]A_h, h = 1, \dots, 8760 \tag{1}$$

The simulation results related to the effect of RTP are based on varying α and ε exogenously. We assume a competitive market structure in generation and in retail. We also assume that both retailers and generators will maximize profits and their profit functions can be written as follows:

$$\pi_R = \sum_{h=1}^{8760}\left[(p_f - w_h)(1-\alpha)D_h(p_f) + (p_r - w_h)\alpha D_h(p_r)\right] \tag{2}$$

$$\pi_G = \sum_{h=1}^{8760}(w_h D_h - cD_h) - rK \tag{3}$$

where R is retail sector, G is generation sector, w is wholesale price, c is marginal generation costs and rK is annual capital costs.

From Equations (1), (2) and (3) we can solve for short-run and long-run equilibrium of the generation and consumption systems. The logic of the simulation system is that the whole generation system is constructed MW by MW under the assumed economic principles (i.e. generators maximize profits according to Equation 3). After the construction of the whole generation system we turn to the retail sector and adjust the flat rate as long as the profits (see Equation 2) for the retail sector are equal to zero. Then we turn again to the generator sector and rebuild the whole system MW by MW with this new flat rate. These two rounds are repeated until we reach the long run equilibrium of the whole system.

4.1 Equilibrium with one Technology

To describe the simulation model and the solution procedure in a simple way assume first that only one capacity type (peaker capacity) without any scale effects is constructed and used. The short run supply curve with an assumed short-run capacity K_{tot}^{SR} is thus an inverted L-shaped curve as depicted in Figure 2.

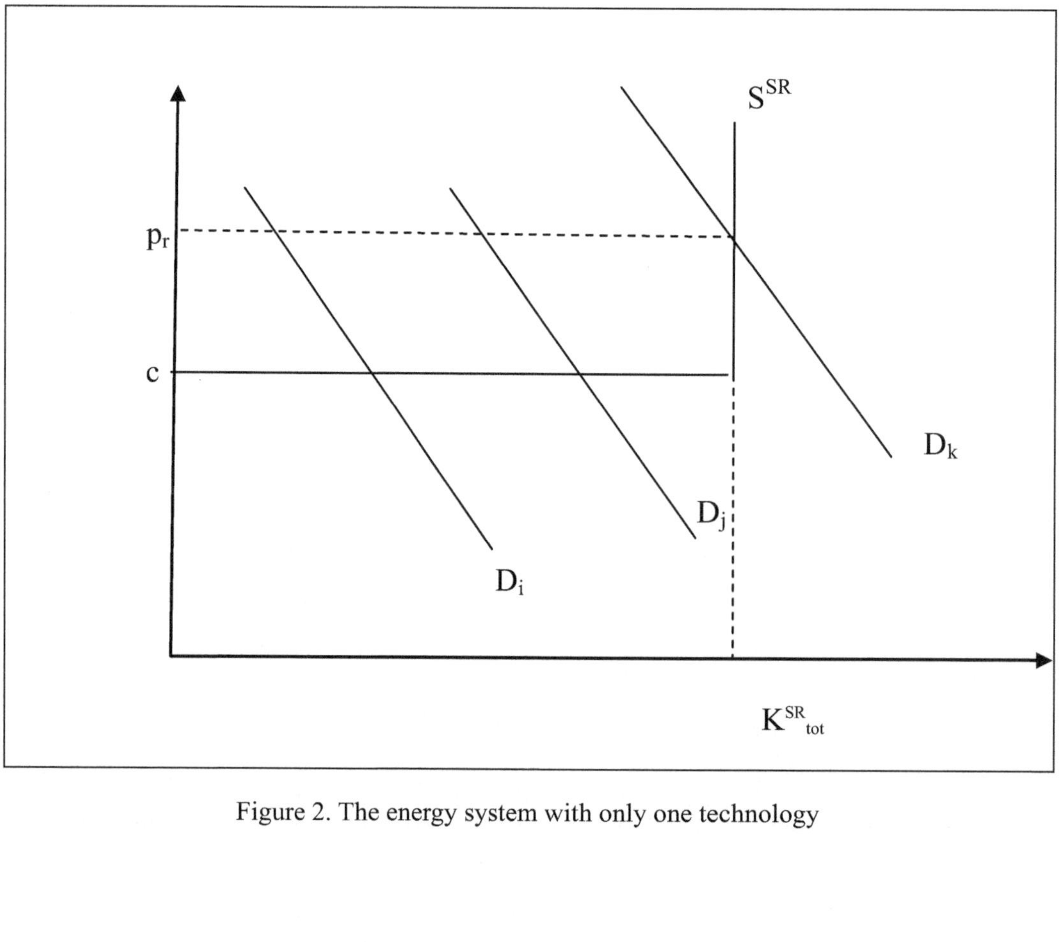

Figure 2. The energy system with only one technology

The short-run profits by hour are maximized by maximizing $\pi_G^{SR} = w_h D_h - cD_h$ with respect to given demand so that we have the known result: $w_h = c$. But this is true only when demand is not greater than the capacity limit, i.e. demands D_i and D_j in Figure 2. For those hours that demand exceeds the short-run capacity limit the real-time price must adjust to take care of market clearing. In equilibrium it has to be that demand equals to supply, i.e.:

$$D_k = \left[\alpha p_r^{\varepsilon} + (1-\alpha)p_f\right]A_h = K_{tot}^{SR} \tag{4}$$

From where we can solve for the price p_r:

$$p_r = \left[\frac{K_{tot}^{SR} - (1-\alpha)p_f A_h}{\alpha A_h}\right]^{-\varepsilon} \tag{5}$$

So the wholesale pricing logic is as follows:

$$w_h = c \text{ when } D_h \leq K_{tot}^{SR} \tag{6}$$

$$w_h = p_r \text{ when } D_h > K_{tot}^{SR} \tag{7}$$

Once we know the demand and wholesale price for each hour we can calculate profits for the generators. Because of competition we assume that capacity is built to the point where profits are equal to zero.

Next we solve for the flat rate. Competition forces the retail sector to zero profits also in the short-run. Abstracting from transmission and retail costs the real-time retail price must always be equal to the wholesale price. Should this not be the case there would be possibilities for undercutting the market price and this will go on as long as the retail price exceeds the wholesale price. So the zero profit condition for the retail sector reduces to:

$$(1-\alpha)\sum_{h=1}^{H}(p_f - w_h)D_h(p_f) = 0 \tag{8}$$

from which we can solve for pf :

$$p_f = \sum_h^H w_h D_h(p_f) \Big/ \sum_h^H D_h(p_f) \tag{9}$$

i.e. the zero profit short-run flat rate is a weighted average of the real-time wholesale price with weights being the relative quantities demanded by customers facing a flat retail price.

In the long run capacity is built to the point where both generators and retailers receive zero profits. As shown in Borenstein-Holland (2005) and Borenstein (2005) this kind of mechanism leads to a unique long-run equilibrium for the total generation capacity, K_{tot}^{LR}.

4.2 Equilibrium with a Technology Mix

We can use the same kind of procedure as explained above to solve for the technology mix. Now however the total technology built consists of different types of technologies with different capital and variable costs. Because of competition we assume that capacity for those technologies which are not capacity constrained is built to the point where profits are equal to zero. Those technologies which are capacity constrained are built to the limit of the capacity constraint. In simulations we first analyze scenarios where peaker and midmerit technologies are optimized and then enlarge the analysis by running simulations where also nuclear power is optimized. Hydro power is assumed to be capacity constrained in every scenario. Wind power is non-dispatchable and its hourly values are subtracted from load duration curve before optimization of the residual curve.

In practice we first solve for short run total capacity K_{tot}^{SR} by using peaker capacity only (just as described above). Then we start to replace peaker capacity with midmerit capacity. We construct the midmerit capacity to the point where the short run profits for midmerit capacity go to zero. If nuclear power is optimized, it is next constructed to replace midmerit capacities to the point where its short run profits are zero. Last we construct the capacity constrained technologies to the point where they reach the limit of their capacity.

After the short run technology mix for generation is solved we turn to the retail market and solve for the short run equilibrium for the retail market. Short run equilibrium for the retail market is reached when the retail sector produces zero profits (see Equation 8). From the zero profit condition we can solve for the new flat rate. With this new flat rate we start to construct again the whole generation technology mix. These two rounds are repeated until the equilibrium conditions for both sectors are simultaneously fulfilled. The long-run energy system equilibrium looks like the system in Figure 3.

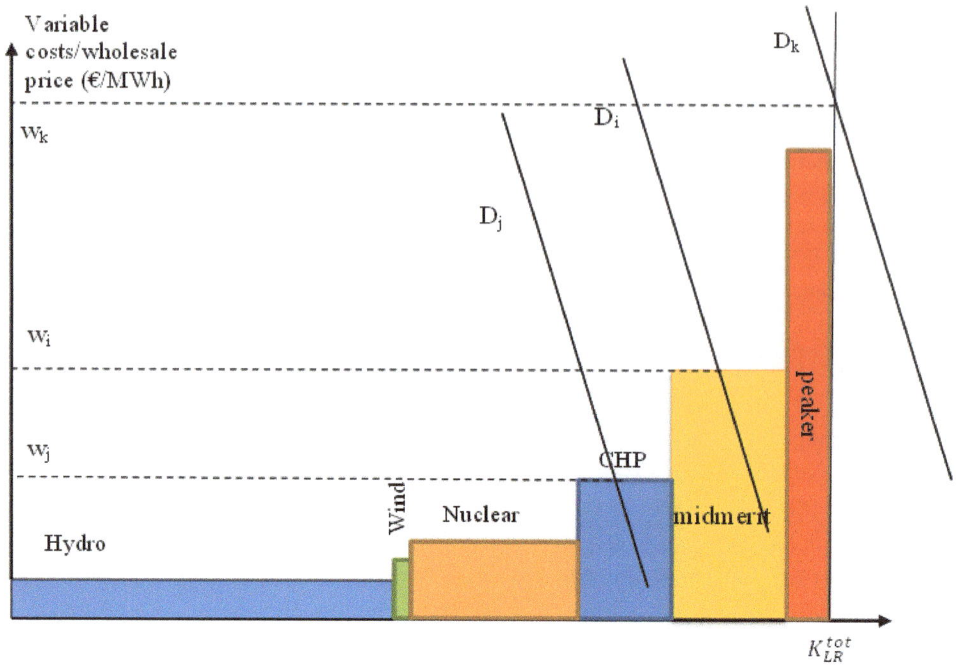

Figure 3. The long-run structure of the energy system

We have explained the basic model logic in reaching the long-run equilibrium with technology mix. In order to clarify the procedure reaching the long-run equilibrium with a complete set of technologies we next give summarization of the iteration procedure used.

Description of the algorithm:

a. Identify hourly based load duration curve. Identify hourly based production values and profile of intermittent generation. Subtract hour by hour intermittent generation from load duration curve to obtain residual load for each hour. Construct residual load duration curve. Specify demand function by using anchor point procedure described in the beginning of section 4.

b. Make an assumption for α and ε. We have assumed that the share of RTP (α) is 33% and value of price elasticity (ε) is -0,05 in every scenarios used in simulations (Note 3, Note 4). Begin with an initial guess for the peaker technology capacity (Note 5). Then expand the quantity of peaker capacity, combine this short run supply function with the hourly demand and calculate short-run profits of the peak power generator. If profits are positive expand the quantity of peaker capacity again, recalculate the profits and so on. Continue the expansion of the capacity until expansion by one more unit causes profits to go negative for peak power generators. The resulting amount of capacity will typically be the equilibrium short run of total amount of all capacities (K_{tot}^{SR}). Note that in this equilibrium where only peak technology is used all profits to the generator are earned when the production is equal to the total amount of peak capacity. In hours where equilibrium quantity is less than K_{tot}^{SR} price is equal to the marginal costs and profits resulting from these hours are equal to zero (see Figure 2).

c. Next, start substituting peaker capacity with midmerit capacity. The midmerit capacity will be lower on the supply function than the peaker capacity and thus it will be used in all hours before any of the peaker capacity. Follow the same procedure as with the peaker capacity, namely: expand midmerit capacity, combine short-run supply function with the demand and calculate profits of the midmerit units. If profits are positive continue expansion until expansion of the midmerit capacity by one more unit will cause profits of all mid-merit units to go negative. We refer to the total amount of midmerit capacity that still results in positive profits by K_{mid}.

d. Next begin substituting mid-merit capacity by the baseload capacity. Follow the same procedure as in stages a. and b. above. Note however, that in those scenarios where baseload capacity is constrained (Note 6) it should be taken into account. Expand the baseload capacity, combine the new short-run supply function to the demand and calculate profits. Again, if profits for nuclear power are positive continue expansion. Expand the capacity up to the point where expansion of capacity by one more unit will causes profits of all nuclear units to go negative or where the expanded capacity reaches the capacity constraint. Refer to the amount of baseload capacity by K_b.

e. Next take the capacity of hydro power generation into account at each hour of the year. Hydro power is first allocated evenly through hours such that total production during the year equals real production from the year where data is received. If then demand for an hour is lower than nuclear and hydropower production together, hydropower is reduced for those hours and reallocated to the later hours such that again yearly production equals real production (Note 7). Check that capacity constraint is not violated. Refer to the amount of hydro power capacity by K_{HP}.

f. Now one has determined the capacities of all types of technologies. One can determine short-run amounts of capacities using the following recursive logic:

i. $Hydro\ power = K_{HP} = K_{HP}$

ii. $Nuclear\ power = K_{NP} = K_b - K_{HP}$

iii. $Midmerit\ power = K_{MP} = K_{mid} - K_{NP} - K_{HP}$

iv. $Peaker\ power = K_{PP} = K_{tot}^{LR} - K_{MP} - K_{NP} - K_{HP}$

g. The final step of this procedure is to note that this equilibrium may not satisfy the retailer breakeven condition. So one must calculate the profits that the retailers earn from flat rate customers in this specific wholesale-producers equilibrium (note that profits for the retailer from RTP customers are always equal to zero). If the retailer profits from flat rate customers are negative, one adjusts the flat rate price up and if profits are positive one adjusts the flat rate price down. After this adjustment we need to re-simulate the capacity so we go back to point a. One has to continue this procedure until we reach equilibrium where retail markets also yield zero profits. This is the long-run unique competitive equilibrium energy system size and structure for a given set of available technologies, technology constraints, and share of customers on RTP and on flat rate.

5. Results

We have simulated the model under different assumptions about the production profile of intermittent generation, about the amount of intermittent generation and about the level of nuclear power. Scenarios A (see Table 2 below) are based on the assumption that nuclear power is constrained to the year 2011 level. In scenarios B we release this assumption and nuclear power is built up to the point where it receives zero profits. In both A and B type of scenarios we run simulations first by using 2011 level of intermittent production and then by assuming that the production is tripled from the 2011 level. The assumption of tripling the wind capacity in Nordic countries is very realistic if we consider the national targets by the year 2020 (Note 8).

We study the impact of the production profile of intermittent generation by assuming three different profiles in the simulations. First, we assume that the profile is similar as it was actually in year 2011 (this is referred as 2011 profile later on). Secondly, we assume that the profile is centered on the beginning of load duration curve, i.e. we organize the hourly wind volumes into a descending order and fit the resulting profile to the load duration curve (this is referred as "beginning" profile later on). Finally we assume that production is centered on the end of the load duration curve, i.e. we organize the hourly wind volumes into an ascending order and fit the resulting profile to the load duration curve (this is referred as "end" profile later on).

The capacity of hydropower is constrained at the level of 2011 in each scenario. Further, we assume that the total amount of electricity produced by using hydropower is the same in each scenario and it is assumed to be at the level of 2011 real production. However, the allocation of hydro power among different hours varies somewhat depending on the amount of wind power generation and nuclear power generation that can remove hydro production from the beginning of the load duration curve to the later hours. The procedure how this is done is reported more detailed in the description of the algorithm.

Table 2. Description of scenarios. Scenarios vary between the assumption made on intermittent generation and nuclear power

Scenario	Intermittent generation	Nuclear power
Baseline scenario A	no	constrained (2011 level)
Scenario A1	2011 production profile	constrained (2011 level)
Scenario A2	"beginning" production profile	constrained (2011 level)
Scenario A3	"end" production profile	constrained (2011 level)
Baseline Scenario B	no	unconstrained
Scenario B1	2011 production profile	unconstrained
Scenario B2	"beginning" production profile	unconstrained
Scenario B3	"end" production profile	unconstrained

5.1 Constrained Nuclear Power

In the Table 3 we present the impact of different scenarios to the total energy consumed, capacities of different technologies, prices and hours when the full amount of production capacity is used. The levels of wind capacity are indicated by the numbers 1 (2011 level) and 2 (triple the level of 2011) after the double dot in each scenario, i.e. A1:1 refers to the scenario where nuclear power is constrained, the wind profile is 2011 profile and the wind power level is at the level of 2011 capacity.

As can be seen from Table 3 and from the Figures 4a, 4b and 4c not only the amount of intermittent generation but also the allocation of this production among different hours has clear impacts on the need for dispatchable generation capacities. These impacts are strengthening as the level of wind capacity increases. If we use the 2011 load profile (scenarios A1) we see that as the wind power capacity increases the total dispatchable capacity decreases. Interesting is, however, that as the amount of midmerit capacity decreases the opposite happens to the peaker capacity. This is because of the location of wind production among different hours on the load duration curve.

Table 3. Production, capacities and prices with constrained nuclear power (Note 9)

Scenario	Total Annual Energy consumed (TWh)	Total dispatchable capacity, MW	Peaker power capacity, MW	Midmerit power capacity MW	Nuclear power capacity	Flat rate €/MWh	Peak price €/MWh	hours at peak quantity (at 8760)
Baseline Scenario A	374.08	62361	10080	20827	9446	65.86	3125.43	87
Scenario A1:1	374.58	60809	10507	18748	9446	65.15	4804.92	83
Scenario A2:1	374.92	62345	10595	19743	9446	64.31	3230.33	87
Scenario A3:1	374.52	57197	7817	17910	9446	64.93	2301.68	99
Scenario A1:2	375.99	58675	12066	14398	9446	62.31	5238.42	79
Scenario A2:2	376.23	62293	11623	16279	9446	62.93	3444.16	82
Scenario A3:2	375.49	47009	3412	12080	9446	62.511	1180.26	182

From the Table 3 and Figures 4a, 4b, and 4c we also see that if wind power is allocated according to “beginning” profile on the load duration curve (scenarios A2), the required amount of total dispatchable capacity changes only slightly. This is because only very little wind is allocated to the peak demand hours which is the main determinant of the total required capacity. We see, however, that also in this scenario wind power replaces midmerit capacity but not peaker capacity, which in fact increases somewhat.

When we look at the results when the wind is allocated according to “end” profile on the load duration curve (scenarios A3) we see that the amount of total dispatchable capacity decreases substantially. Because wind is now located more on peak demand hours it replaces a significant amount of peak capacity but has only little impact to the amount of midmerit capacity. The changes in the capacity mixture indicate that the wind profile might have significant impact on the system costs and also to the emissions. The flat rate decreases somewhat as the amount of wind capacity increases. It seems, however, that the flat rate is not very sensitive to the production profile of intermittent generation. Peak price is clearly lowest under the scenario where intermittent generation is located more on the end of the load duration curve. This is because now the hours at peak quantity increase and the costs of peaker technology are recovered by more hours than under other profile scenarios.

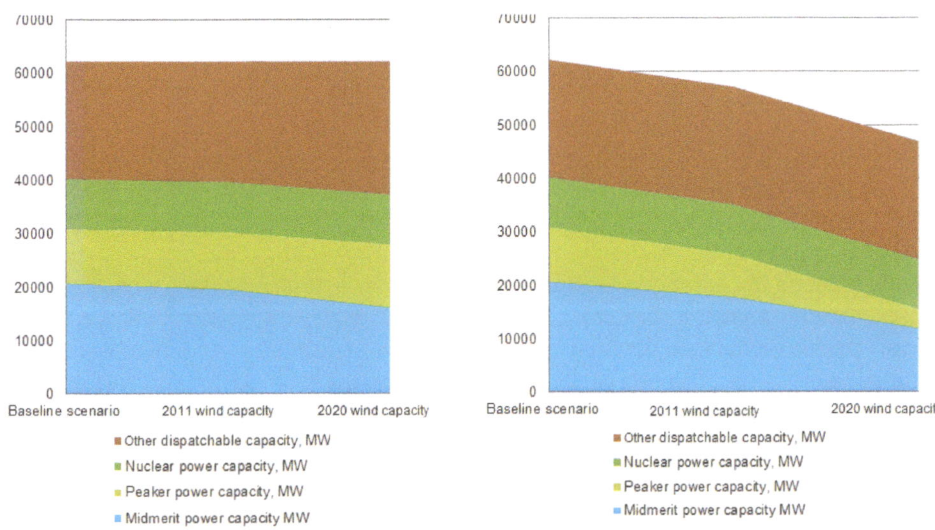

Figure 4. Capacities with constrained nuclear power

Next in Table 4 we present the impacts of different scenarios to production costs and profits. According to our assumptions only constrained (or intermittent) technologies can be able to make profits since the model assumes that midmerit and peaker technologies are optimized under zero profit condition.

Table 4. Production costs and profits under constrained nuclear power

Scenario	Total production costs million € (with wind in brackets)	Costs €/MWh (with wind in brackets)	Profits million €, hydro power	Profits, million €, nuclear power	Profits, million €, wind power	total profits, million €	Total profits €/Mwh
Baseline scenario A	17975.65	48.05	5937.80	539.48		6477.24	17.31
Scenario A1:1	17082.81 (17807.13)	47.45 (47.54)	5732.73	437.96	137.45	6308.14	16.84
Scenario A2:1	17204.73 (17929.05)	47.75 (47.82)	5489.59	276.26	-124.59	5641.26	15.05
Scenario A3:1	16810.05 (17534.37	46.71 (46.82)	5882.31	513.44	478.19	6873.94	18.35
Scenario A1:2	15357.32 (17530.28)	46.23 (46.62)	5057.53	71.72	52.2	5181.54	13.78
Scenario A2:2	15563.38 (17736.34)	46.82 (47.14)	5000.20	-178.51	-690.38	4131.31	10.98
Scenario A3:2	14491.77 (16664.73)	43.69 (44.38)	5689.59	424.50	1401.35	7515.49	20.16

From the results we see that generation profile of intermittent power has clear impact on the production costs as well as on the profits. Interesting is that total production costs as well as production costs per MWh decrease in every scenario compared to the baseline scenario. One should note also that as the share of wind power increases the system costs slightly decrease. System costs are lowest when the intermittent generation is allocated according to "end" profile on the load duration curve.

The profitability of the producers is dependent, among other things, on the production profile and amount of intermittent generation. The highest profits are received by hydro power producers. Three observations to note are that wind power producers are making positive profits in four scenarios (without any support mechanism), that nuclear power producers are making negative profits under one scenario and that profits of dispatchable technology decreases as the amount of wind capacity increases. Total profits are highest when the intermittent generation is located according to "end" profile on the load duration curve (scenario A3) and it is increasing with the level of wind power generation. Total profits per MWh are clearly highest under scenario A3:2 i.e. when the amount of wind capacity is tripled from the level of 2011 and production profile is allocated according to "end" profile on the load duration curve.

From Table 5 we see that total emissions (t/CO_2) are decreasing clearly as the share of intermittent generation increases. Interesting is that opposite to the results related to capacity, costs and profits, emissions are not very sensitive to the production profile of intermittent generation. They seem to, however, increase somewhat if the intermittent generation is located more on peak demand hours (scenarios A3). This is because now the intermittent generation replaces more heavily peaker power production than midmerit production which has higher emissions per produced MWh. Midmerit power is assumed to use coal and peat, whereas peaker power is using oil and gas as a fuel. We have used the following fuel based emission multipliers (tCO_2/MWh) in calculations (Note 10): coal (0.341), peat (0.382), gas (0.202) and oil (0.267).

Table 5. Emissions (constrained nuclear power)

Scenario	Emissions peaker power, million tCO_2	Emissions midmerit power, million tCO_2	Total emissions, million tCO_2	Total emissions t/MWh
Baseline scenario A	2.30	82.28	84.58	0.226
Scenario A1:1	2.43	69.65	72.08	0.192
Scenario A2:1	2.42	70.01	72.43	0.193
Scenario A3:1	1.84	70.85	72.69	0.194
Scenario A1:2	2.86	44.53	47.39	0.126
Scenario A2:2	2.67	45.92	48.60	0.129
Scenario A3:2	0.91	48.11	49.01	0.131

We have analysed the sensitivity of model results to changes in shares of real-time price customers (parameter α) and price elasticities of demand (parameter □). The model results related to total capacity and total costs are strengthened (i.e. total capacity and total costs both decrease more compared to the baseline) as the share of real-time price customers increases or demand becomes more prices elastic. The results related to emissions are not sensitive to changes in parameter values of α and □.

5.2 Unconstrained Nuclear Power

Next we present our simulation results under the situation where the capacity of nuclear power is not constrained. The model is now simulated in such a way that nuclear power is built under the same zero profit condition as peaker and midmerit power capacities. From Table 6 we can see the impact of different scenarios to production, capacities and price. We see that in the baseline scenario where we assume that there is no intermittent generation the optimal capacity level of nuclear power is 17746 MW. After the introduction of intermittent generation the optimal level of nuclear power is decreasing. Notable is that intermittent generation is clearly replacing also nuclear power in each of the scenarios. The level of nuclear power is smallest under scenario B2:2 i.e. when intermittent production is at the triple level of 2011 and it is allocated according to "beginning" profile to the load duration curve. It is also almost as low under the scenario B1:2 which uses the 2011 production profile. This is a significant result since it indicates that if (and as) the amount of intermittent generation is increasing it should be taken into account when planning new nuclear power investments. According to our results it is possible that further investments in nuclear power are not economically efficient if done simultaneously with large intermittent investments.

From the results we also see that when nuclear power is not constrained the production profile of intermittent generation becomes even more important. The difference on total dispatchable capacity among "beginning" and "end" profile is significant. One further implication of unconstrained nuclear power is that now the flat rate is clearly lower in every scenario compared to the scenarios where nuclear power is constrained.

Table 6. Production, capacities and prices with unconstrained nuclear power (Note 11)

Scenario	Total Annual Energy consumed (TWh)	Total dispatchable capacity, MW	Peaker power capacity, MW	Midmerit power capacity MW	Nuclear power capacity	Flat rate €/MWh	peak price €/MWh	hours at peak quantity (at 8760)
Baseline Scenario B	378.26	62745	10158	10108	17746	55.23	3137.02	87
Scenario B1:1	378.22	61153	10580	9857	16151	55.45	4841.70	83
Scenario B2:1	377.95	62638	10655	10797	15843	56.22	3248.20	87
Scenario B3:1	378.49	57551	7877	8513	16697	54.37	2316.67	99
Scenario B1:2	378.10	58885	12111	9360	12849	56.26	5267.55	79
Scenario B2:2	377.43	62419	11653	12165	12034	58.45	3451.47	82
Scenario B3:2	379.04	47294	3438	5328	14597	52.45	1187.29	99

From Figures 5a, 5b and 5c we see the impacts of different scenarios to the capacity mixture and total dispatchable capacity. If the intermittent production is allocated according to "beginning" profile on the load duration curve (scenarios B2) the total dispatchable capacity does not change almost at all. As the amount of intermittent generation increases it replaces only nuclear power capacity and actually the amount of peaker and midmerit capacities increase. This indicates possibilities for higher costs and emissions. The very opposite is true if the intermittent generation is located according to "end" profile (scenarios B3). Now, it replaces heavily both peaker and midmerit capacity and also decreases strongly the total amount of dispatchable generation.

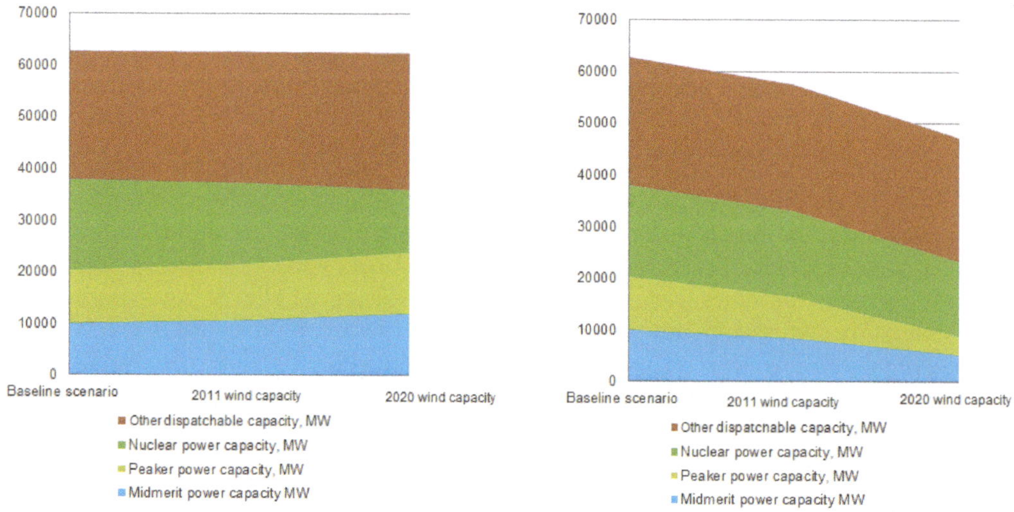

Figure 5. Capacities with unconstrained nuclear power

In Table 7 we present the results related to system costs and profits under the assumption of unconstrained nuclear power. Total production costs are decreased under every scenario compared to the scenarios when nuclear power is constrained. The same is true for the production costs per MWh. The costs are lowest under the scenario B3:2 i.e. when the amount of wind capacity is tripled and the production is located according to "end" profile on the load duration curve. Releasing the assumption of constrained nuclear power has a clear impact to the profits of the industry. Now, as assumed, the nuclear power is making zero profits. The profits of hydro power are reduced and consequently also the total profits are reduced. Under the "end" profile scenarios wind power is able to make positive profits, otherwise its profits are negative.

Table 7. Production costs and profits with unconstrained nuclear power

Scenario	Total production costs million € (with wind in brackets)	Costs €/MWh (with wind in brackets)	Profits million €, hydro power,	Profits, million €, nuclear power	Profits, million €, wind power	total profits, million €	Total profits €/Mwh
Baseline	16508.82	43.65	4161.50	0	0	4161.50	11.00
Scenario B							
Scenario B1:1	15814.51 (16538.83)	43.39 (43.73)	4124.72	0	-13.84	4111.01	10.87
Scenario B2:1	15925.65 (16649.97)	43.83 (44.05)	4294.64		-276.88	4017.76	10.63
Scenario B3:1	15485.87 (16210.19)	42.56 (42.83)	4102.86	0	386.66	4489.71	11.86
Scenario B1:2	14466.02 (16638.99)	43.27 (44.01)	4124.90	0	-234.79	3890.31	10.29
Scenario B2:2	14756.32 (16929.29)	44.23 (44.85)	4562.90	0	-830.81	3732.11	9.89
Scenario B3:2	13443.73 (15616.69)	40.10 (41.20)	3986.04	0	1139.11	5125.28	13.52

In Table 8 we present the results related to emissions from different scenarios and assuming optimized nuclear capacity. We see that total emissions decrease significantly as the nuclear power is not constrained. Emissions per MWh are lowest under the scenario B3:2 i.e. when the level of intermittent capacity is tripled and the production is located according to "end" profile i.e. more on the peak demand hours. The result was expected as we noted the strong change in capacity mixture discussed above.

Table 8. Emissions (unconstrained nuclear power)

Scenario	Emissions peaker power, million tCO_2	Emissions midmerit power, million tCO_2	Total emissions, million tCO_2	Total emissions t/MWh
Baseline scenario B	2.31	23.47	25.78	0.068
Scenario B1:1	2.44	22.66	25.11	0.066
Scenario B2:1	2.43	25.17	27.60	0.073
Scenario B3:1	1.85	19.83	21.68	0.057
Scenario B1:2	2.87	21.37	24.24	0.064
Scenario B2:2	2.68	28.55	31.23	0.082
Scenario B3:2	0.91	12.56	13.47	0.036

Similarly as with respect to results related to constrained nuclear power scenarios we have analysed the sensitivity of model results also here. The results related to total capacity and total costs are also in this case strengthened as the share of real-time price customers increases or demand becomes more prices elastic. Again the results related to emissions are not sensitive to changes in parameter values.

6. Conclusions

We have utilized a long-run oriented simulation model in order to analyze the economic value of intermittent electricity generation in the Nordic power markets. We have simulated the model under different assumptions about the production profile of intermittent generation, about the amount of intermittent generation and about the level of nuclear power. First we assume that nuclear is constrained to the year 2011 level but later we release this assumption and nuclear power is built up to the point where it receives zero profits. Our simulations show that generation profile of intermittent power has clear impacts on the capacity structure, production costs as well as on the profits.

If higher amount of intermittent generation occurs at lower demand hours it replaces midmerit technology, but total dispatchable capacity changes only slightly. This is because only very little wind is allocated to the peak demand hours which is the main determinants of the total required capacity. The reverse is true if wind is allocated more heavily to the end of the load duration curve. In that case total dispatchable capacity decreases substantially. Because wind is now located more on peak demand hours it replaces a significant amount of peak capacity but has only little impact on the amount of midmerit capacity. The changes in the capacity mixture indicate that the wind profile might have significant impact on the system costs and also to the emissions. Interesting is that total production costs as well as production costs per MWh decrease in every scenario with intermittent generation compared to the baseline scenario when nuclear power is constrained. It is often argued that the integration of renewable energy decreases the cost efficiency of the power industry. According to our results this seems not to be the case. System costs are lowest when the intermittent generations is allocated according to the "end" profile i.e. more on the peak demand hours. Notable is that profits of dispatchable technologies decrease as the amount of wind capacity increases regardless of the production profile of intermittent generation. Production profile of intermittent generation has, however, significant effect on the profitability of wind generators and if the intermittent generation is located according to the "end" profile wind producers make clear profits. Consequently, even though the profits of dispatchable generators have decreased, total profits are highest when the level of wind capacity is tripled and it is allocated according to the "end" profile to the load duration curve.

Total emissions (t/CO_2) are decreasing clearly as the share of intermittent generation increases. Interesting is that opposite to the results related to capacity, costs and profits, emissions are not very sensitive to the production profile of intermittent generation in the case when nuclear power is constrained. Emissions seem to, however, increase somewhat if the intermittent generation is located more on peak demand hours. This is because now the intermittent generation replaces more heavily peaker power production than midmerit production which has higher emissions per produced MWh.

When we look at the results when nuclear power is not constrained the production profile of intermittent generation becomes even more important. The difference on total dispatchable capacity among the "beginning" and the "end" profile is significant. Notable is that intermittent generation is clearly replacing also nuclear power in each of the scenarios. The level of nuclear power is smallest under scenario when intermittent production is at the triple level of 2011 and it is allocated according to "beginning" profile on the load duration curve. It is also almost at low under the scenario which uses the 2011 production profile. This is a significant result since it indicates that if (and as) the amount of intermittent generation is increasing it should be taken into account when planning new nuclear power investments. According to our results it is possible that further investments in nuclear power are not economically efficient if done simultaneously with large intermittent investments.

One further implication of unconstrained nuclear power is that now the flat rate is clearly lower in every scenario compared to the scenarios where nuclear power is constrained. We see also that total emissions decrease significantly as the capacity of nuclear power is not constrained but optimized. Contrary to the constrained nuclear power case, also the production profile of intermittent generation has clear impact on the emissions. Emissions are clearly lowest when the intermittent generation is allocated according to the “end” profile.

Acknowledgements

We thank the participants of the following conferences for their valuable comments: the 11th International Industrial Organization Conference (Boston, 2013), the International Energy Workshop (Paris, 2013), the 20th Annual Conference of European Association of Environmental and Resource Economists (Toulouse, 2013) and the Annual Conference of the Finnish Economic Society (Mariehamn, 2013). We thank Yrjö Jahnsson foundation for funding.

References

Borenstein, S. (2005). The Long-Run Efficiency of Real-Time Electricity Pricing. *The Energy Journal, 26*(3), 93-116. http://dx.doi.org/10.5547/ISSN0195-6574-EJ-Vol26-No3-5

Borenstein, S. (2007). Wealth Transfers from Implementing Real-Time-Retail Electricity Pricing. *The Energy Journal, 28*(2), 131-49. http://dx.doi.org/10.5547/ISSN0195-6574-EJ-Vol28-No2-6

Borenstein, S., & Holland, S. (2005). On the Efficiency of Competitive Electricity Markets with Time-Invariant Retail Prices. *RAND Journal of Economics, 36*(3), 469-93.

Bye, T., & Hansen, P. V. (2008). How do spot prices affect aggregate electricity demand? Discussion Paper 527, Research Department of Statistics Norway.

Chalkiadakis, G., Robu, V., Kota, R., Rogers, A., & Jennings, N. R. (2011). Cooperatives of distributed resources for efficient virtual power plants. Proc. of 10th Int. Conf. on Autonomous Agents and Multiagent Systems - Innovative Applications Track (AAMAS 2011), Tuner, Yolum, Sonenberg and Stone (Eds.).

Chen, J., Mount, T. D., Thorp, J. S., & Thomas, R. J. (2005). Location based scheduling and pricing for energy and reserves: a responsive reserve market proposal. *Decision Support Systems, 40*, 563-577. http://dx.doi.org/10.1016/j.dss.2004.09.006

Damsgaard, N. (2003). Deregulation and regulation of electricity markets (Unpublished doctoral dissertation). Stockholm School of Economics, Sweden.

EC (European Commission). (2007). EnergyPortal. Retrieved November 11, 2012, from ww.energy.eu/publications

Eurelectric. (2012). Power Statistics & Trends, 2012 Edition. Retrieved February 30, 2013, from www.eurelectric.org/powerstats2012

Green, R., & Vasilakos, N. (2010). Market behavior with large amounts of intermittent generation. *Energy Policy, 38*, 3211-3220. http://dx.doi.org/10.1016/j.enpol.2009.07.038.

Johnsen, T. A. (2001). Demand, generation and price in the Norwegian market for electric power. *Energy Economics, 23*, 227-251. http://dx.doi.org/10.1016/S0140-9883(00)00052-9

Joskow, P. L. (2011). Comparing the costs of intermittent and dispatchable electricity generating technologies. EUI Working Papers RSCAS 2011/45.

Joskow, P. L. (2012). Creating a Smarter U.S. Electricity Grid. *Journal of Economic Perspectives, 26*, 29-48. http://dx.doi.org/10.1257/jep.26.1.29.

Kopsakangas-Savolainen, M., & Svento R. (2013). Promotion of market access for renewable energy in the Nordic Power markets. *Environmental and Resource Economics, 54*, 549-569. http://dx.doi.org/10.1007/s10640-012-9605-z

Kopsakangas-Savolainen, M., & Svento, R. (2012a). *Modern Energy Markets; Real-Time Pricing, Renewable Resources and Efficient Distribution*. Springer UK, London.

Kopsakangas-Savolainen, M., & Svento, R. (2012b). Real-Time Pricing in the Nordic Power Markets. *Energy*

Economics, 34(4), 1131-1142. http://dx.doi.org/10.1016/j.eneco.2011.10.006

Lamont, A. D. (2008). Assessing the long-term system value of intermittent electric generation technologies. *Energy Economics, 30*, 1208-1231. http://dx.doi.org/10.1016/j.eneco.2007.02.007

Li, N., Chen, L. & Low, S. H. (2011). *Optimal Demand Response Based on Utility Maximizing in Power Networks. IEEE* 978-1-4577-1002-5/11. Retrieved from http://dx.doi.org/10.1109/PES.2011.6039082

Mount, T., Maneevitjit, S., Lamadrid, A. J., Zimmerman, R. D., Thomas, R. J. (2011). *The hidden system costs of wind generation in a deregulated electricity market.* WP 2011-01, Cornell University, New York.

Saad, W., Han Z., Poo,r H V., & Basar, T. (2012). Game-Theoretic Methods for the Smart Grid. An overview of microgrid systems, demand-side management, and communications. *IEEE* Signal Processing Magazine 2. http://dx.doi.org/10.1109/MSP.2012.2186410

Saad, W., Han, Z., Poor, H. V. (2011). Coalitional Game Theory for Cooperative Micro-Grid Distribution Networks. *IEEE* 978-1-61284-955-3/11. Retrieved from http://dx.doi.org/10.1109/iccw.2011.5963577

Tarjanne, R., & Kivistö, A. (2008). Comparison of electricity generation costs, Research report EN A-56 Lappeenranta University of Technology, Finland. Retrieved from http://urn.fi/URN:ISBN:978-952-214-588-8

Vainio, T. (2011). Sähkön tuotantokustannusvertailu, Työpaperit, Tiede+teknologia, Aalto University, Helsinki. Retrieved from http://lib.tkk.fi/TIEDE_TEKNOLOGIA/2011/isbn9789526041353.pdf

Notes

Note 1. We use wind as a representative technology of intermittent generations because due to weather conditions the role of solar power is, at least at the moment, very modest in the Nordic Power Markets.

Note 2. Estonia joined the common Nordic Power market in 2010. Because, however, hourly consumption data from 2011 (which we use in simulations) is not available from Estonia, it is excluded from this study.

Note 3. For the model sensitivity of the share of real-time price customers and price elasticity of demand see Kopsakangas-Savolainen and Svento (2012b).

Note 4. Several studies support the view of inelastic electricity demand in Nordic countries (see e.g. Johnsen (2001), Damsgaard (2003) and Bye and Hansen (2008)). In most of the studies estimates for elasticity values varies from -0,35 to -0,79.

Note 5. We begin with peaker technology because if it used at all it will be used in the highest demand hour.

Note 6. Even though in final equilibrium baseload capacity refers to the nuclear power, for modeling reasons we first, assume that the baseload constraint is equal to the capacity of nuclear power and hydro power together.

Note 7. By utilizing hydro power this way, the results from simulations can be generalized also to such markets where production technologies do not include hydro power.

Note 8. The aggregate wind power capacity of Finland, Sweden, Denmark and Norway was 6553 MW in 2011 and the target is to reach 25600 MW by the year 2020. We have excluded the wind capacity of Norway (441 MW) because of missing hourly production data and because at least Finland is behind the targeted investment schedule, we use only tripled level (18336 MW) of 2011 capacity instead of level 25600 MW.

Note 9. Total dispatchable capacity includes only the average amount of hydropower capacity in use. This assumption does not change the results because we assume that hydropower capacity and its amount of production are constrained to the level of year 2011.

Note 10. Note that these multipliers have to be used to the amount of primary energy used in production.

Note 11. We make similar assumption on hydro power as in scenarios with constrained nuclear power.

7

Powering Ahead: The Reform of the Electricity Sector in Uganda

Joseph Mawejje[1], Ezra Munyambonera[1] & Lawrence Bategeka[1]

[1] Economic Policy Research Centre, Makerere University Campus, Kampala, Uganda

Correspondence: Joseph Mawejje, Economic Policy Research Centre, Makerere University Campus, Kampala, Uganda. E-mail: jmawejje@eprc.or.ug

Abstract

Uganda is among the very few countries in Africa that have fully embraced reforms in the electricity sector. These reforms that included unbundling, privatization, and the creation of an independent regulator were intended to improve sector performance and efficiency. The key feature of Uganda's model is that it sought privatization in the form of concessions and encouraged competition in the form of independent power producers on long term contracts. By focusing on developments in the generation and distribution segments, this paper gives a recount of the reforms and evaluates the progress made. Despite the post reform crises that rocked the electricity sector in 2005–2006 and again in 2011–2012 occurrences of load shedding have reduced greatly; the crisis periods have been replaced by promising developments. This has been on account of additional generation capacity and the Umeme license review that set new operational targets in distribution. We argue that the renewed Government commitment in the sector is set to lead to improved sector performance. In addition we propose some policy and regulatory measures that could lead to further improvements in the sector.

Keywords: Uganda, electricity, reforms

1. Introduction

During the past three decades, many countries, including Uganda, have undertaken reforms in their electricity sectors. Such reforms mainly entailed the transfer of ownership from Government to the private sector and usually involved the unbundling of Government utilities into different business centers for generation, transmission and distribution. Such reforms were motivated by a quest for improved overall sector efficiency (World Bank, 1993).

In Uganda, just like in many other developing countries, the reforms were driven by the need to reduce sector dependence on Government subsidies and thus free resources for other critical development needs. In addition, the reforms were intended to attract private sector participation, accelerate developments in generation and to improve overall electrification coverage (Engurait, 2005).

During the period leading to the reforms, the Uganda electricity sector was faced with numerous challenges among which were excessive power outages, limited investments, and very high technical and non technical losses all of which made for poor commercial performance (UETCL, 2008). With advice and financial support from the International Monetary Fund and the World Bank, the Government undertook major institutional restructuring measures to address these challenges. It was then that a strategic decision was reached to split the vertically integrated Uganda Electricity Board into different commercial segments in distribution, transmission and generation with the private sector expected to lead investments in the sector.

It is against this background that this paper seeks to evaluate the progress with regard to the developments in the sector since the reforms. Specifically, we focus on developments in generation and distribution and argue that renewed Government commitment will be critical in firming the achievements in the sector. In addition, Government has to decidedly deal with corruption in the procurement processes that have in the past delayed project start-off. The rest of the paper is organized as follows: section two reviews the history of the electricity sector in Uganda; section three discusses the framework for electricity reforms; section four presents the data and methods; section five discusses the results; section six discusses the renewed Government commitment in the sector and lastly section seven provides the conclusions.

2. The Historical Context of Uganda's Electricity Sector

The development of Uganda's electricity sector dates back to 1905 when Sir Winston Churchill identified the potential of the Nile for electricity generation. However, the plans for setting up the electricity sector were delayed due to the perceived lack of an effective market for electricity at the time (Gore, 2009). It was therefore not until 1936 that the development of the electricity sector in Uganda got some headway when the head of the Kenya based East African Power and Lighting Company (EAP & L) and Governor Philip Mitchell held discussions about the opportunities for electricity generation and distribution in the Uganda protectorate. In the same year EAP&L was awarded a license for thermal generation and distribution in three major towns of Uganda. Electricity services on a commercial basis started in 1938 in three towns of Kampala, Jinja, and Entebbe (Mwaura, 2012).

The period after the world war was characterized by great economic change and growth. The 1946 Uganda development plan had placed much emphasis on the development of the electricity sector to drive an export-oriented and agricultural-based industrial economy (Gore, 2009). The plan was to start on a large hydro-electricity project and the creation of a quasi-independent body to oversee developments in generation, transmission and distribution of electricity. Thus in 1948, the Uganda Electricity Board was created as a vertically integrated natural monopoly, initially with a vision to supply electricity to the wider East African region (World Bank, 2000).

The electricity sector in Uganda was planned around exploiting the immense potential of the Nile. The prioritized dam projects were the Owen Falls Dam, the Bujagali falls and the Ripon falls. The Owen Falls project was deemed superior to the two other potential locations on technical grounds (Gore, 2009). The Owen Falls project started operations in December 1953 and was officially commissioned in April 1954, initially with a capacity of 150 MW that was later expanded to 180 MW. It was widely expected that the electricity project would spur economic and social transformation. However, in the immediate years following the completion of the Owen falls dam the anticipated benefits were not realized because of the financial and technical constraints that faced the UEB (Gore, 2012). The major challenge that faced the UEB was the lack of finances required to expand the grid to the rural areas, with the result that new customer numbers flattened out quickly and that the sector relied heavily on Government subsidies (UNECA & UNEP, 2007). Despite these challenges, the UEB set out on an ambitious plan to expand generation capacity. The Bujagali project was first proposed in 1957 and Government approval was eventually granted in 1965 (Gore, 2012).

The political unrest that followed the post-independence years, however, coupled with the waning interest from major project financiers, including the World Bank, meant that the project did not take off on time. What followed was a near collapse of the electricity sector in Uganda. The generation infrastructure was rundown, capacity fell to a mere 60 MW and customer numbers fell by 25 percent (Engurait, 2005).

It was not until 1986 that interest in developing the Uganda electricity sector was re-awakened. In 1993, the Government of Uganda with the support of the World Bank under the public enterprise reform and divesture (PERD) programme undertook major reforms that led to the divesture of many public enterprises (Kuteesa et al., 2010). Many state-owned companies were put up for privatisation, but interestingly UEB was not initially among them (Tangri & Mwenda, 2001). Six years later, the decision not to privatize the UEB would be reversed. A new legislation, the 1999 Electricity Act, was passed and this formed the basis for the unbundling of the monopoly into three separate segments for generation, transmission and distribution. While the generation and distribution segments have been opened to private sector participation, the transmission segment is still wholly Government owned.

3. The Framework for Electricity Reforms

Developing countries were encouraged to privatise their electricity sectors following earlier successes in countries such as Chile, Argentina, UK, Wales and Norway (Gratwick & Eberhard, 2008). The early reformers were usually driven by the desire to embrace a more private sector led approach that would improve efficiency across the sector, through increased competition and regulation.

Such reformers followed a standard prescription that involved eight steps that were the recommended course of action for full liberalisation of the electricity markets. These steps as laid out by Gratwick and Eberhard (2008:3952) are as follows: corporatization - commercialisation - legislation - regulation - introduce Independent Power Producers - unbundle - divesture of generation and distribution assets and lastly - introduce competition. The reforms were usually promoted the World Bank and the International Monetary Fund and were geared towards attracting private sector investments. However, as Gratwick and Eberhard (2008:3957) point out, only a handful of countries have adopted the standard restructuring model of the electricity markets. Instead a new hybrid structure, which is characterised by dominant state-owned actor with the entry of IPPs permitted on the margins of the sector, is emerging.

The emergence of the hybrid model could be due to the mismatch between the agendas of the state and of the private companies. The mismatched agendas, among other factors, could have resulted into limited investments in generation and the limited willingness of private sector players to extend distribution lines to more distant and sparsely populated rural areas where costs and losses are higher (Pineau, 2002; Wamukonya, 2003). Thus what we see is a return to business by the state. In many countries therefore, the state still plays an active and significant role in the electricity markets.

3.1 Overview of Uganda Electricity Reforms in Uganda

The reforms in Uganda's electricity sector sought to make the sector commercially viable; reduce dependence on the Government Subsidies; improve access to electricity including rural electrification; improve reliability and quality of supply; attract private investors and finally take advantage of the regional markets (UNECA and UNEP, 2007). A series of events that included the development of a strategic plan that would provide the roadmap for sector transformation; and the introduction of a new legislation that provided for the creation of a sector regulator followed. All this was aimed at improving private sector participation that would enable increased supply of reasonably priced and reliable electricity (Karekezi et al., 2004).

The reforms necessitated the formulation of the Electricity Act which was passed in 1999 to provide for the legal and regulatory framework. In addition, the Rural Electrification Trust Fund was created to promote pro-poor electricity access. In 2001, the UEB was unbundled into three companies namely: the Uganda Electricity Generation Company Ltd (UEGCL); the Uganda Electricity Transmission Company Ltd (UETCL); and, the Uganda Electricity Distribution Company Ltd (UEDCL). The generation concession was awarded to Eskom Enterprises in 2002, while in May 2004 Umeme Ltd, the largest distribution company in Uganda, signed a 20-year concession agreement to manage and operate the UEB distribution network. The transmission infrastructure remained Government owned.

The legal and regulatory framework for the Uganda energy sector was complimented by a policy framework that includes the energy policy, which was completed in 2002 and the renewable energy policy which was completed in 2007. Other organs such as the Electricity Disputes Tribunal, the Rural Electrification Agency (REA) and Board (REB) were established to provide guidelines for resolution of sector disputes, and promote, support and provide for rural electrification programs respectively.

When the reforms started being implemented only one hydro power plant at Nalubaale (180 MW) was generating electricity for the national grid (ERA & MFPED, 2008). Subsequent efforts led to the completion of the 200 MW Kiira plant in 2003. In 2003 the assets of the two plants were leased out to Eskom Uganda Ltd, a subsidiary of Eskom South Africa, a public utility company. Two years later Eskom strengthened its stake in the Uganda electricity by assuming joint control of the distribution business with UK based Globeleq under the trading name Umeme. However in 2006 Eskom exited the distribution business and Globeleq assumed full control of Umeme. However, the sequencing of the reforms seemed to have been flawed; the sector was liberalized without a clear policy framework. The energy policy was approved three years after the Electricity Act in 2002 and the Renewable Energy Policy eight years later in 2007. Ideally, the law should follow policy such that the guidelines to regulate the sector are reflected in the policy.

4. Data and Methods

The data used in this study were sourced from three major sources: the electricity regulatory authority, the ministry of energy and mineral development (MEMD), and the Uganda bureau of statistics. Additional data was sourced through a comprehensive review of the literature. Data from the electricity regulatory authority (ERA) covers the time period 1990–2011, while data from the Uganda bureau of statistics is for the time periods 1992, 1999, 2002, 2006, and 2009 corresponding to the Uganda National Household Surveys. The data from the Uganda bureau of statistics helped us provide an in-depth analysis of electricity access and utilization, while data from the ERA and MEMD contained important variables used in this study that include: generation, demand, customer numbers, subsidies, losses, and price data. In addition, we used data from the literature. Specifically papers by Karekezi et al. (2004) and the United Nations Commission for Africa (2007) provided useful data particularly on the performance of the electricity sector before the reforms.

The data was analyzed using simple descriptive techniques as well a comprehensive review of the literature. This allowed us to assess the current status of the electricity sector against the reform objectives. Specifically, we investigated the outcomes of the reforms in two segments were competition was opened up, that is generation and distribution. In generation, we examine the growth in installed capacity and the events that led to the crisis that resulted into reduced generation capacity in 2006 and the subsequent corrective Government action and commitments. With regard to developments in distribution, we examine the entry of new distribution companies and their current capacities. In addition we investigated electricity access, customer growth, distribution efficiency and Umeme's sector investments.

5. Results and Discussions

5.1 Developments in Generation

Until 2005, hydro sources provided most of Uganda's electricity generation capacity. The joint capacity of the Nalubaale and Kiira dams was 380 MW with the former contributing 180 MW and the latter 200 MW. However, the actual combined output from the two dams was always below capacity. Therefore, following the liberalization of the electricity sector Government sought to attract private investments in generation to add to the existing capacity. Bujagali was the project of first priority. However, due to allegations of corruption in the procurement processes, negotiations with potential financiers that also included the World Bank became controversial and lengthy. These procurement challenges resulted into delayed project take-off and were partly the reason Uganda faced an energy crisis between 2005 and 2006. Details of the procurement and corruption challenges that also involved top Government officials and project contractors can be found in Kapika and Eberhard (2013) and Gore (2009). In addition some civil society groups expressed important environmental concerns that had to be dealt with before the project would continue. All these factors led to the delayed construction works of the Bujagali project.

In the meantime, actual electricity generation from both the Nalubaale and Kiira plants declined in 2006 as water levels in Lake Victoria dropped due to hydrological constraints in the Lake Victoria basin at the time (see Figure 1). The combined electricity generation from the two plants fell to about 140 MW resulting in severe electricity deficits (Rugumayo, 2006).

The receding water levels (Figure 1) created so much debate (for details see Apunyo, 2007; Adam, 2006; Kull, 2006) and most of the analysis therein seemed to suggest that development of electricity projects on the Nile and the poorly conceived designs caused excessive water outflows from the Lake Victoria Basin with dire consequences for electricity generation. On Friday February 17, 2006 two civil society groups led by Lori Pottinger of International Rivers and Frank Muramuzi of the National Association of Professional Environmentalists (NAPE) wrote a letter to Mr. Gobind Nankani, the World Bank vice president for Africa (International Rivers, 2006). In the letter the World Bank, which provided financial support for the projects was requested to take full responsibility and institute adequate measure to address the issues at hand.

Such debates and actions by the civil society have been challenged by among others Tickodri-Togboa (2011) as misinformed and based on poor analysis. Moreover a paper by Yousef et al. (2000) provides a more plausible explanation to the receding water levels as the effects of the "Wolf-Gleissberg Solar Cycles". This paper was able to predict in 2000 that Lake Victoria water levels would recede in $2009 \pm 2-3$ years and goes ahead to predict similar expected occurrences in $2021 \pm 2-3$ and $2033 \pm 2-3$. Yousef et al. (2000:222) then suggested what would have been a solution that would have ensured sustained electricity generation thus:

"The problem of electricity production at drought conditions perhaps can be solved by an engineering technique that can use something like varying water level in two basins as in the Panama Canal."

However, such advice was not heeded. We therefore conclude that the crisis was due to poor planning and failure to use available knowledge and information. As a matter of fact by 2009 the combined available generation capacity for the two facilities was estimated to have risen to 175 MW (Kapika & Eberhard, 2013), and has since risen again to averages of 200–250 MW[1] suggesting that the problem was temporally. This however is not to discount the possible effects of human activity including the development of dams on the River Nile.

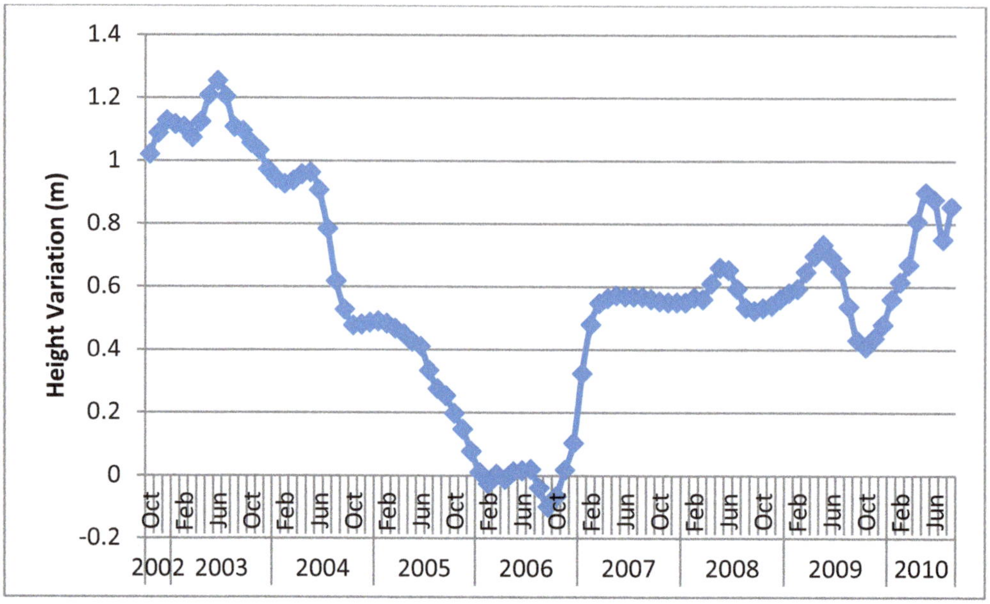

Figure 1. Lake Victoria height variation with respect to nine-year mean levels in meters

Source of data: USDA Foreign Agricultural Service, at http://www.pecad.fas.usda.gov/lakes/images/lake0314.N.1.4.smooth.txt

The severe energy shortfalls that were occasioned by the constrained water levels led to massive electricity deficits (Figure 2). To mitigate the deficits in electricity supply, Government procured three thermal plants each with a capacity of 50 MW. Two of these plants were operated by Aggreko; one located at Mutundwe and the other at Lugogo and the other operated by Jacobsen (U) Ltd located at Namanve. In 2008 the Eletromaxx thermal plant located at Tororo provided a further 20 MW bringing the total thermal capacity to 170 MW.

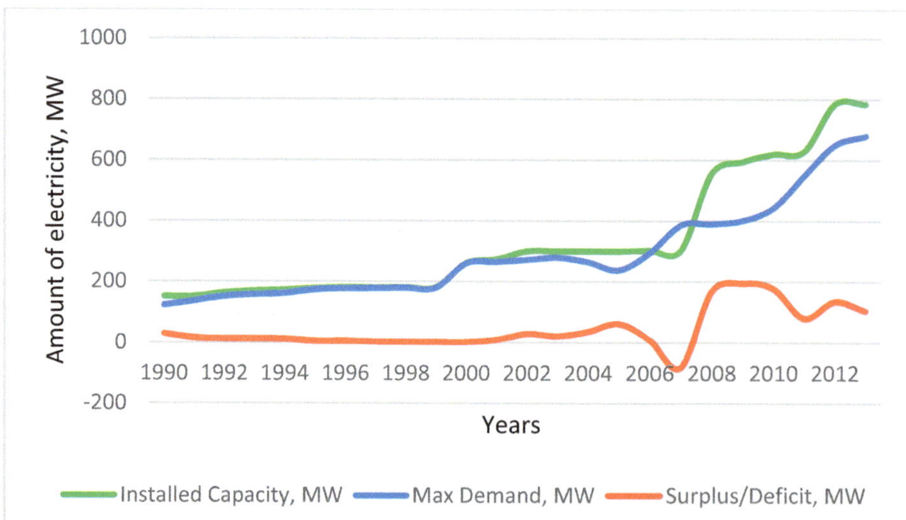

Figure 2. Installed capacity and demand

Sources of data: ERA, UBOS, MEMD.

The increased reliance on emergency thermal generation and its costly nature necessitated Government to increase subsidies to the sector so as to cushion the electricity consumer from the price increments. By 2011 the electricity subsidies had grown to UGX 447 billion, about 37 percent of the entire budget for the ministry of energy and mineral development (Electricity Regulatory Authority, 2011). By early 2010, thermal generation accounted for 41 percent of total generation up from 23 percent by 2006 (Table 2). By 2012 thermal costs accounted for 85 percent of total generation costs. The resultant effect was price increases of 35 percent and 41 percent in June and November 2006 respectively. However, the price increases were not sufficient to meet the financing gap. By 2010 the domestic electricity prices were UGX 385.6 per kWh against a cost reflective tariff of UGX 828 per kWh, the difference being met from Government budget by way of subsidies.

Table 1. Electricity subsidies, UShs, Billion

	Years					
	2006	2007	2008	2009	2010	2011
Subsidies in billion UGX	107	78	169	229	351	447

Source: Electricity Regulatory Authority.

The subsidies to the sector became unsustainable and in 2011 Government defaulted on its obligations to the thermal IPPs resulting in their temporary cessation of generation (Kapika & Eberhard, 2013). The closure of some thermal plants were compounded by a fire at the Aggreko plant based at Mutundwe. The resultant massive electricity shortfalls had big implications for doing business in Uganda. Electricity rationing was started on a daily 12-hour schedule. At the peak of the crisis, a section of some Ugandan traders held massive demonstrations across the country protesting the expensive and usually unavailable electricity.

Table 2. Electricity generation mix

Generation in GWh	Time periods					
	2006	2007	2008	2009	2010	2011
Hydro	1,160	1,264	1,374	1,230	1,255	1,308
Renewable	79	91	132	171	209	347
Thermal	370	539	590	896	1022	1029
Total Generated	1,609	1,894	2,096	2,298	2,486	2,684
% thermal in total generated	23	29	28	39	41	38

Source: Electricity Regulatory Authority.

The completion of the 250 MW Bujagali plant, the 9.5 MW plant at Buseruka and the 3.5 MW plant at Nyagak has provided great reprieve to the Uganda electricity sector and for the first time in over three decades Uganda has enough electricity to satisfy peak demand. This has led to the decommissioning of two of the thermal plants. In addition the 1.7 MW bagasse co-generation project at Lugazi Sugar Cooperation of Uganda Limited (SCOUL) was completed raising total installed capacity from about 540 MW in 2005 to 800 MW by March 2013.

In January 2012, with a view of using the scarce available resources to secure future electricity generation and supply, Government scrapped all subsidies to the electricity sector. Generation projects such as Karuma and Ayago (both 600 MW) have been earmarked for immediate construction.

Table 3. Installed electricity capacity in MW 2008–2013

Plant Name	2008	2009	2010	2011	2012	2013
Hydro Electricity						
Nalubale	180	180	180	180	180	180
Kiira	200	200	200	200	200	200
Kasese Cobalt Company Ltd	10	10	10	10	10	10
Kilembe Mines	5.4	5.4	5.4	5.4	5.4	5.4
Bugoye Tronder	-	13	13	13	13	13
Mpanga	-	-	18	28	18	18
Ishasha	-	-	6.5	6.5	6.5	6.5
Buseruka	-	-	-	-	9.5	9.5
Bujagali	-	-	-	-	250	250
Kisizi	0.3	0.3	0.3	0.3	0.3	0.3
Nyagak	-	-	-	-	3.5	3.5
Thermal Generation						
Jacobsen Namanve	50	50	50	50	50	50
Electromaxx Tororo	-	20	20	20	20	20
Aggreko Kiira	50	50	50	50	-	-
Aggreko Mutundwe	50	50	50	50	-	-
Bagasse cogeneration						
Kakira sugar works	12	12	12	12	12	12
Kinyara Sugar works	-	5	5	5	5	5
Lugazi SCOUL	-	-	-	-	1.7	1.7

Source: Electricity Regulatory Authority and Uganda Bureau of Statistics.

In addition to the operational electricity projects highlighted above, the Government has earmarked the following projects as priority for development in the near future: Karuma, 600 MW; Isimba, 188 MW; and Ayago, 600 MW. Other potential projects include Ariang 400 MW and thermal generation from the oil refinery that is expected to start at 100 MW but eventually increase to 700 MW.

5.2 Developments in Distribution

Following the reform process, the distribution network that belonged to UEB was concessioned to Umeme under a 20-year lease contract. The state-owned UEDCL continues to own the distribution network at 33kV and below in the areas where UEB used to operate with a few additions made by REA and Umeme. Additional but smaller distributional companies include The West Nile Rural Electrification Company (WENRECO) and Ferdsult Engineering Services Ltd. In a bid to speed up rural electrification two cooperative societies namely; Pader - Abim Energy Cooperative Society and Bundibugyo Energy Cooperative Society were granted licenses to distribute electricity in some of the most remote areas of Uganda.

Table 4. Distribution companies in Uganda

Distribution company	Status
Umeme	Umeme inherited customers that were once served by UEB and was leased the UEDCL assets under a 20 year concessional arrangement and controls 97 percent of the districbution market in Uganda. Umeme had a total of 5000,000 2012
Ferdsult	Ferdsult operates and maintains a rural electricity distribution network concessionaire under a 10 year agreement with the Rural Eelectrification Agency. Areas of operation include the districts of: Kibaale, Kyenjojo, Rukungiri, Kanungu, Ntugamo, Isingiro, Rakai, and Masaka. Ferduslt pioneered the pre-paid metering system in Uganda and currently serves about 10,000 consumers.
West Nile Electricity Company (WENRECO)	WENRECO operates an off-grid distribution network in the Northwestern districts of Arua, Paidha, Nebbi, Koboko, Maracha, Zombo and Yumbe. The company operated the 3.5MW Nyagaka HPP and served about 4000 customers by March 2013.
Bundibugyo Energy Coop. Society (BECS)	BECS runs the distribution concessionaire in Bundibugyo district since 2009. Accordingly, BECS took charge of electricity distribution, grid maintenance and managing the revenue from power consumers. Currently BECS serves about 1500 customers.
Pader - Abim Energy cooperative	Serves about 1500 customers in Pader, Abim and Agago districts
Kilembe Investments Limited (KIL)	KIL runs a 10-year concessionaire to distribute and sell electricity in the Districts of Kasese, Rubirizi and surrounding areas. The license runs for ten years and is renewable. Currently KIL serves about 2000 customers on the pre-payment system. KIL intends to introduce solar energy for users in isolated areas.

5.2.1 Electricity Access and Customer Growth

With an energy consumption per capita level of 75 kWh and electrification rates of 11 percent, Uganda has one of the lowest electricity penetration rates in Sub-Saharan Africa (GOU, 2010). The constraints to electrification have been identified as the slow pace of developments in generation, vandalism of distribution equipment, and the limited effective demand especially among the rural areas (Ezor, 2009). Another challenge is that of the long distances and sparsely populated rural areas make it unattractive for the profit minded private distribution companies and therefore electrification, especially in the rural areas has been slow (Pineau, 2002). In this paper we investigated the evolution of electricity access in Uganda since 1992 using the Uganda Bureau of Statistics (UBOS) Uganda National Household Survey (UNHS) data. Results indicate significant increment in electricity access, between 1992 and, with overall electricity access rates rising from 7 percent in 1992 to 11 percent in 2009 (Figure 3).

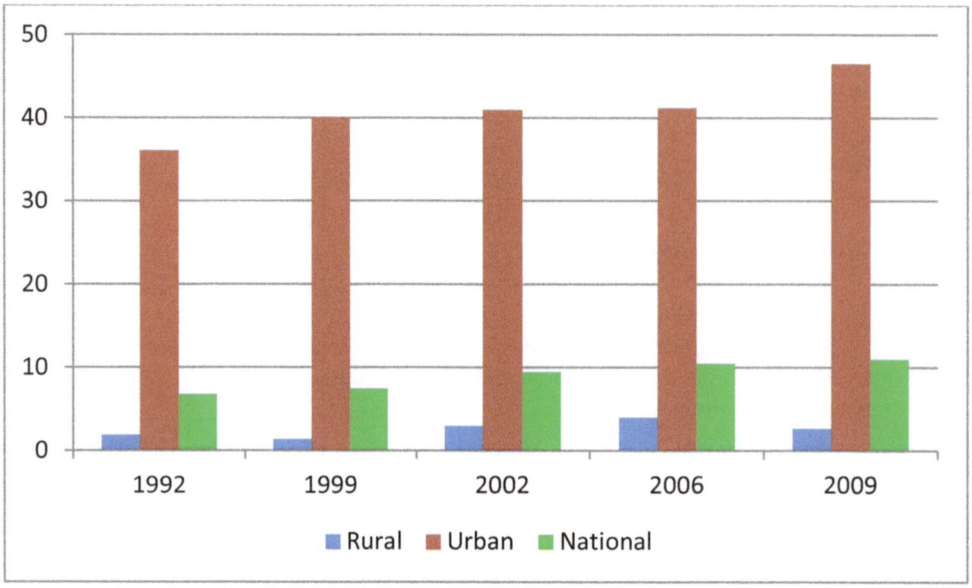

Figure 3. Electrification rates in Uganda, percentage

Source: Author Computations from UBOS UNHS data.

What we see is that electrification rates in Uganda are slowly growing. The completion of the Bujagali project and other priority projects is likely to spur further growth in access as is evidenced in the accelerated number of connections across all customer segments for Umeme, highlight the extent to which constrained electricity generation had suppressed demand as shown in Figure 4 below:

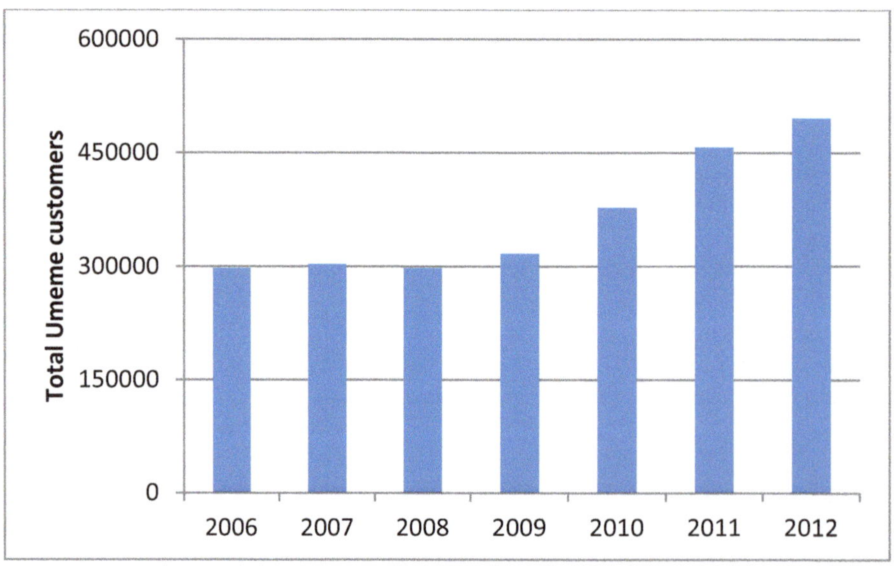

Figure 4. Umeme customers

Source: Electricity regulatory Authority.

5.2.2 Distribution Efficiency

Historically, the distribution losses have been very high in the 30–40 percent range, although they seem to have settled below the 35 percent mark from 2005. According to targets set in its concession agreement, the private distribution company was expected to reduce losses to about 30 percent by 2010 and to about 28 percent by 2011. The distribution company has done well in reducing the losses from about 38 percent in 2002 to about 27 percent in 2012. In 2012 a review of the Umeme concession agreement was carried out by the Electricity Regulatory

Authority and a loss reduction trajectory that will ensure distribution losses reduce to 13.25 percent by 2018 was agreed to. In addition quality of customer service has improved.

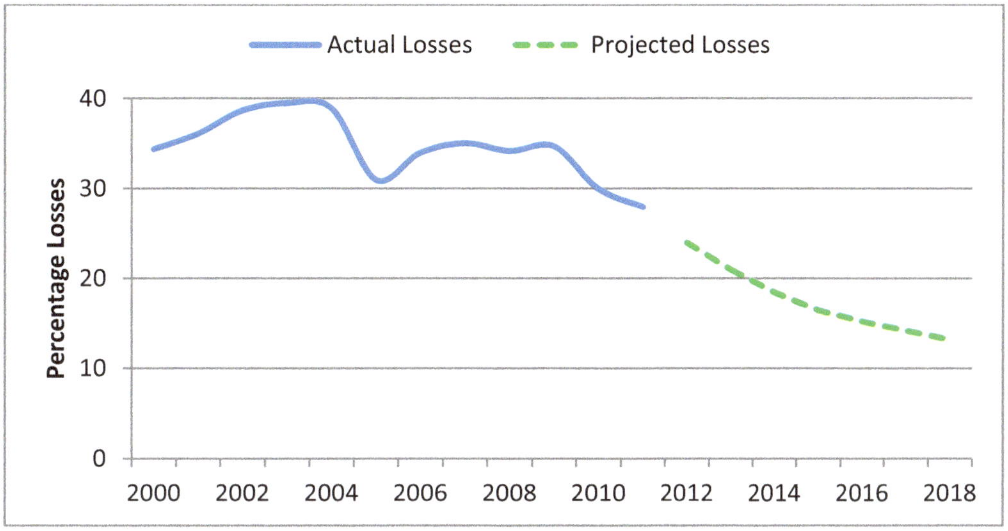

Figure 5. Electricity distribution efficiency

Source: Electricity Regulatory Authority.

In line with sector best practices, the distribution companies in Uganda have started rolling out pre-payment metering systems with a view of reducing electricity losses due to power theft and non-payment. Ferdsult and Wenreco have been running the pre-paid metering systems. Umeme Ltd started piloting the pre-paid meter system in 2012. All other smaller distributors use prepaid meters. It is expected that all electricity users in Uganda will have pre-paid meters within two years as a new measure to check power distribution losses due to electricity theft.

5.2.3 Sector Investments

The concession agreement between the Government of Uganda (UEDCL) and Umeme requires Umeme Ltd to have invested US$65 million over a period covering the first five years of operation. By December 2011, Umeme had more than doubled the required target having invested US$134 million (Umeme, 2012). In 2012 alone Umeme had invested over US$36 million (UGX 93 billion) that was largely spent on new connections, feeders and electricity substations.

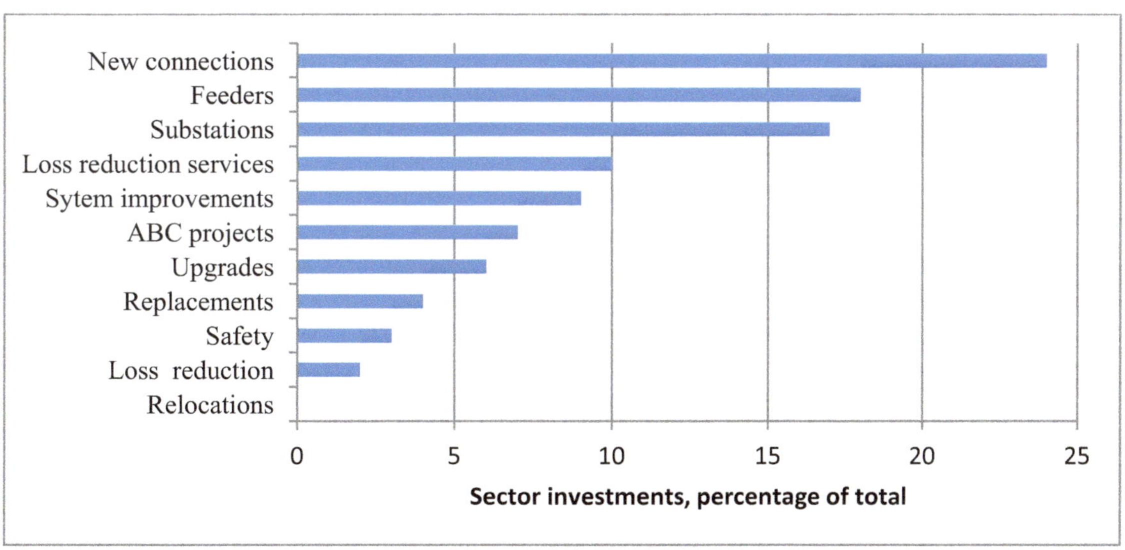

Figure 6. Distribution of 2012 Umeme sector investments

Source: Electricity Regulatory Authority.

In November 2012, Umeme cross listed on the Uganda Stock Exchange (USE) thus tapping into a larger pool of resources. The IPO was expected to raise US$69 million and part of this money intended to clear its shareholder loans so as to improve its balance sheet and attract more funding for investments in network expansion, upgrading and loss reduction.

6. Renewed Government Commitment to the Sector

Perhaps the most important factor in the development of the electricity sector in Uganda is the renewed commitment by Government. Following the challenges that led to the slow investments in the electricity sector, particularly in generation, and the crisis that befell the industry leading to massive blackouts in 2005–2006 and again in 2011–2012, the Government of Uganda decided to renew interest in the sector. Government policy is now focused on infrastructure development, especially in the provision of electricity and roads. This is highlighted in the five-year National Development Plan (NDP) that was completed in 2010 and the subsequent national budget speeches. Government's plan is to increase consumption per capita from 75 kWh in 2010 to 3668 kWh by 2040 and installed capacity from 425 MW to 41,738 MW under the same time period (GOU, 2010).

Table 5. Electricity generation and consumption projections

Year	2010	2015	2020	2025	2030	2035	2040
Capacity MW	425	3,885	8,601	14,670	22,222	31,252	41,738
Consumption (kWh/ capita)	75	674	1,273	1,872	2,470	3,069	3,668

Source: National Development Plan for Uganda.

After the completion and commissioning of the 250 MW Bujagali plant, Government has expressed its commitment to provide partial funding the construction of the 600 MW Karuma plant without external help. Going forward Government has committed to take lead in future energy projects. To this effect the Government set up an energy fund through the Uganda Energy Credit Capitalization Company (UECC) that will provide financing for private sector energy investments especially for small projects with a capacity of up to 20 MW and also to support rural electrification projects in Uganda.

Renewed Government commitment has also been expressed in the distribution segment. The new Rural Electrification Plan for the Period 2013–2022 will among others ensure that Government assumes "greater responsibility for planning, financing and overall management of the rural electrification sector" (GOU, 2013). The renewed interest in rural electrification has been occasioned by the reluctance of the major electricity distributor to extend electricity to rural areas.

7. Conclusions

Good progress has been achieved in Uganda's electricity sector, and this has been partly attributed to renewed Government interest in the sector. However, the electricity sector is still faced with some challenges especially in generation and distribution. In generation, there have been challenges of attracting private sector investments exacerbated by corruption in the procurement and tendering processes that have delayed project implementation. In distribution, the energy losses are still high and electrification is hampered by the reluctance of private profit-seeking companies to extend power lines to distant sparsely populated neighbor hoods.

To confront the challenges in generation Government has abolished subsidies in the sector with a view of prioritizing interventions that would secure future supply security. In distribution, the review of the distributor's license in 2012 has helped to set new operational targets that will result into a systematic loss reduction trajectory. The new rural electrification Plan for the Period 2013-2022 will ensure that among others Government assumes greater responsibility in pro-poor electrification. All these developments augur well for improved performance in the Uganda electricity sector. The post reform crises have been replaced by promising developments on account of improved Government involvement in the sector.

We make the point that this renewed Government commitment is set to lead to improved sector performance, especially in generation and distribution. Our arguments have important implications for Government policy. Specifically, the renewed Government commitment in the sector should be complimented by a renewed will to fight corruption that has affected project take-off and implementation in the past. In addition, the distribution companies should be supported with strong legal and regulatory framework that criminalizes electricity theft so as to reduce the non-technical losses due to electricity theft.

References

Adam, D. (2006). *Uganda accused of 'pulling plug' on disappearing waters of Lake Victoria.* The Guardian. Retrieved from http://www.guardian.co.uk/world/2006/feb/09/uganda.davidadam

Electricity Regulatory Authority. (2008). Constraints to Investment in Uganda's Electricity Generation Industry.

Electricity Regulatory Authority. (2011). Study on distribution system losses and collection rates by Umeme Ltd. Parsons Brinckerhoff Africa (PTY) Ltd (PB).

Engurait, S. P. (2005). Power sector reforms in Uganda: status and achievements. *International Journal of Global Energy Issues, 23*(2), 110-118. http://dx.doi.org/10.1504/IJGEI.2005.006873

Ezor, Z. (2009). *Power to the people: rural electrification in Uganda. School for International Training.* Retrieved from http://digitalcollections.sit.edu/cgi/viewcontent.cgi?article=1675&context=isp_collection

Gore, C. (2009). Electricity and Privatisation in Uganda: The origins of the crisis and problems with the response. In McDonald David (Ed.), *Electric Capitalism: Recolonising Africa on the Power Grid*(pp. 359-399). HSRC Press, Capetown.

Government of Uganda. (2000). *Rural electrification strategy and plan: covering the period 2001 to 2010.* Ministry of Energy and Mineral Development.

Government of Uganda. (2012). *Rural Electrification Strategy and Plan 2013-2022.* Ministry of Energy and Mineral Development.

Gratwick, K. N., & Eberhard, A. (2008). Demise of the standard model for power sector reform and the emergence of hybrid power markets. *Energy Policy, 36*, 3948-3960. http://dx.doi.org/10.1016/j.enpol.2008.07.021

International Rivers. (2006). *Letter to the World Bank Regarding Owen Falls Dam and Lake Victoria.* Retrieved from http://www.internationalrivers.org/resources/letter-to-the-world-bank-regarding-owen-falls-dams-and-lake-victoria-3304

Kapika, J., & Eberhand, A. (2013). *Power sector reform and regulation in Africa: Lessons from Kenya, Tanzania, Uganda, Zambia, Namibia and Ghana.* Human Sciences Research Council, Cape Town.

Karekezi, S., Kimani, J., Mutiga, A., & Amenya, S. (2004). Energy Services for the Poor in Eastern Africa Sub Region: Energy Access Study of East Africa. Global Network on Energy for Sustainable Development, African Energy Research Network.

Kull, D. (2006). Connections between recent water level drops in Lake Victoria, Dam Operations and Drought. *International Rivers*. Retrieved from http://www.internationalrivers.org/files/attached-files/full_report_pdf.pdf

Kuteesa, F., Tumusiime-Mutebile, E., Whitworth, A., & Williamson, T. (2010). *Uganda's Economic Reforms: Insider Accounts*. Oxford University Press.

Luyombo, S. N. (2008). The Transmitter. Uganda Electricity Transmission Company Limited (UETCL). Kampala.

Mwaura, F. M. (2012). Adopting electricity prepayment billing system to reduce non-technical losses in Uganda: Lessons from Rwanda. *Utilities Policy, 23*, 72-79. http://dx.doi.org/10.1016/j.jup.2012.05.004

Pineau, P. O. (2002). Electricity sector reform in Cameroon: is privatisation the solution? *Energy Policy, 30*, 999-1012. http://dx.doi.org/10.1016/S0301-4215(02)00054-X

Rugumayo, A. (2006). *The Electricity Supply Situation in Uganda and Future Direction*. Ministry of Energy and Mineral Development.

Tangri, R., & Mwenda, A. (2001). Corruption and Cronyism in Uganda's Privatisation in the 1990's. *African Affairs, 100*, 117-133. http://dx.doi.org/10.1093/afraf/100.398.117

Tickodri-Togboa, S. S. (2011). *On the contribution of Victoria Nile river discharge to the hydrological performance of East Africa's Lake Victoria. Second International Conference on Advances in Engineering and Technology*. Retrieved from http://mak.ac.ug/documents/Makfiles/aet2011/Tickodri-Togboa.pdf

Umeme. (2012). *Annual Report 2012*. Retrieved from http://umeme.co.ug/downloads/umeme_annual_report_2012.pdf

United Nations Economic Commission for Africa (UNECA) and United Nations Environmental Programme. (2007). Making Africa's Power Sector Sustainable: An Analysis of Power Sector Reforms in Africa. *A Joint Report*.

Wamukonya, N. (2003). Power sector reform in developing countries: mismatched agendas. *Energy Policy, 31*, 1273-1289. http://dx.doi.org/10.1016/S0301-4215(02)00187-8

World Bank. (1993). *The World Bank's Role in the electric power sector*. Retrieved from http://www-wds.worldbank.org/external/default/WDSContentServer/WDSP/IB/1999/09/17/000178830_98101911183588/Rendered/PDF/multi_page.pdf

World Bank. (2000). *Uganda Fourth Power Project*. Retrieved from http://www-wds.worldbank.org/servlet/WDSContentServer/WDSP/IB/2000/07/07/000094946_00062905314733/Rendered/PDF/multi0page.pdf

Yousef, S. M., & Amer, M. M. (2000, September). The sharp rise of Lake Victoria, a positive indicator to solar Wolf-Gleissberg cycles turning points. *International Conference on the Environmental Hazards Mitigation*. Conferences of Virtual Academia. Retrieved from www.virtuacademia.com/conferences

Apunyo, R. (2007, September). Dam Construction Flaws and the Receding Lake Victoria-Uganda. *10th International River symposium and Environmental Flows Conference*. Brisbane, Australia.

Government of Uganda. (2013). *National Development Plan 2010/11 – 2014/15*. National Planning Authority.

Note

Note 1. Electricity Regulatory Authority, personal communication.

8

Investigating Impact of Flow Rate and Wettability on Residual Trapping in CO_2 Storage in Saline Aquifers through Relative Permeability Experiments

Mansour Soroush[1], Dag Wessel-Berg[2], Ole Torsaeter[1] & Jon Kleppe[1]

[1] Norwegian University of Science and Technology- NTNU, Trondheim, Norway

[2] SINTEF, NO-7465, Trondheim, Norway

Correspondence: Mansour Soroush, Department of Petroleum Engineering and Applied Geophysics, NTNU, Trondheim, Norway. E-mail: mansour.soroush@ntnu.no

Abstract

In order to investigate effects of injection rate and aquifer influx in imbibition processes and also wettability behaviour for CO_2 storage in aquifers, two representative fluids are chosen for relative permeability measurements. These two fluids represent CO_2 and brine at the reservoir conditions. The first set of experiments is done by n-heptane and a mixture of glycerol and water, flowing in a glass beads porous medium. The density difference and viscosity ratio are designed to be in the range of CO_2-brine systems normally found at reservoir conditions. Another set of experiments is designed based on dodecane and a mixture of glycerol and water. The second mixture is chosen so that the same ratios of density differences and viscosity ratios are maintained. Interfacial tension and contact angles are measured for both cases. By this set up, two cases of strongly water-wet and water-wet systems are designed. The purpose of this study is to quantify the impact of wettability and flow rate through relative permeability experiments. Results show that the relative permeability is sensitive to both rate and wettability, and after interpreting the data from experiments in a history matching process, based on the effects of drainage and imbibition rates and also the effect of wettability, correlations are developed to predict the amount of CO_2 trapped in the pores. This newly developed method will be useful in obtaining good estimations of real case trapping volume in CO_2 storage processes. Scaling analysis of the experiment shows that the tests are well designed in the range of real reservoir conditions.

Keywords: CO_2 storage, residual trapping, relative permeability, imbibition, drainage, CO_2-brine systems, hysteresis, wettability, rate, history matching, scaling

Nomenclature

d Glass beads diameter, μm

GBs Glass beads

CGR Capillary to gravity ratio

k Absolute permeability, md

k_r Relative permeability

S Saturation, fraction

S_w Water saturation, fraction

S_g Gas (CO_2) saturation, fraction

P_c Capillary pressure, mbar

L, E, T, A and B Constants in kr models

M Mobility ratio

N Correy parameter

N_c Capillary number

Q flux, m/s

U Characteristics velocity, m/s

φ Porosity, %

ρ Density, gr/cc

α Dipping angle, degree

μ Dynamic viscosity, cP

λ Pore size distribution index

γ Interfacial tension, mN/m

Subscripts

g Gas (CO_2)

w Water

i initial

ir irreducible

r residual

Superscript

0 End point value

* Normalized value

1. Introduction

A good option for reduction of the emission of CO_2 to the atmosphere is CO_2 capture and storage (Bennion & Bachu, 2005). CO_2 storage in geological formations is considered to be an important solution for preventing CO_2 emissions to the atmosphere; however, it is vital to ensure safe and secure storage projects. Success of the project is highly dependent on understanding the process in the reservoirs and different geological characteristics. There are many potential geological storage possibilities including depleted oil and gas reservoirs and coal beds, but the main attraction of brine formations as suitable storage sites for CO_2 is the availability and possibly larger volumes (Bennion & Bachu, 2005).

A successful CO_2 storage project in the aquifers requires accurate modeling of the CO_2 behaviour in the subsurface formation. One of the important issues in the modeling process is the relative permeability (the ability of flow of a fluid through a porous material in presence of other fluid) of the CO_2-brine system. Relative permeabilities of these systems are influenced by different reservoir parameters such as absolute permeability (the ability of flow of a fluid through a porous material), temperature, wettability, hysteresis effect, interfacial tension and displacement rates. Many factors have been targeted for the studies in previous petroleum researches, but not many specific research projects on CO_2-brine systems are reported (Perrin et al., 2009; Pini et al., 2011; Krever et al., 2011).

There are some relative permeability measurements reported for CO_2-brine systems. Bennion et al. (2008) studied the impact of interfacial tension and pore size distribution and capillary pressure on CO_2 relative permeability at reservoir conditions. They did experiments on three sandstone and carbonate samples that are representing potential sequestration zones. They observed that the database of measured relative permeability for different zones allows appropriate selection of CO_2-brine sites. Later, the database has been applied in simulation studies.

Recent investigations of CO_2-brine systems, including experimental capillary pressure and relative permeability studies, have been reported by the Benson research team at Stanford. Perrin et al. (2009) reported two sets of relative permeability measurements at different flooding rates. They have also reported slight difference in the end points and corresponding relative permeability values due to the injection rate. Pini et al. (2011) have reported experiments on capillary pressure and heterogeneity of the CO_2-brine systems in sandstone rocks at reservoir conditions. They have used an experimental configuration in addition to a medical X-ray computer tomography (CT) scanner to determine the saturations. They have also observed the rate dependency in the capillary pressure measurements. Kua et al. (2011) reported simulation studies of the effects of flow rate and small-scale heterogeneities on multiphase flow of CO_2 and brine. They have simulated steady-state core scale multiphase flow experiments at wide range of fractional flows and concluded that when the rates are large enough or low enough gravity effect is minimized. Krever et al. (2011) have measured relative permeability and completed studies on trapping of CO_2 in sandstone rocks at reservoir conditions. They have measured drainage relative permeability and residual gas saturations at reservoir conditions. They have reported that the relative permeability in four samples was consistent with general characteristics of drainage in strongly water-wet rocks. They observed that residual trapping can play a major role in the immobilization of CO_2 injected into the subsurface and they have suggested that further investigations should be performed to establish the wetting properties of rocks. Krause et al. (2011) did laboratory core flooding experiments coupled with CT scanning. They have used the results to measure core properties such as absolute and relative permeability, and also to determine sub-core scale saturation and porosity distributions. In that study by doing experiments on the subcore scale, fluid distributions during the displacement process, displacement efficiency of CO_2-brine drainage processes, residual trapping and fluid saturation at the millimeter to sub-millimeter scale were studied.

The effect of gravitational forces on capillary trapping during CO_2 storage has been studied by Bandara et al. (2011). They have used pore scale models to provide sensitivity analysis of the trapping volume to the gravity number. Their results show that, for instance, for large gravity numbers, must of the injected CO_2 reached to the caprock. Another work by Suekane et al. (2011) also clarifies the effect of gravity through imaging the trapped air bobbles in Berea sandstone after spontaneous imbibition at atmospheric pressure. They reported that trapped N_2 bubbles are stable against the water flow rate corresponding to a capillary number in the range of 1.0×10^{-4}.

There are also some researches in theoretical and analytical aspects of the residual trapping problem. Spiteri et al. (2008) developed a new trapping model based on pore scale modeling. They believe that hysteresis is the most important factor in the prediction of CO_2 trapping.

Wettability and flow rate in both drainage and imbibition processes can affect the residual trapping mechanism and this is quantified through this study. The aim of this study is to target this problem through relative permeability experiments and to find correlations to apply the results in real storage sites. Scaling analysis of the experiment admits that the tests are designed in the range of real reservoir conditions. The correlations give good estimation of the trapped volume based on injection rate and wettability characteristics.

2. Methods

2.1 Relative Permeability Measurement Methods

Steady-state and unsteady-state methods are two procedures for measuring relative permeability. In steady-state method, fixed ratios of two immiscible fluids are injected at the same time into the porous media until saturation and pressure equilibrium. In unsteady-state method one fluid phase is displaced by another immiscible fluid phase (Honarpour et al., 1986). In this study unsteady-state method is used and relative permeabilities are measured from recorded pressure difference and production during drainage and imbibition. Either explicit or implicit methods can be used for calculating the relative permeability from the experiment dada. The widely used explicit method is JBN technique (Jonhson et al., 1959). The implicit method uses numerical simulations for history matching (Wang et al., 2006). The history matching of the data in this study is done by Sendra software. In that a two phase one dimensional black oil simulation model is used for interpreting the flooding data. Through this process the experimental data matched by adjusting the relative permeability models. There are different correlations that may be used to match the relative permeability from the experiment data. In the Sendra simulator these correlations are described as follows. The normalized water saturation is used in all correlations:

$$S_w^* = \frac{S_w - S_{wi}}{1 - S_{wi} - S_{gr}} \tag{1}$$

For CO_2-brine systems, the formulations are given for gas-water; but they are similar for oil-water and oil-gas systems (Sendra user guide, 2012). Burdine (1953) introduced the relative permeability values as below:

$$k_{rw} = k_{rw}^0 (S_w^*)^{\frac{2+3\lambda}{\lambda}} \tag{2}$$

$$k_{rg} = k_{rg}^0 (1 - S_w^*)^2 \left[1 - (1 - S_w^*)^{\frac{2+\lambda}{\lambda}}\right] \tag{3}$$

In that, the superscript 0 is the end point value, superscript * is normalized value for water saturation from equation (1) and λ is the pore size distribution index (Burdine, 1953). Later, Corey (1954) proposed a new correlation that has been widely used:

$$k_{rw} = k_{rw}^0 (S_w^*)^{N_w} \tag{4}$$

$$k_{rg} = k_{rg}^0 (1 - S_w^*)^{N_g} \tag{5}$$

N_w and N_g are the water and gas Corey parameters. These parameters define the curvature of water and gas relative permeability plots. Sigmund and McCaffery (1972) modified the Corey correlation:

$$k_{rw} = k_{rw}^0 \frac{(S_w^*)^{N_w} + A S_w^*}{1 + A} \tag{6}$$

$$k_{rg} = k_{rg}^0 \frac{(1 - S_w^*)^{N_g} + B(1 - S_w^*)}{1 + B} \tag{7}$$

Here N_w and N_g are the same as Corey parameters. The constants A and B are small values for linearizing the curves when relative permeability values approach zero. These equations are the same as Corey equation if the constants A and B are zero. Later, Chierici (1984) introduced new correlation:

$$k_{rw} = k_{rw}(S_{gr}) e^{-BR_w^{-M}} \tag{8}$$

$$k_{rg} = k_{rg}(S_{wi}) e^{-AR_w^{L}} \tag{9}$$

$$R_w(S_w) = \frac{S_w - S_{wi}}{1 - S_{gr} - S_w} \tag{10}$$

Here, instead of normalized water saturation terms, R_w is used (Chierici, 1984). LET Correlation is developed by Lomeland et al. (2005):

$$k_{rw} = k_{rw}^0 \frac{(S_w^*)^{L_w}}{(S_w^*)^{L_w} + E_w (1 - S_w^*)^{T_w}} \tag{11}$$

$$k_{rg} = k_{rg}^0 \frac{(1 - S_w^*)^{L_g}}{(1 - S_w^*)^{L_g} + E_g (S_w^*)^{T_g}} \tag{12}$$

The parameter L describes the shape of the curve in the lower parts, while the parameter T is related to the top of the curve (Sendra user guide, 2012) (Lomeland et al., 2005). In this study, all the above-mentioned correlations have been examined to find the best possible history matching of the experimental data and in most of the cases the best fit was obtained by using the Sigmund and McCaffery correlation.

Relative permeability models are also developed to predict residual values in flowing systems. These models are usually developed empirically and recently are matched through pore scale network modeling. Land (1971) has developed an imbibition relative permeability relation for two and three phases. In that relation he assumed that trapping volume is function of the non-wetting saturation. Land proposed a widely used empirical correlation that is still the base for some modified models (Soroush et al., 2013).

$$S_{gr}^{*} = \frac{S_{gi}^{*}}{1+CS_{gi}^{*}} \quad (13)$$

$$S_{gr}^{*} = \frac{S_{gi}^{*}}{1+CS_{gi}^{*}} \quad (14)$$

$$C = \frac{1}{S_{gr}^{*max}} - 1 \quad (15)$$

In the Land correlation, S_{gr}^{*} is residual saturation, S_{gi}^{*} is initial saturation and S_{gr}^{*max} is maximum residual saturation. Killough (1976) developed Land's model by introducing an interpolation method to find scanning curves that is an intermediate curve to predict the relative permeability curve and the end points. In this method relative permeability falls between two endpoint saturations. In Killough's method trapped critical saturation for each value of reached saturation *S* is:

$$S_{gr} = S_{gr}^{max} + \frac{S-S_{gi}}{1+C(S-S_{gi})} \quad (16)$$

$$C = \frac{1}{S_{gr}^{max}-S_{gi}} - \frac{1}{S_{g}^{max}-S_{gi}} \quad (17)$$

Kleppe et al. (1997) questioned the validity of applying Land's trapping relationship. The experiments were based on artificial core analysis (Aeroolithe 10), and did not match Land's results. They proposed a correlation based on the maximum residual saturation at the end of complete imbibition:

$$S_{gr} = \frac{S_{gi}}{S_{g}^{max}} S_{gr}^{max} \quad (18)$$

Carlson (1981) developed a method that produces a scanning curve that is parallel to the imbibition curve. The idea is shifting the imbibition curve horizontally until it intersects the drainage curve (Schlumberger Eclipse technical description, 2012).

2.2 Analytical Description

For the purpose of analytical description of the residual trapping mechanism during CO_2 storage in the saline aquifers, two phenomena are considered; injection of CO_2 into the subsurface formations and the behaviour of the CO_2 after injection. Then, further development of the model is obtained by adding a dip angle to the formation layers.

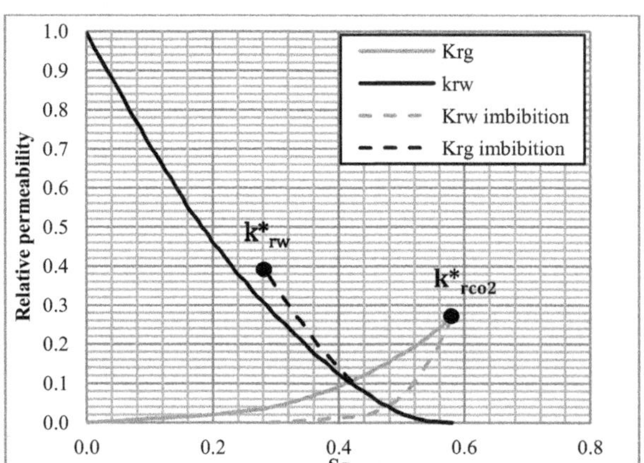

Figure 1. Typical relative permeability curve for CO_2-brine systems, drainage and imbibition

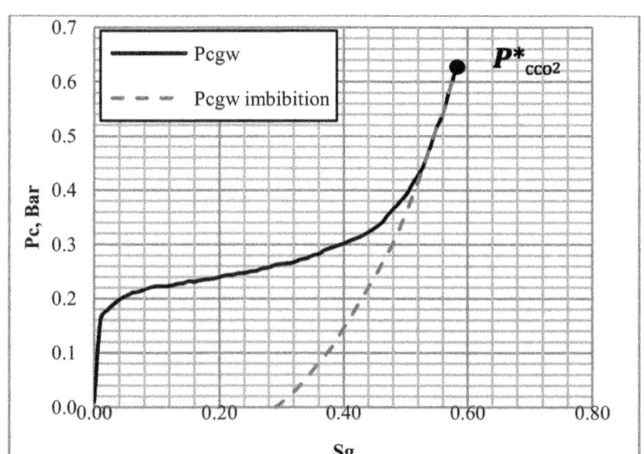

Figure 2. Typical capillary pressure curve for CO_2-brine systems, drainage and imbibition

Here, a one dimensional, two phase system for immiscible and incompressible flow of CO_2 in a narrow porous channel filled with brine has been considered. By introducing dimensionless parameters it is possible to find the scaling parameters for the problem. The medium is initially filled with the brine at $S_w = 1$ and it is designed to be very long compared to the thickness of the reservoir, and it has a dip angle of α. Two typical relative permeability curves with the end point relative permeability values of $k^*_{rCO_2}, k^*_{rw}$ and a capillary pressure curve with the maximum pressure of $P^*_{cCO_2}$ are selected for this problem, as shown in Figures 1 and 2 (Soroush et al., 2013). To develop the governing equation, conservation of mass and the Darcy equation have been normalized based on dimensionless length, time, porosity and flux. By simplifying the equation and rearranging, three main dimensionless groups are derived (Soroush et al., 2013). The benefit of this scaling investigation is a relationship between field scale CO_2 storage and experimental results.

$$N_{cv} = \frac{P^*_c k^*_{rCO_2} \bar{k}}{QL\mu_{CO_2}} \tag{19}$$

$$N_{gr} = \frac{\Delta\rho g sin\alpha k^*_{rCO_2} \bar{k}}{Q\mu_{CO_2}} \tag{20}$$

$$M = \frac{k^*_{rw}/\mu_w}{k^*_{rCO_2}/\mu_{CO_2}} \tag{21}$$

Here, Q is injection flux (m/s) and $\bar{k}$ is average permeability (m^2) and L is length of the model (m). Gupta (1979) has introduced the following capillary number for small-scale environment:

$$N_c = \frac{U\mu}{\gamma} \tag{22}$$

Here γ is interfacial tension between two fluids and U is characteristics velocity. The capillary-to-gravity ratio (CGR) has been also defined to provide quantification of the relative effect of gravity in the presence of capillary pressure. CGR definition is first given by Iffly et al. (1997) as follows:

$$N_{gr} = \frac{2\gamma}{\Delta\rho g h \sqrt{\frac{k}{\varphi}}} \tag{23}$$

2.3 Experiment Set-Up and Procedure

2.3.1 Fluid Properties

In this study, two sets of fluids are chosen for experiments at laboratory conditions. The fluid selection is based on scaling analysis. This means that density difference and viscosity ratio are chosen in the range of CO_2-brine conditions at the reservoir. The second fluid set has the same scaling ratios but the difference is the wettability behaviour to the porous material. The viscosities of the fluid sets are measured by Brookfield LVDV-II+Pro viscometer, and are measured at atmospheric pressure and 22 °C (Table1). The DV-II+Pro consist of a rotating spindle that is placed in the chamber full of the testing fluid. After calibrating the spring, the viscous drag of the fluid against the spindle is measured by the spring deflection.

Table 1. Fluid properties at laboratory condition

	Fluid set 1		Fluid set 2	
Fluid name	n-heptane	glycerol (50%) fresh water (50%)*	dodecane	glycerol (66%) fresh water (34%)*
Molecular formula	C_7H_{16}	$C_3H_8O_3+H_2O$	$C_{12}H_{26}$	$C_3H_8O_3+H_2O$
Densities (gr/cc)	0.68	1.1	0.78	1.2
Density difference (gr/cc)	0.42		0.42	
Viscosities (cp)	0.67	5.8	1.68	14.5
Viscosity ratio ($\frac{\mu_{Hep.}}{\mu_{Gly.+water}}, \frac{\mu_{Dode.}}{\mu_{Gly.+water}}$)	0.1159		0.1155	

*based on weight %.

Ranges of fluid properties for the CO_2-brine systems are different for various reservoir conditions but the overall ranges are reported in the literature. CO_2 and brine densities and viscosities are strongly dependent on the reservoir conditions including depth, temperature and pressure. Salinity of the brine also affects the density and viscosity of it (Nordbotten et al. 2005). Nordbotten et al. (2005) have categorized the reservoir conditions to cold basin with temperature gradient of 25 °C /km, with a surface temperature of around 10 °C, and warm basin with temperature gradient of 45 °C /km, and surface temperature of about 20 °C. Shallow depth is considered to be around 1000 m and deep formations assumed to have around 3000 m depth. Hydrostatic pressure gradient of 10.5 MPa/km is also considered for brine formations. According to these assumptions, Table 2 shows the range of different fluid properties for CO_2-brine systems. Comparing the fluid properties of the designed sets and the typical ranges of density and viscosity of the CO_2 storage sites, good agreement is found between the fluid model designs and deep and cold storage sites.

Table 2. Fluid properties of CO_2 and brine taken from Nordbotten et al. (2005)

	Cold Shallow	Cold Deep	Warm Shallow	Warm Deep
Density of CO_2 ($\boldsymbol{kg/m^3}$)	714	733	266	479
Density of brine($\boldsymbol{kg/m^3}$)	1012-1230	995-1202	998-1210	945-1145
Viscosity of CO_2 (mPa.s)	0.0577	0.0611	0.0395	0.0395
Viscosity of brine (mPa.s)	0.795-1.58	0.378-0.664	0.491-0.883	0.195-0.312
Density difference ($\boldsymbol{kg/m^3}$)	298-516	262-469	732-944	466-666
Viscosity ratio	0.037-0.073	0.095-0.162	0.026-0.127	0.127-0.203

2.3.2 Absolute and Relative Permeability Measurements

Stainless steel chamber is used for having glass beads inside and inserting that in the core holder. The chamber is filled with glass beads through a continuous shaking procedure. After at least one hour of shaking, the glass beads are packed. It is tried to shake the chamber in different directions to have a homogeneous glass beads pattern. Nylon filters (60 μm) in both ends of the chamber prevent glass beads production. The helium porosimeter method has been used for porosity determination. The helium porosimeter utilizes the principle of gas expansion that is described by Boyle's law. To have a known volume of helium gas, the chamber (empty, but with filter papers inside) is filled with helium at a known pressure. Then the helium is expanded into the chamber that is filled with glass beads. The second pressure record depends on the volume of the chamber minus the glass beads volume, so we have the porosity. In the next stage, the chamber is fully saturated with the wetting phase (glycerol mixture with water) by use of vacuum cell. The chamber is placed in the vacuum and after a complete process of vacuuming the mixture opened to the chamber. Figure 3 shows the set up and Figure 4 is the chamber and separator. Schematics of the set-up and connections can be found in Figure 5.

An absolute permeability measurement of the porous media in the chamber is done by injecting a glycerol-water mixture in different rates and recording the pressure drop.

The unsteady-state relative permeability method is used and the chamber is initially filled with a glycerol-water mixture and it is placed into the core holder. n-heptane is vertically injected from the top in the first section of runs (test number 1 to 4 in Table 5) and in the other set of experiments (test numbers 9 to12 in Table 5) dodecane is injected into the chamber. After a complete drainage (more than 1.2 pore volume injection) imbibition has been performed and the non-wetting phase has been injected into the chamber again vertically from bottom. The test is replicated by injecting glycerol mixtures from each fluid set at the same experimental condition and different injection rates (test number 5 to 8 and 13 to16 in Table 5).

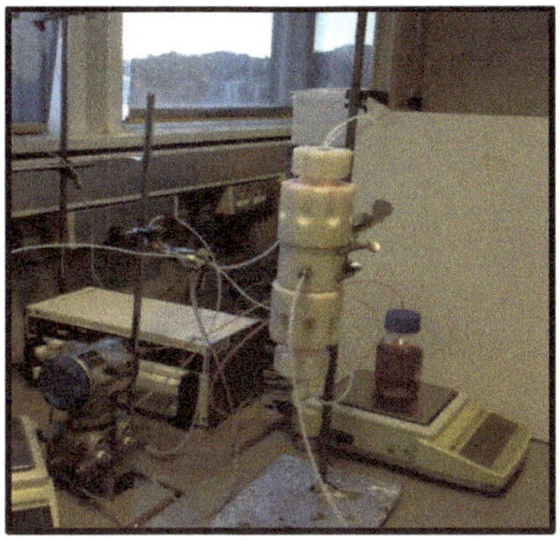

Figure 3. Set-up picture

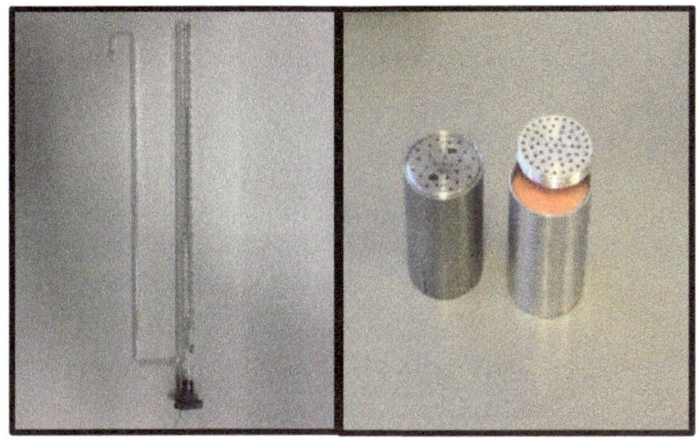

Figure 4. Separator and glass beads chamber

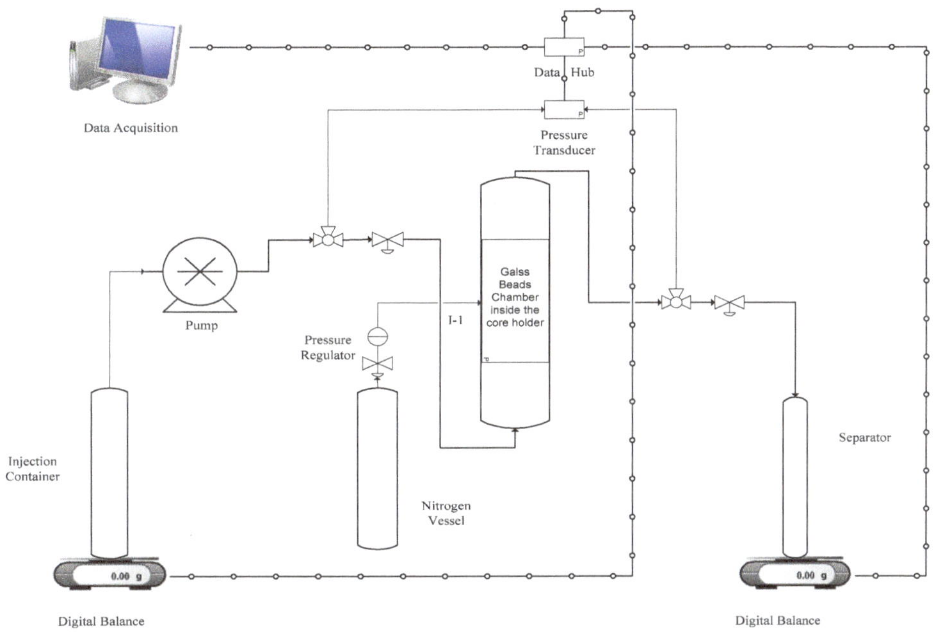

Figure 5. Schematic of the relative permeability set-up

2.3.3 Interfacial Tension (IFT) Measurement

Interfacial tension between two sets of fluids has been measured with pendant drop method. Minimum one hour is considered to let the drop to be stable at atmospheric pressure and 22 °C. Figures 6 and 7 show the droplet after stabilization and the curvature is fitted to obtain the IFT for two sets of the fluids. Figure 8 compares the results of the IFT measurements. Both tests resulted in the same range of IFT and in stabilized period they matched together. The IFT estimations from the experiments are in the range of $26\ mN/m$. Hebach et al. (2002) studied CO_2-brine IFT at the range of reservoir pressure and reported values is around $30\ mN/m$. Chiquet et al. (2006) studies also confirmed the former measurement.

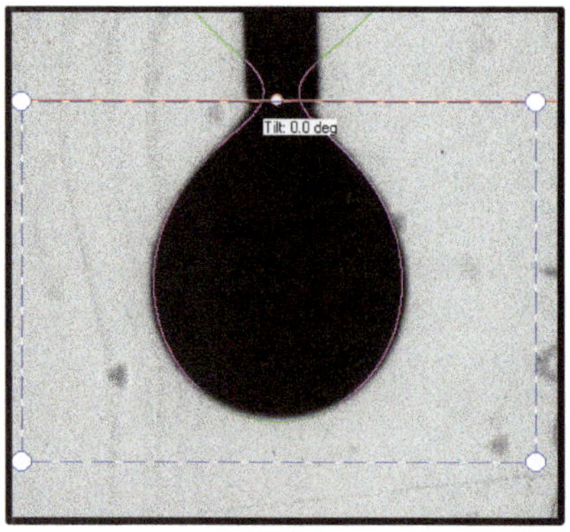

Figure 6. A droplet of heptane in glycerol and water mixture, fluid set 1

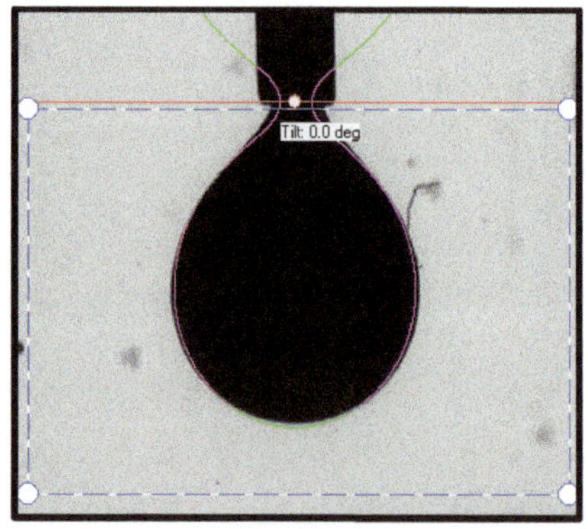

Figure 7. A droplet of dodecane in glycerol and water mixture, fluid set 2

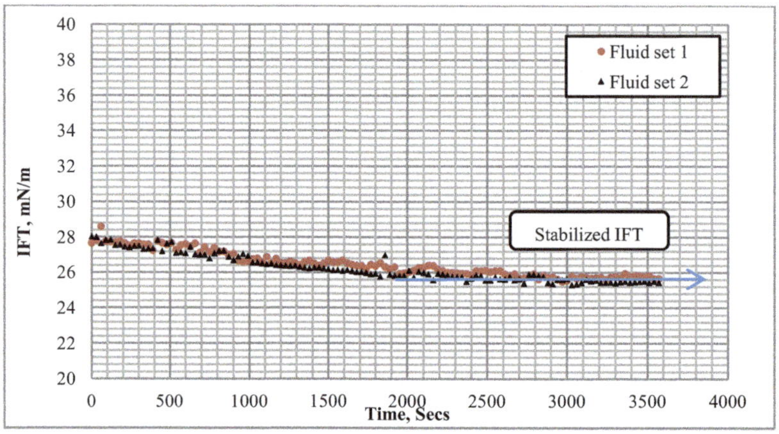

Figure 8. IFT measurement results

2.3.4 Contact Angle Measurements

An imaging method is used for measuring the contact angle for two sets of fluids. Practically, in porous materials it is complex to determine the real contact angle because of the pore shapes and complex mineralogy, but in order to obtain an estimation of the range of contact angle and knowledge about the wetting characteristics of each fluid set, contact angle is measured in presence of fluids and a solid surface of glass that is cleaned and polished. The chamber is filled with the non-wetting phase in each fluid set (n-heptane or dodecane) and a drop of the wetting phase is placed on the glass surface and an enlarged image of the drop is obtained by photography. The experiment continued for minimum one hour to let the droplet to be stable. Very small changes in the angle are observed after one hour of stabilization period. Figures 9 and 10 are droplet photos and matching curves. Figure 11 compares the contact angles obtained from the experiment. Contact angles are calculated for both left and right sides of the drop and almost the same values are predicted, but the final result is the average of right and left measurements. About 16 degrees difference is observed in two experiments. According to conventional wettability criteria wetting phase in fluid set 1 is categorized as strongly wet to the media, while fluid set 2 is considered wet. The wettability characteristic is validated by the result of Amott cell wettability measurements.

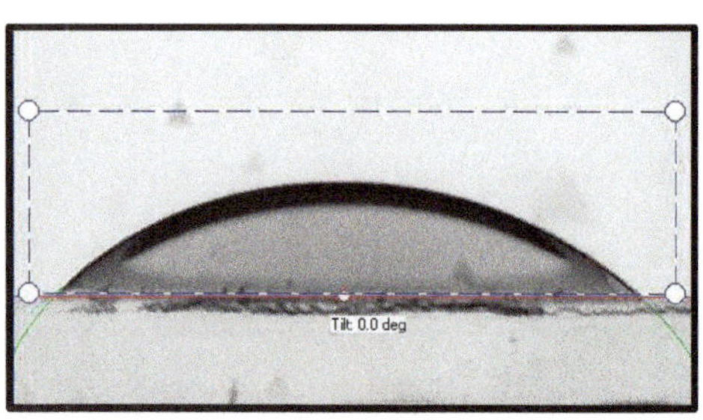

Figure 9. Contact angle measurements, fluid set 1

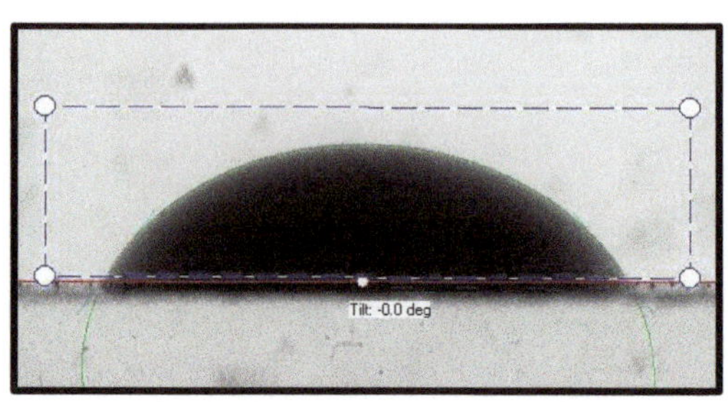

Figure 10. Contact angle measurements, fluid set 2

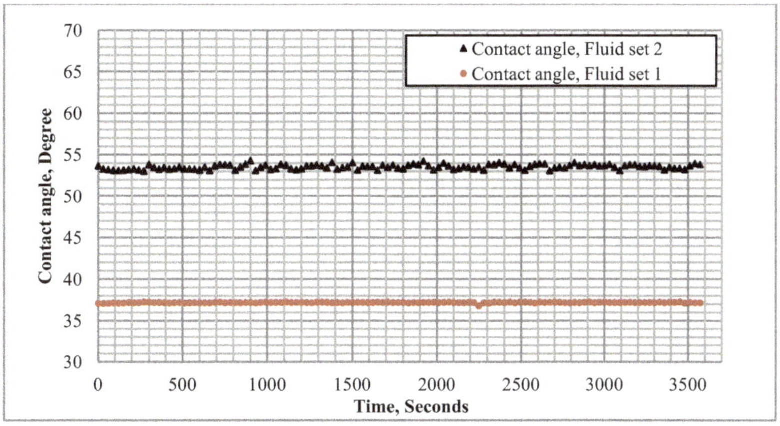

Figure 11. Contact angle measurements for two sets of fluids

2.3.5 Amott Cell Imbibition Test

An Amott cell is used to measure the wettability index of the fluid systems. Only an imbibition process is completed to quantify the residual values of the non-wetting in this test. In this experiment the non-wetting phase will be produced only by the gravitational force and will be accumulated at top of the Amott cell. The chamber that is fully saturated with the non-wetting phase (n-heptane or dodecane) is placed in the imbibition cell surrounded by wetting phase (glycerol and water mixtures). The wetting phase starts to penetrate into the chamber and the non-wetting phase displaces the wetting phase out of the chamber until equilibrium is reached. The mechanism of production can be both spontaneous and gravitational because of having heavier fluid at top of the lighter one in the chamber. The production volume is measured in a scaled section at the top of the cell. The non-wet production gives an estimation of the effect of wettability when repeating the test with fluid set 2 in the same porous condition. As it is shown in Figure 12, production rate is higher in the early stages of the experiment, but it is stabilized at late times.

Table 3. Amott cell experiments design

Test No	fluid	Imbibition test total time, sec	Porosity, %	Distance from top of the chamber to top of the Amott cell, cm	PV, cc	Residual fluid, fraction
1	fluid set 1	30000	39.7	28	15.156	0.173
2	fluid set 2	30000	39.6	28	15.127	0.313

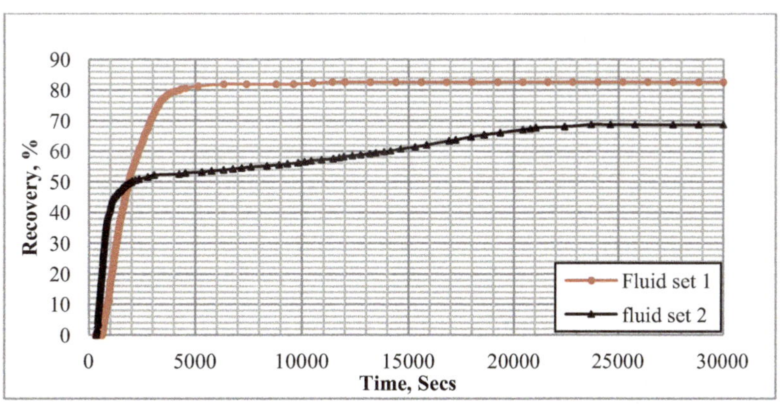

Figure 12. Amott cell imbibition cell results

3. Results

3.1 Absolute Permeability Results

The absolute permeability is used as an adjusting parameter in simulations by the Sendra software to match the experimental pressure drop observed in the relative permeability measurements. The absolute permeability calculation is based on injection of wetting fluid of fluid set 1 by different injection rates. The pressure drop created by the injection is plotted in Figure 13. The stabilized pressure in each step gives a point in the pressure versus rate plot, and by use of the Darcy equation absolute permeability is obtained.

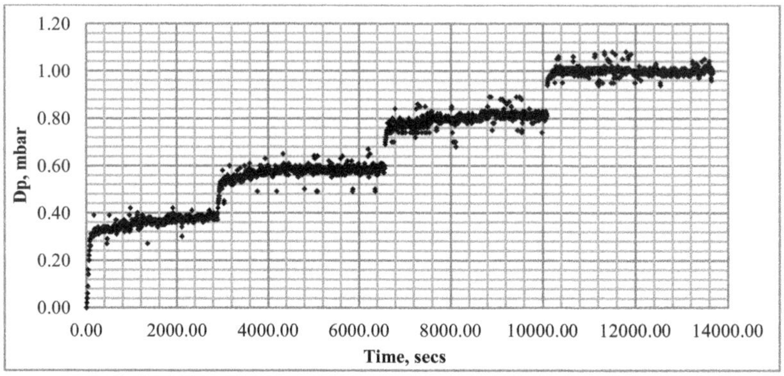

Figure 13. Pressure difference for different rate in injection

Table 4. Stabilized pressure for each injection rate

Rate, cc/min	Rate, cc/sec	Stabilized Dp, mbar	Stabilized Dp, atm
0	0	0	0
0.0667	0.001111	0.37	0.00036519
0.1000	0.001667	0.59	0.00058233
0.1333	0.002222	0.80	0.00078960
0.1667	0.002778	1.00	0.00098700

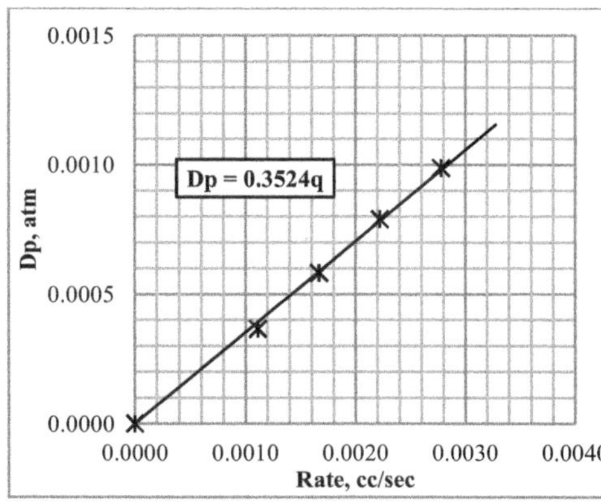

Figure 14. Pressure difference versus injection rate

The results are in good agreement with the correlation given by Rump and Gupta (1975) that is for estimating the absolute permeability in glass bead packs.

$$k = \frac{\varphi^{5.5}}{5.6} d^2 \tag{24}$$

Here φ is porosity and d is glass bead diameter in μm and permeability will be estimated in md. Table 5 shows experiment condition and table 6 shows the absolute permeability values from experiment and correlation.

Table 5. Experimental conditions

Items	Properties
Injection rate	Variable in each set
System pressure (bar)	Atmospheric pressure
Overburden pressure (bar)	10
Experiment temperature (°C)	22
Chamber Length (cm)	5.56
Chamber Diameter (cm)	4.18

Table 6. Absolute permeability measurements

Calculated k, md	Minimum GBs size **70 μm**	Maximum GBs size **110 μm**	Average
From equation 24	5437	13426	9431
From experiment			7980

3.2 Relative Permeability Results

Table 7 gives overall information about the successful tests done in this study. All the experiments are carried out for porosity of 39.7 % and calculated pore volume is 15.6 cc.

Table 7. Experiments in this study

Experiment type	Test No	Fluids	Injection rate, cc/min	Comments
Drainage	1	fluid set 1	0.1	
	2	fluid set 1	0.2	
	3	fluid set 1	1	
	4	fluid set 1	5	
Imbibition	5	fluid set 1	0.1	
	6	fluid set 1	0.2	
	7	fluid set 1	1	
	8	fluid set 1	5	end points values
Drainage	9	fluid set 2	0.1	
	10	fluid set 2	0.2	
	11	fluid set 2	1	
	12	fluid set 2	5	
Imbibition	13	fluid set 2	0.1	
	14	fluid set 2	0.2	
	15	fluid set 2	1	
	16	fluid set 2	5	end points values

3.2.1 Results for Fluid Set 1

According to the measured absolute permeability and by using history matching of pressure drop and production, the best relative permeability model is matched to the data. Figures 15 and 16 show the history matching results for a very low injection rate test and Figure 17 is a plot of relative permeability obtained from this test. The tests arecarried out in different injection rates and the results are presented in Figure 18. Both saturation end-points and corresponding relative permeabilities are affected by the rate. Imbibition tests are completed after the drainage test with the highest rate. Test number 4 is replicated, and afterwards imbibition tests are carried out with different rates. Four different rates of imbibition are examined and three reasonable pressure drops obtained. Very high imbibition rates lead to unstable pressure response but it was still possible to measure the end point values. Figure 21 compares the relative permeability measurements for the imbibition test. The results are less sensitive to the rate, comparing to the drainage tests. Table 8 shows the correlation and parameters used for history matching and end point values. For simplicity, the k_{rg} and k_{rw} terms used in the plots, that are actually relative permeabilities to the non-wetting and wetting phases.

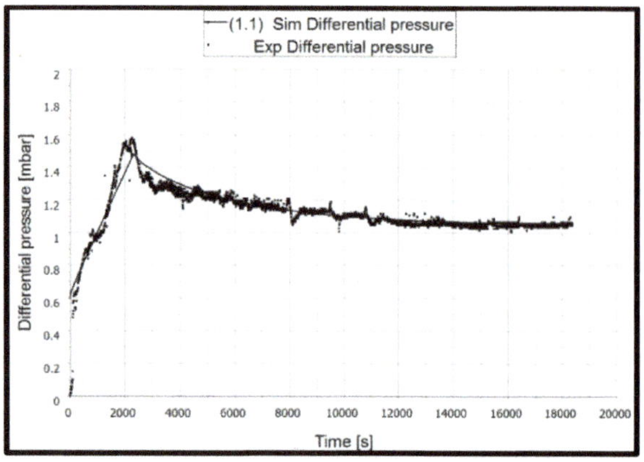

Figure15. Drainage test, history matching of differential pressure, injection rate of 0.1 cc/min

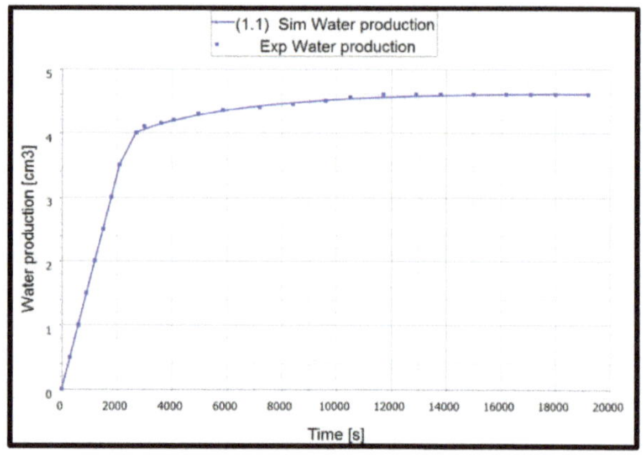

Figure 16. Drainage test, history matching of water production, injection rate of 0.1 cc/min

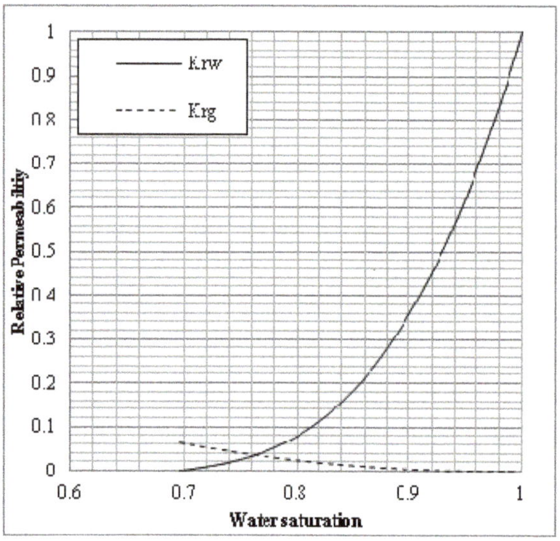

Figure17. Drainage test, injection rate=0.1 cc/min

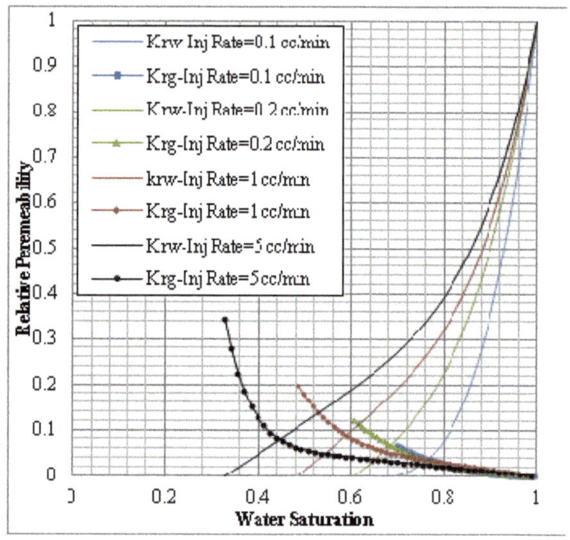

Figure 18. Drainage tests, effect of injection rate

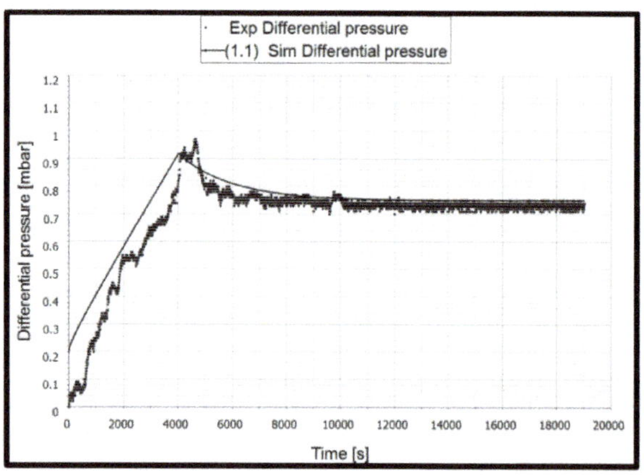

Figure19. Imbibition test, history matching of differential pressure, injection rate of 0.1 cc/min

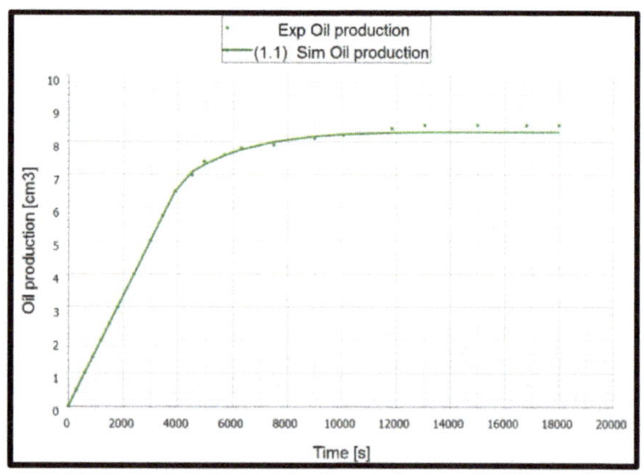

Figure 20. Imbibition test, history matching of water production, injection rate of 0.1 cc/min

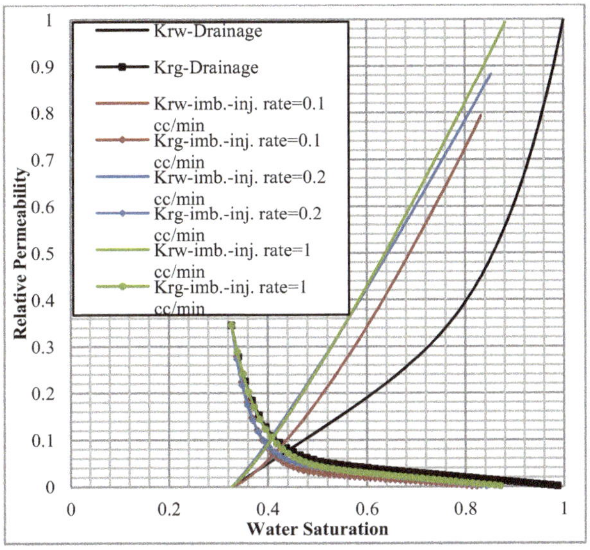

Figure 21. Imbibition tests, effects of injection rate

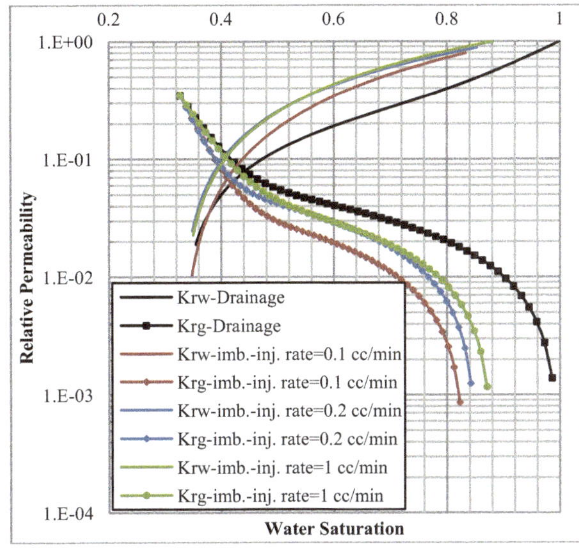

Figure 22. Imbibition tests, effects of injection rate – logarithmic

Table 8. History matching data - fluid set 1

Test type	Correlation	Inj. rate, cc/min	Capillary Number	Nw	Ng	A	B	Krw(Sgi)	Krg(Swir)	Swir	Sgr
Drainage	Sigmund & McCaffery	0.1	$\mathbf{1.956 \times 10^{-9}}$	2.842	2.317	0.116	0.006	1.000	0.065	0.696	0.000
	Sigmund & McCaffery	0.2	$\mathbf{3.912 \times 10^{-9}}$	3.245	3.136	0.479	0.221	1.000	0.121	0.604	0.000
	Sigmund & McCaffery	1	$\mathbf{1.956 \times 10^{-8}}$	4.823	6.631	0.794	0.600	1.000	0.198	0.485	0.000
	Sigmund & McCaffery	5	$\mathbf{9.781 \times 10^{-8}}$	5.765	8.680	0.816	0.245	1.000	0.346	0.327	0.000
Imbibition	Sigmund & McCaffery	0.1	$\mathbf{1.693 \times 10^{-7}}$	1.382	12.110	0.035	0.138	0.793	0.346	0.327	0.167
	Sigmund & McCaffery	0.2	$\mathbf{3.867 \times 10^{-7}}$	1.128	14.188	0.001	0.214	0.882	0.346	0.327	0.147
	Sigmund & McCaffery	1	$\mathbf{1.693 \times 10^{-6}}$	1.184	10.525	0.001	0.197	0.992	0.346	0.327	0.118

3.2.2 Results for Fluid Set 2

The same procedure carried out for fluid set 2. The trend of rate effect is almost the same in the new set of tests but end points and corresponding relative permeabilities are affected due to wettability change. The only variable parameter in this set of experiment is wettability characteristics of the fluids and the porous material. Figures 23 to 28 show the result. Table 9 shows different correlations used to match the results.

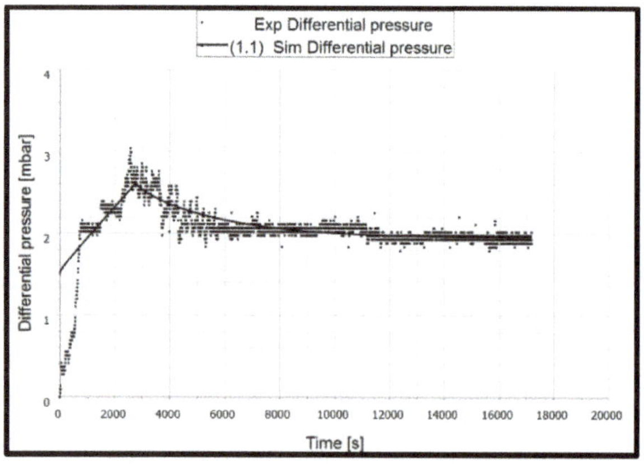

Figure 23. Drainage test, history matching of differential pressure, injection rate=0.1 cc/min

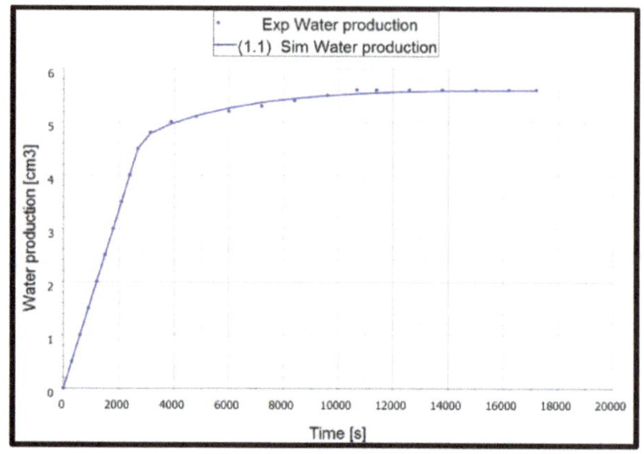

Figure 24. Drainage test, history matching of water production, injection rate=0.1 cc/min

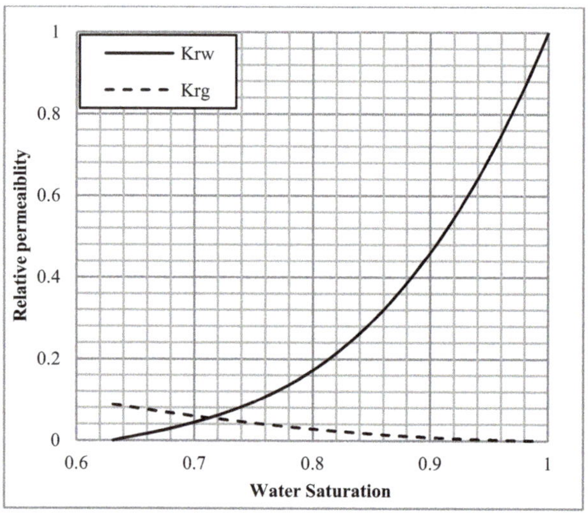

Figure 25. Drainage test, injection rate=0.1 cc/min

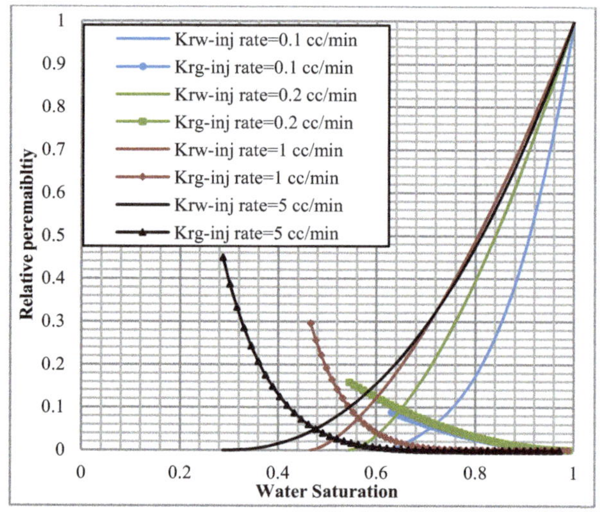

Figure 26. Drainage tests, injection rate=0.1 cc/min

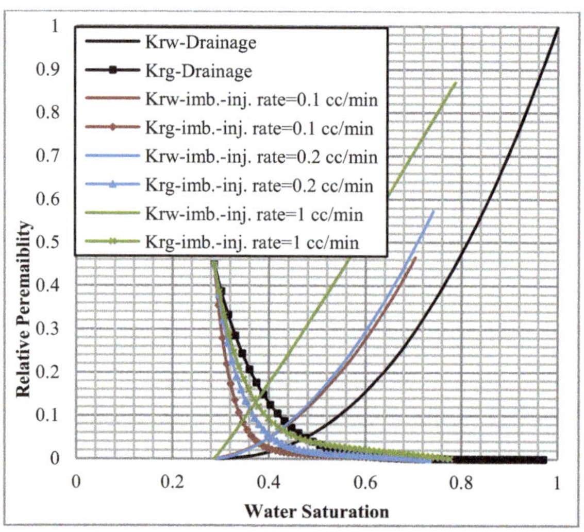

Figure 27. Imbibition tests, effect of injection rate

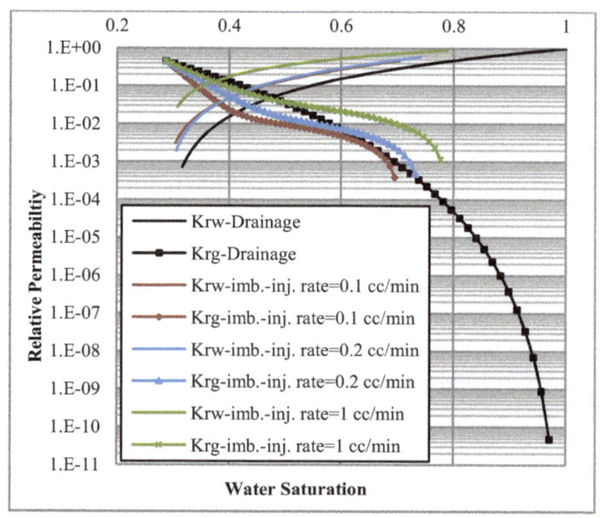

Figure 28. Imbibition tests, effect of injection rate-logarithmic

Table 9. History matching data - fluid set 2

Test type	Correlation	Inj. rate, cc/min	Capillary Number	Nw	Ng	A	B	Krw(Sgi)	Krg(Swir)	Swir	Sgr
Drainage	Sigmund & McCaffery	0.1	$\mathbf{4.905 \times 10^{-8}}$	2.973	1.894	0.260	0.028	1.000	0.088	0.631	0.000
	Sigmund & McCaffery	0.2	$\mathbf{9.810 \times 10^{-7}}$	1.607	1.988	0.001	0.015	1.000	0.158	0.545	0.000
	Sigmund & McCaffery	1	$\mathbf{4.905 \times 10^{-7}}$	1.527	6.852	-	-	1.000	0.295	0.466	0.000
	Sigmund & McCaffery	5	$\mathbf{2.454 \times 10^{-6}}$	2.266	3.542	-	-	1.000	0.449	0.287	0.000
Imbibition	Sigmund & McCaffery	0.1	$\mathbf{4.233 \times 10^{-7}}$	1.928	12.010	0.127	0.044	0.465	0.449	0.287	0.296
	Sigmund & McCaffery	0.2	$\mathbf{8.467 \times 10^{-7}}$	1.788	8.602	0.001	0.057	0.573	0.449	0.287	0.259
	Sigmund & McCaffery	1	$\mathbf{4.233 \times 10^{-6}}$	1.088	8.193	0.049	0.142	0.869	0.449	0.287	0.213

4. Discussions

The nature of flooding brine with CO_2 and imbibition process afterwards makes it to be difficult to control. Very low viscosity of CO_2 and high density difference results in unfavourable mobility ratio. Viscous fingering seems to be an evitable process in such systems. Some former studies are done at conditions that lead to onset of viscous fingering in displacement tests. Peters and Flock (1981) have presented a dimensionless instability number to estimate the onset of viscous fingering in immiscible displacement processes:

$$I_{sc} = \frac{(M-1)(v-v_c)\mu_w D^2}{C^* \sigma k_{wor}} \tag{25}$$

In this equation I_{sc} is a dimensionless stability number, M is mobility ratio, v is superficial velocity and v_c is characteristics velocity, D is core diameter, C^* is wettability number determined experimentally and k_{wor} is permeability to water. They observed that dimensionless instability numbers below 13.56 is considered as flooding with presence of viscous fingering. In this study, according to the stability criteria for core flooding, during drainage test and injecting the non-wetting phase with very high viscosity ratio between fluids, the process should be done at very low displacement velocities (very low injection rates) to achieve a stable front. The problem with very low displacement rate is the creation of capillary end-effects. This means that the injection rate should be high enough to avoid capillary end-effect and low enough to avoid viscous fingering problems. This is a major challenge.

In order to overcome this challenge, results of the experiments with different rates are used to define correlations to fit to the parameters in the stabilized period. Figure 29 shows the effects of rate on the relative permeability end points values. By increasing the rate, these values increase and it start to stabilize in high injection rate. Following equation is fitted to the results:

$$k_r = a_1(1 - e^{-a_2 q}) \tag{26}$$

Where a_1 and a_2 are matching parameters. A similar method has been suggested by Bennion et al. (1991) in the case of rate dependent flooding experiments. Stabilized flow conditions are ideally obtained when considering infinite flow rates that lead to relative permeability values of $k_r = a_1$. The same procedure is used to fit the residual and initial saturation values. In this case the trend is decreasing at higher injection rates.

$$S_{wir} = a_1(1 + e^{-a_2 q}) \tag{27}$$

$$S_{CO2r} = a_1(1 + e^{-a_2 q}) \tag{28}$$

Here S_{wir} is irreducible water saturation after drainage from curve fitting and S_{CO2r} is the residual CO_2 saturation. Again, stabilized flow conditions are considered in the case of infinite flow rates that lead to $S_{wir} = a_1$, $S_{CO2r} = a_1$. After fitting the correlations and determining the stabilized values, we have one set of relative permeability curves for each fluid set. The modifications show that the results are close to the highest injection rate tests. Curve fitting values from correlation 26 and 27 and 28 give the end points, while the curvatures of the final relative permeability models are obtained from the highest injection rate imbibition and drainage tests. Table 10 shows the fitting parameter and Figures 33 and 34 are relative permeabilities obtained after modification. In order to scale the results to the real storage reservoir conditions, the capillary number is calculated for the experiments. Figure 32 shows the residual CO_2 saturation obtained from the experiment and is plotted against capillary number. Former simulation studies of Utsira brine formation of Sleipner and Viking formation (Canadian brine formation) revealed the range of capillary number using inter-block velocity in simulation model (Soroush et al., 2013).

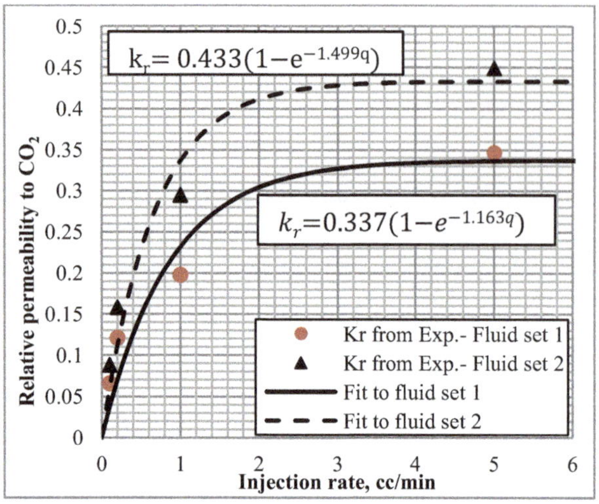

Figure 29. Relative permeability to CO_2, effect of imbibition rate and wettability

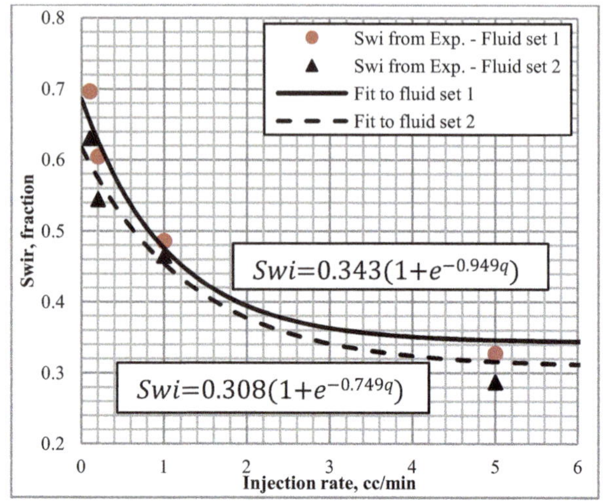

Figure 30. Swi, effect of drainage rate and wettability

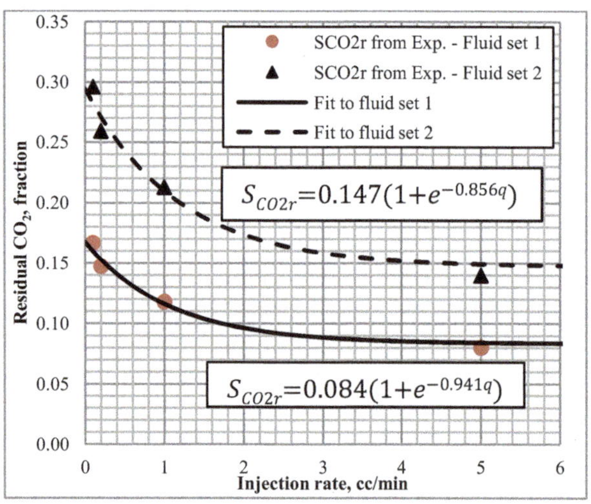

Figure 31. Residual CO_2, effect of imbibition rate and wettability

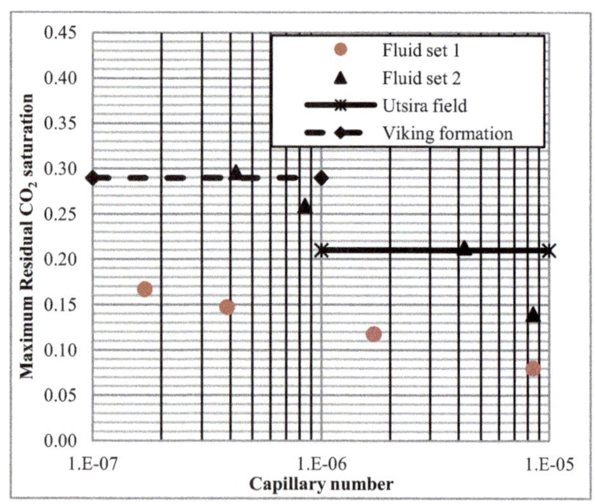

Figure 32. Residual CO_2 saturation effect of imbibition rate and wettability

Table 10. History matching data, eliminating effect of rate, modified relative permeability

Test type	Correlation	Fluid set	Nw	Ng	A	B	Krw(Sgi)	Krg(Swir)	Swir	Sgr
Drainage	Sigmund & McCaffery	1	5.765	8.680	0.816	0.245	1.000	0.337	0.343	0.000
	Corry	2	2.266	3.542	-	-	1.000	0.433	0.308	0.000
Imbibition	Sigmund & McCaffery	1	1.184	10.240	0.001	0.197	0.992	0.337	0.343	0.084
	Sigmund & McCaffery	2	1.088	8.193	0.049	0.142	0.869	0.433	0.308	0.147

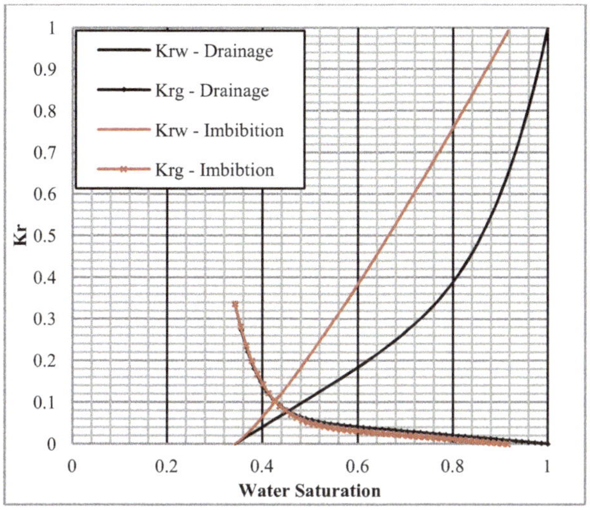

Figure 33. Modified relative permeability, fluid set 1

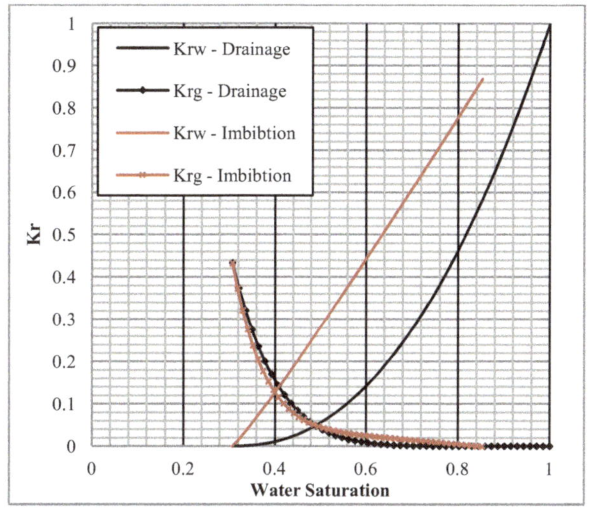

Figure 34. Modified relative permeability, fluid set 2

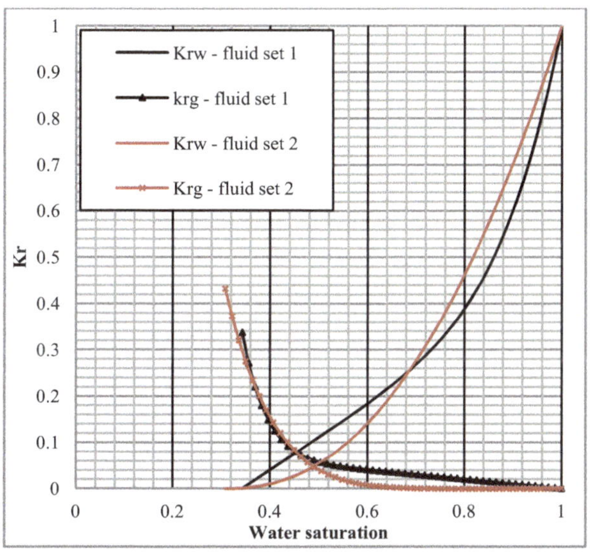

Figure 35. Drainage relative permeability, fluid set 1 and 2, effect of wettability

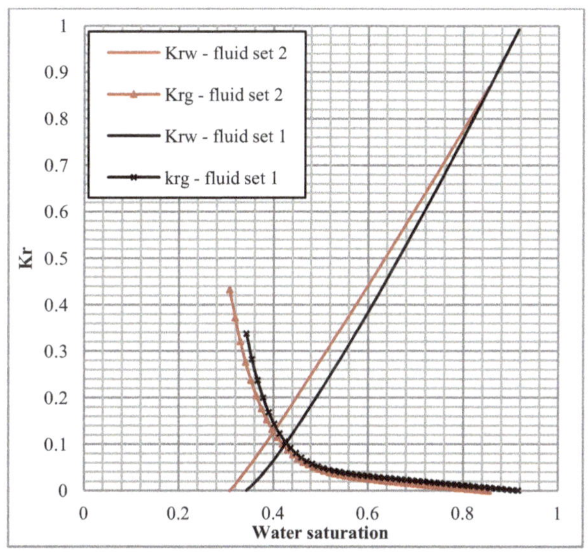

Figure 36. Imbibition relative permeability, fluid set 1 and 2, effect of wettability

In the experiments with fluid set 1 (higher wettability of wetting phase to the medium) residual CO_2 saturation is increased from 8% to 16.7% (8.7% grow) because of lowering the imbibition rate. On the other hand, in fluid set 2 (lower wettability of wetting phase) it is increased from 14% to 29.6% (15.6% grow). Using fluid set 2, the relative permeability to the CO_2 (non-wetting phase in the representative fluid) is increased by 9.6% while $Swir$ is decreased only by 3.5% because of rate effect.

By eliminating the effect of rate and comparing the results of the imbibition tests, the influence of wettability is analysed. Using fluid set 2, the relative permeability to the CO_2 (non-wetting phase in the representative fluid) is increased by 9.6% while Swir is decreased only by 3.5%. Doing imbibition with fluid set 2 results in 12.3% decrease in relative permeability to water at the end point and CO_2 residual saturation increased from 8.4% to 14.7% (6.3% increase) and this is just because of different wettability characteristics of the system. Although the percentage of the change due to wettability is limited to this study, it can show the overall trend.

4.1 Validity of the Results

Relative permeability calculations are based on pressure drop readings from pressure transducer and volumetric and material balance and visual readings of the production. Although the pressure drops are in low ranges, the accuracy of the Fuji electronics pressure drop measurements that is used in this study is in the order of 0.01 mbar, which is acceptable. The production recording is double-checked by both volumetric and visual readings from the separator, and a minimum error is observed. 60 μm nylon filters are used at the chamber both ends to prevent glass beads production. Filter size has been optimized so that it is not being too fine to create errors in pressure drop and not too coarse to allow glass beads production. The pressure drop created by this filters is considered to be very small in comparison to the glass beads column pressure drop. There are also other laboratory conditions and simulations errors, but overall range of the error is acceptable.

5. Conclusions

In this study, laboratory experiments are carried out to investigate the effects of flooding rates in both drainage and imbibition processes in CO_2-brine systems. The wettability behaviour also created some interesting results in these systems. Results may be used in both new experimental activities (2D model set up in connection with this work) and in simulation studies. The following conclusions are offered:

(1) Results of this study verify that using two sets of fluids representing the CO_2-brine systems in the reservoir can facilitate the research. Using a liquid-liquid system with the same density difference and viscosity ratio may not be exactly representative of the CO_2 storage at reservoir conditions; however, scaling the results shows that the designs are pretty close to the real conditions (figure 32).

(2) Correlations are developed by fitting the residual CO_2 saturation and corresponding relative permeability in different rates and wettability conditions. Using these correlations, it is possible to estimate the CO_2 residual values in different aquifer influxes. This method is useful in obtaining estimations of CO_2 residuals in real cases.

(3) The results show that in the case of less wettability of the wetting phase to the porous media, residual CO_2 trapping is more sensitive to the imbibition rate. This is clear from trend of CO_2 trapping volume versus capillary number in figure 32. For both sets of fluids, more CO_2 is trapped in lower capillary number (lower imbibition rate).

(4) Comparison of the analysis of drainage and imbibition tests for two fluid sets shows the effect of wettability. In conclusion, using fluid set 2 (lower wettability of the wetting phase to the porous material) leads to increasing relative permeability to CO_2 and decreasing $Swir$. Residual CO_2 saturation after imbibition process is increased because of using fluid set 2. Wettability affected the relative permeability values at the end points more than other parameters. The weakest effect is observed on the irreducible water saturation after drainage.

Acknowledgements

This publication has been produced with support from the BIGCCS Centre, performed under the Norwegian research program Centres for Environment-friendly Energy Research (FME). The authors acknowledge the following partners for their contributions: Conoco Phillips, Gassco, Shell, Statoil, TOTAL, GDF SUEZ and the Research Council of Norway (193816/S60). Special thanks to Erik Lindeberg, Torleif Holt, Øyvind Haave, Bård Bjørkvik, Szczepan Polak and Svein Arlid for invaluable technical supports, help and advice in SINTEF petroleum research.

References

Bandara, U. C., Tartakovsky, A. M., & Palmer, J. (2011). Pore-scale study of capillary mechanism during CO_2 injection in geological formations. *International Journal of Greenhouse Gas Control 5,* 1566-1577. http://dx.doi.org/10.1016/j.ijggc.2011.08.014

Bennion, B., & Bachu, S. (2005). Relative permeability characteristics for supercritical CO_2 displacing water in variety of potential sequestration zones in the western Canada sedimentary basin. SPE 95547, *SPE annual technical conference, Dallas, Texas, US.* http://dx.doi.org/10.2118/95547-MS

Bennion, B., & Bachu, S. (2008). Drainage and imbibition relative permeability relationships for supercritical CO2/brine and H2S/brine systems in intergranular sandstone, carbonate, shale, and anhydrite rocks. *SPE Reservoir Evaluation & Engineering, 11*(3), 487-496. http://dx.doi.org/10.2118/99326-PA

Bennion, B., & Thomas, F. B. (1991). Recent improvement in experimental and analytical techniques for determination of relative permeability from unsteady state flow experiments. Prepared for presentation at SPE 10th Technical conference in Port of Spain, Trinidad. Retrieved from http://www.weatherfordlabs.com/media/34746/recent_improvements_in_experimental-rel_perm.pdf

Burdine, N. T. (1953). Relative permeability calculations from pore size distribution data. *Journal of Petroleum Technology, 5*(3), 71-78. http://dx.doi.org/10.2118/225-G

Carlson, F. M. (1981). Simulation of relative permeability hysteresis to the non-wetting phase. SPE10157-MS, *Soc. of Pet. Eng., Richardson, TX.* http://dx.doi.org/ 10.2118/10157-MS

Chiquet, P., Daridon, J. L., Broseta, D., & Thibeau, S. (2007). CO< sub> 2</sub>/water interfacial tensions under pressure and temperature conditions of CO< sub> 2</sub> geological storage. *Energy Conversion and Management*, *48*(3), 736-744. http://dx.doi.org/ 10.1016/j.enconman.2006.09.011

Chierici, G. (1984). Novel relations for drainage and imbibition relative permeabilities. *Old SPE Journal, 24*(3), 275-276. http://dx.doi.org/10.2118/10165-PA

Corey, A. T. (1954). The interrelation between gas and oil relative permeabilities. *Prod., 19*(1), 38-41.

Gupta, S., & Trushenski, S. (1979). Micellar flooding-compositional effects on oil displacement. *Old SPE Journal, 19*(2), 116-128. http://dx.doi.org/ 10.2118/7063-PA

Hebach, A., Oberhof, A., Dahmen, N., Kögel, A., Ederer, H., & Dinjus, E. (2002). Interfacial tension at elevated pressures measurements and correlations in the water+ carbon dioxide system. *Journal of Chemical & Engineering Data, 47*(6), 1540-1546. http://dx.doi.org/ 10.1021/je025569p

Honarpour, M., Koederitz, L., & Herbert, H. A. (1986). *Relative permeability of petroleum reservoirs.* C.R.C Press Inc.

Iffly, R., Rousselet, D. C., & Vermeulen, J. L. (1972, October). Fundamental study of imbibition in fissured oil fields. In *Fall Meeting of the Society of Petroleum Engineers of AIME.* http://dx.doi.org/10.2118/4102-MS

Johnson, E. F., Bossler, D. P., & Naumann, V. O. (1959). Calculation of relative permeability from displacement experiments. *Trans. AIME, 216*, 370-372.

Kaviany, M. (1995). *Principles of heat transfer in porous media* (2nd Ed.). New York: Springer.

Killough, J. E. (1976). Reservoir simulation with history-dependent saturation functions. *Old SPE Journal, 16*(1), 37-48. http://dx.doi.org/ 10.2118/5106-PA

Kleppe, J., Delaplace, P., Lenormand, R., Hamon, G., & Chaput, E. (1997, October). Representation of capillary pressure hysteresis in reservoir simulation. In *SPE Annual Technical Conference and Exhibition.* http://dx.doi.org/10.2118/38899-MS

Krause, M., Perrin, J. C., & Benson S. M. (2011). Recent progress in predicting permeability distribution for history matching core flooding experiments. *Energy Procedia, 4*, 4354-4361. http://dx.doi.org/10.1016/j.egypro.2011.02.387

Krever, S. C. M., Pini, R., Zou, L., & Benson, S. M. (2011). Relative permeability and trapping of CO_2 and water in sandstone rocks at reservoir conditions. *Water resources research, 48.* http://dx.doi.org/10.1029/2011WR010859

Kua, C. W., Perrin, J. C., & Benson S. M. (2011). Simulation studies of effects of flow rate and small scale heterogeneity on multiphase flow of CO_2 and brine. *Energy Procedia 4,* 4516-4523. http://dx.doi.org/10.1016/j.egypro.2011.02.408

LAND, C. (1971). Comparison of calculated with experimental imbibition relative permeability. *Old SPE Journal, 11*(4), 419-425. http://dx.doi.org/10.2118/3360-PA

Lomeland, F., Ebeltoft, E., & Thomas, W. H. (2005, August). A new versatile relative permeability correlation. In *international symposium of the society of core analysts held in Toronto, Canada.* Retrieved from http://www.ri.reslab.no/sendra-papers.aspx

Nordbotten, J. M., Celia, M. A., & Bachu, S. (2005). Injection and storage of CO_2 in deep saline aquifers: Analytical solution for CO2 plume evolution during injection. *Transport in Porous media, 58*(3), 339-360. http://dx.doi.org/10.1007/s11242-004-0670-9

Perrin, J. C., & Benson, S. (2010). An experimental study on the influence of sub-core scale heterogeneities on CO2 distribution in reservoir rocks. *Transport in porous media, 82*(1), 93-109. http://dx.doi.org/10.1007/s11242-009-9426-x

Peters, E., & Flock, D. (1981). The onset of instability during two-phase immiscible displacement in porous media. *Old SPE Journal, 21*(2), 249-258. http://dx.doi.org/10.2118/8371-PA

Pini, R., Krevor, S. C. M., & Benson, S. M. (2012). Capillary pressure and heterogeneity for the CO_2-brine system in sandstone rocks at reservoir conditions. *Advances in Water Resources 38,* 48-59. http://dx.doi.org/10.1016/j.advwatres.2011.12.007

Rumpf, H., & Gupte, A. R. (1975). The influence of porosity and grain size distribution on the permeability equation of porous flow. *NASA STI/Recon Technical Report N, 75*, 30484.

Schlumberger eclipse technical description. (2012). Retrieved from http://www.slb.com

Sendra user guide. (2012). Retrieved from http://www.sendra.no

Sigmund, P. M., & McCaffery, F. G. (1979). An Improved Unsteady-State Procedure for Determining the Relative-Permeability Characteristics of Heterogeneous Porous Media (includes associated papers 8028 and 8777). *Old SPE Journal, 19*(1), 15-28. http://dx.doi.org/ 10.2118/6720-PA

Soroush, M., Wessel-Berg, D., & Kleppe J. (2013), Effects of wetting behaviour on residual trapping in CO_2-brine systems. 165334-MS, *SPE Western Regional & AAPG Pacific Section Meeting, Joint Technical Conference, Monterey, CA, USA.* http://dx.doi.org/ 10.2118/165334-MS

Spiteri, E., Juanes, R., Blunt, M., & Orr, F. (2008). A new model of trapping and relative permeability hysteresis for all wettability characteristics. *Spe Journal, 13*(3), 277-288. http://dx.doi.org/10.2118/96448-PA

Suekane, T., Zhou, N., Hosokawa, T., & Matsumoto, T. (2010). Direct observation of trapped gas bubbles by capillarity in sandy porous media. *Transport in porous media, 82*(1), 111-122. http://dx.doi.org/10.1007/s11242-009-9439-5

Wang, J., Dong, M., & Asghari, K. (2006, April). Effect of Oil Viscosity on Heavy Oil-Water Relative Permeability Curves. In *SPE/DOE Symposium on Improved Oil Recovery.* http://dx.doi.org/10.2118/99763-M

Wessel-Berg, D., Kleppe, J., & Soroush, M. (2013, June). Investigating Residual Trapping Mechanism in CO_2 Storage in the Brine through Analytical and Numerical Simulation Methods. In *75th EAGE Conference & Exhibition incorporating SPE EUROPEC 2013.* http://dx.doi.org/ 10.2118/164808-MS

9

Carbon Market Funding for the Development of Areas of Pastures: The Case of Small Farms in the State of Rio de Janeiro, Brazil

Fabiano Oliveira Chagas[1], Osvaldo Luiz Gonçalves Quelhas[1], Roberto Guimarães Pereira[1] & Valdir Lameira[2]

[1] Universidade Federal Fluminense, Rio de Janeiro, Brasil

[2] University of Coimbra, Coimbra, Portugal

Correspondence: Fabiano Oliveira Chagas, Universidade Federal Fluminense, Rio de Janeiro, Brasil. E-mail: fabiano.chagas@gmail.com

Abstract

This study aims to examine, within the Local Development, the reforestation with eucalyptus on land used for grazing. Uses the Market Certified Emission Reductions (CER) as a source for funding, through projects of Clean Development Mechanism (CDM). The study analyzes the cash flow of small-scale beef cattle operation and simulates the eucalyptus plantation on the farms. The area selected for this study covers the family-structured farms of small and medium extension on the beef cattle industry in north-central state of Rio de Janeiro, Brazil. The research method used was a literature review, application of the methodology recommended by the UN and cash flow analysis. Analysis of Net Present Value (NPV) considering the eucalyptus plantations for energy purposes, has brought positive values. The study confirms that the implementation of forestry along these models is itself a profitable activity. This study applied the methodology AR-AMS0001, recommended by the executive board of the United Nations on Climate Exchange.

Keywords: sustainable development, CDM for reforestation, engineering sustainability, local development

1. Introduction

The farms all over the state of Rio de Janeiro, whose source of income is agriculture and livestock, according to agricultural census of the Brazilian Institute of Geography and Statistics (CIDE, 2008) covers 1.6 million hectares. The cattle ranches basically depend on grazing and sugar cane, for the livestock maintenance and in some cases, a small farming focused not much more than subsistence. Many of these farms are maintained by families who hold possession of these areas because they are heirs of ancient estates, as presented in the study Nozoe (2006).

The work presented here aims to answer the following main question:

Is it financially and economically feasible to replace the beef cattle and milk production by small-scale eucalyptus plantations for siderurgical charcoal production and obtain financing through CDM forestry projects, after obtaining obtaining funding through forestry CDM projects?

The problem was investigated within the following economic and financial conditions observed in the Brazilian environment:

- Farmers who economically depends on the cattle on a small scale production have little chance to prosper economically;
- The charcoal supply for the steel industry does not fully address the necessary demand from steel producers, and the planned investments to increase forest for this purpose does not follow the same proportion of growth investments in steel production;
- The implementation of CDM forestry projects in Brazil for energy co-generation is a strategic alternative energy source to replace fossil fuels.

This paper aims to analyze the CDM implementation on forestry projects as a strategic alternative to the Local Development on underused areas focused so far only for grazing.

Derive the following specific objectives:

- Run the economic and financial analysis of beef cattle and milk in the area where the research was performed;
- Studying the cash flow of a eucalyptus plantation in reforestation;
- Simulate an economic and financial planning to plant eucalyptus trees and cut trees on a 20 years plan to meet the siderurgical charcoal market demand.

The object of study are the farms of Rio de Janeiro's north-central area, which comprises Macae, Conceição de Macabu, Quissamã and Carapebus, family-owned small and medium extensions, between 25 and 60 hectares, and who use the cattle beef and milk as a source of income.

2. Research Method

We use a comparative study between the average income of livestock farms and the income from the quantification of Certified Emission Reduction (REC) from reforestation with eucalyptus sales to the charcoal market during 20 years.

The project CARBON FIX-TERRA BOA developed in Guararapes-Sao Paulo, Brazil, used this method in your project to calculate the reducing emissions of greenhouse gases through carbon sequestration by restoration of native forest. The area where the project was developed had been degraded due to its use as pasture area.

3. Literature Review

Green and Unruh (2010) in their study report that the implantation of CDM reforestation, such as the region studied here, generate positive impacts on local agriculture and increase income to the farmer, improving soil quality, increased capacity hydrographic investments in regional infrastructure, transfer of technology applied to agriculture, among others. However, the authors alert that there are potential risks of negative impacts that can limit the project implementation, such as the destruction of existing vegetation resources, increase threats of watter damage or even the decrease of watter availability.

The possible impacts caused on this type of implementation is not part of this document. On Brazil, by December 2010 only one forest carbon credits project won approval by the Authority designated as told by Puentes (2010). The project developed by Plantar S/A, entitled "Reforestation as Renewable Source of Wood Supply for Industrial Use in Brazil", aims to reduce GHG emissions in Brazilian metal industry and has an area of 11,700 hectares, located in the state of Minas Gerais, providing removal of 2.27 million tons of CO_2 over a period of 30 years. Puentes (2010) continues, reporting that two other projects were under development in Brazil: the first is the AES Tiete to reforest areas for permanent preservation around the reservoirs of the Tietê River, an area of 8000 hectares can capture 2.7 million hectares of tonnes of CO_2 in 30 years. The second is the Agroforestry Soroteca with 2319 hectares, removing 6.5 million tons of CO_2 in 24 years.

Reforestation CDM projects worldwide are using the methodology AR-AMS0001 to support their estimates. Puentes (2010) cites in his dissertation projects in India, with reductions of 11.5 million tonnes of CO_2 in Vietnam with reductions of 2.66 million tons of CO_2, and in Bolivia, Paraguay and Uganda minors scales.

3.1 Estimating Carbon Sequestration

According to the Brazilian Agricultural Research Corporation-EMBRAPA (2002), it is known that the ability of fixation and carbon uptake by trees is directly dependent on the species, considering the growth rate, longevity, planted area, the climate and other variables.

The study conducted here does not consider the costs of hiring regulated firms to analyze the feasibility of implementing the CDM project, the implementation of the project itself, and also does not consider the costs of monitoring and audits to be conducted during the 21 years of project implementation.

According to the Brazilian Agricultural Research Corporation-EMBRAPA (2002), it is known that the ability of fixation and carbon uptake by trees is directly dependent on the species, considering the growth rate, longevity, planted area, the climate and other variables.

The study conducted here does not consider the costs of hiring regulated firms to analyze the feasibility of implementing the CDM project, the implementation of the project itself, and also does not consider the costs of monitoring and audits to be conducted during the 21 years of project implementation.

3.1.1 Calculation Methodology Adopted

In the proposed case study, the AR-AMS0001 methodology will be used, which becomes the "Methodology simplified baseline and monitoring of afforestation and reforestation small-scale project activities under the clean development mechanism implemented on grasslands". This methodology is approved by IPCC-NGGIP, as part of ONU, to evaluate the economic potential of the use of CDM projects as a source of funding.

The methodology presents a series of equations for calculation, divided into two sections: the first is the baseline calculation and the second section is reserved for the monitoring processes.

The methodology provides equations to calculate:

- Carbon stock baseline;
- The above ground biomass estimates;
- Underground biomass estimates;
- Carbon stock below ground estimates;
- Baseline removal of Greenhouse gases within the project area;
- Calculation of identified leakage;
- RCE.

At the Monitoring section, there are equations to perform the following calculations:

- Carbon stocks of above-ground biomass;
- Carbon stocks in below-ground biomass;
- Measuring leakages;
- Net GHG removal calculation;
- Monitoring frequency.

3.2 Financial Analysis and Projections

To perform a detailed analysis of the feasibility of a project some mathematical equations were used-financial, where part of a baseline and projections based on statistical simulations.

In these sections it applies methods of mathematical and financial for the following scenarios:

- Small-scale farms;
- Wood market for charcoal from eucalyptus reforestation covering 11/12 of the pasture area;
- Implementation of CDM project to carbon credits market in the reforested area.

4. Data Collection and Analysis, and Proposed Use of Carbon Credit Market or Local Development Growth

4.1 Estimate of Economic and Financial Beef Cattle and Milk in the Region

Schier (2005), on his research, uses the principles of full absorption costing, which seeks to allocate the entire cost of producing the products.

The method described by Schier (2005) to calculate the full absorption costing considers the following variables:

- Expenses related to production (fixed and variable), which are treated as "sales cost";
- Net sales less cost of goods sold, resulting in gross profit for the period;
- Gross profit less the expenses of the period resulting in profit before applying taxes and social contribution on net income;
- When the production period is not totally sold, products in stock so go to the next period.

In this study, the following assumptions were made for the parameterization of the calculations for the determination of the cost:

- For the milk market, the entire production is purchased by the dairy cooperatives in the city;
- Regarding to beef cattle market, when a farmer has availability to sell, there are never problems in abattoirs and / or butchers cities in buying the squad;
- To consider these two items, the research will not be considered the remaining stock or it was not sold in the period, the milk is harvested or weight of cattle for slaughter;
- Since the raw material is nothing else than the amount of productive livestock that the farmer has, and knowing

that we are dealing with two distinct markets with the same ingredient (beef and milk), the cost of raw materials will result from the amount of productive animals and their depreciation;

• As characteristic of the region, because the flock feed predominantly on pasture, the cost of feeding livestock are derisory;

• The reproduction takes place by natural procedures, excluding the need to carry out artificial insemination. Therefore, there is no cost involved with insemination.

4.1.1 Accounting and Financial Analysis of Beef Cattle and Milk

For this study, the sample was taken for a property in the county of Macabuzinho in the city of Conceição de Macabu which has 72.5 hectares. Of these, 3 hectares for planting sugarcane, 1 hectare to the farmhouse and barn and all the rest, 67.5 acres for grazing. The property has 139 head of cattle crossbred with 98 matrices, 1 and 40 breeding males for fattening and subsequent sale for slaughter. This is scenario of livestock for the dairy industry used in this research.

4.1.1.1 Roster Maintenance

Using the method of calculating the cost of milk production Segala and Silva (2007), we have in Table 1 the distribution of cows for a period of 4 months.

Table 1. Distribution of the squad for a period of four months

Number of cows selected for reproduction	nov/09	Dec/09	jan/10	Feb/10
In lactation	61	58	58	62
Dry	18	19	20	19
Heifers and heifer calves	19	21	19	18
I cannot find precise English equivalent	98	98	97	99

The squad has an average of 59 cows with this characteristic and applied to the purchase price of the matrices is $ 2,000.00 already in milk production. Table 2 shows the monthly depreciation based on the presented data.

Table 2. Depreciation of matrices for the milk production

Matrizes	Valor de compra	Valor total	Depreciaçã o por ano	% depreciação por mês	Valor depreciado por mês
59 in the production phase - Nov-09	2000	R$ 122.000,00	10%	0,83%	R$ 1.012,60
58 in the production phase - Dec-09	2000	R$ 116.000,00	10%	0,83%	R$ 962,80
58 in the production phase - Jan-10	2000	R$ 116.000,00	10%	0,83%	R$ 962,80
62 in the production phase - Feb-10	2000	R$ 124.000,00	10%	0,83%	R$ 1.029,20

Source: Research data.

About the energy costs were considered using a water pump used to clean and supply water tank, the equipment that keeps the refrigerated tank, forage harvester, tractor, milking equipment of the cane shredder (forage). At the time of getting vaccinated animals, buying up a lot for all the flock, and the cutting of milk plus a margin of 10%, so there is not a control vaccination per individual. On the farm, working directly with cattle in milk production the owner and one employee.

Table 3 shows the relationship of equipment and machinery used directly or indirectly for the milk production. Similarly the research used by Segala and Silva (2007), data were parameterized to contemplate the reality of the farm, and therefore the region.

Table 3. Monthly Depreciation of equipment

Item	Purchase amount	Purchase year	Annual depreciation	Percentage of use	Percentage of use in other activities	Monthly depreciation amount
Forage	R$ 9.700,00	2001	14,28%	100%	0%	R$ 115,43
Milking	R$ 30.000,00	2004	10,00%	100%	0%	R$ 250,00
Cooler	R$ 28.000,00	2005	6,67%	100%	0%	R$ 155,63
Mowing	R$ 2.800,00	2003	10,00%	50%	50%	R$ 23,33
Feed truck	R$ 30.000,00	2002	10,00%	20%	80%	R$ 250,00
Water pump	R$ 380,00	2000	5,00%	60%	40%	R$ 1,58
Equipment's monthly depreciation						R$ 795,98

Source: Research data, adapted to Segala and Silva (2007).

Table 4 below, is presented the calculation of depreciation of buildings and facilities that are designed to produce milk.

Table 4. Estimated monthly depreciation for facilities

Item	Investment value	Purchase year	Annual depreciation	Utilization percentage	Utilization percentage in other activities	Monthly depreciation
Covered corral	R$ 70.000,0 0	2001	4%	100%	0%	R$ 233,33
Machinery shed	R$ 25.000,0 0	2001	4%	33%	67%	R$ 83,33
Employee housing	R$ 20.000,0 0	2001	4%	100%	0%	R$ 66,67
Fences	R$ 10.000,0 0	2000	10%	100%	0%	R$ 83,33
Power grid	R$ 10.000,0 0	1999	10%	100%	0%	R$ 83,33
Water/plumbing network	R$ 10.000,0 0	1999	10%	20%	80%	R$ 83,33
Breeding bull	R$ 5.000,00	1999	10%	100%	0%	R$ 41,67
Monthly depreciation of facilities						R$ 675,00

Source: Research data, adapted to Segala and Silva (2007).

4.1.1.2 Overall Production Costs

From these surveys covering up the front approaches and adaptations to research Segala and Silva (2007), Table 5, below was generated, and presents a summary of the realized values, containing the calculation of the monthly average cost of milk production. On not considering the depreciation, the value of production would decrease a lot, and it is shown in Table 6 below.

Table 5. Overall cost production

Cost of Milk Production	November 2010	Vertical analysis	December 2010	Vertical analysis	January 2010	Vertical analysis	February 2010	Vertical analysis
Fuel	R$ 400,00	5,72%	R$ 400,00	5,73%	R$ 400,00	5,76%	R$ 400,00	5,70%
Light consumption	R$ 800,00	11,43%	R$ 800,00	11,46%	R$ 800,00	11,52%	R$ 800,00	11,41%
Facility depreciation	R$ 675,00	9,65%	R$ 675,00	9,67%	R$ 675,00	9,72%	R$ 675,00	9,63%
Depreciation of matrices	R$ 1.012,60	14,47%	R$ 962,80	13,79%	R$ 962,80	13,86%	R$ 1.029,20	14,68%
Depreciation of machinery and equipment	R$ 795,98	11,37%	R$ 795,98	11,40%	R$ 795,98	11,46%	R$ 795,98	11,35%
Service, maintenance and cleaning	R$ 45,00	0,64%	R$ 45,00	0,64%	R$ 45,00	0,65%	R$ 45,00	0,64%
Medicine, vaccines, and veterinary	R$ 114,50	1,64%	R$ 148,00	2,12%	R$ 112,00	1,61%	R$ 112,00	1,60%
Salaries (including charges)	R$ 3.154,60	45,08%	R$ 3.154,60	45,19%	R$ 3.154,60	45,42%	R$ 3.154,60	44,99%
Total costs	R$ 6.997,68	100%	R$ 6.981,38	100%	R$ 6.945,38	100%	R$ 7.011,78	100%
Monthly milk production	21960	litros	20880	litros	20880	litros	22320	litros
Cost per liter of milk	R$ 0,32		R$ 0,33		R$ 0,33		R$ 0,31	
Selling price per liter	R$ 0,65		R$ 0,65		R$ 0,65		R$ 0,65	
Profit or loss per liter of milk produced	R$ 0,33		R$ 0,32		R$ 0,32		R$ 0,34	
Monthly revenue	R$ 7.276,32		R$ 6.590,62		R$ 6.626,62		R$ 7.496,22	

Source: Research data, adapted to Segala and Silva (2007).

Table 6. Overall cost production excluding depreciation

Cost of Milk Production	November 2010	Vertical analysis	December 2010	Vertical analysis	January 2010	Vertical analysis	February 2010	Vertical analysis
Fuel	R$ 400,00	8,86%	R$ 400,00	8,80%	R$ 400,00	8,87%	R$ 400,00	8,87%
Light consumption	R$ 800,00	17,72%	R$ 800,00	17,59%	R$ 800,00	17,73%	R$ 800,00	17,73%
Service, maintenance, and cleaning	R$ 45,00	1,00%	R$ 45,00	0,99%	R$ 45,00	1,00%	R$ 45,00	1,00%
Medicine, vaccines, and veterinary	R$ 114,50	2,54%	R$ 148,00	3,25%	R$ 112,00	2,48%	R$ 112,00	2,48%
Salaries (including charges)	R$ 3.154,60	69,88%	R$ 3.154,60	69,37%	R$ 3.154,60	69,92%	R$ 3.154,60	69,92%
Total costs	R$ 4.514,10	100%	R$ 4.547,60	100%	R$ 4.511,60	100%	R$ 4.511,60	100%
Monthly Milk Production	21960	litros	20880	litros	20880	litros	22320	litros
Cost per liter of milk	R$ 0,21		R$ 0,22		R$ 0,22		R$ 0,20	
Selling price per liter	R$ 0,65		R$ 0,65		R$ 0,65		R$ 0,65	
Profit or loss per liter of milk produced	R$ 0,44		R$ 0,43		R$ 0,43		R$ 0,45	
Monthly revenue	R$ 9.759,90		R$ 9.024,40		R$ 9.060,40		R$ 9.996,40	

Source: Data from the survey, adapted the Segala and Silva (2007).

4.1.1.3 Values from the Slaughter Sales Operation

On the farm taken per sample, born approximately 45 males and 45 females annually, which represents 6-7 animals per month. Females are focused exclusively milk production and thus replacing the animals leaving the productive period and headed to slaughter. Table 7 shows the estimated annual births in the farm and the amount of animals that are fattening and subsequent slaughter. Table 8 shows the projected monthly revenue considering the income with cattle for slaughter. The price for the animal is published on an electronic online service market for cattle in Brazil (PECUARIA, 2012).It was taken the value of cattle for slaughter by R $ 1,170.00 and R $ 880.00 for the female.

Table 7. Estimated monthly revenue with cattle

Number of animals born	Per year	Females To Reset Milk Production	Annual surplus	Monthly for slaughter	Price of the animal for slaughter	Value calculated on sale
Males	42	0	42	3	R$ 1.170,00	R$ 3.510,00
Females	42	8	34	2	R$ 880,00	R$ 1.760,00
Total						R$ 5.270,00

Table 8. Forecast monthly and annual revenue of ranching family

Without Depreciation	Month 1	Month 2	Month 3	Month 4	Average monthly value	Average annual value
Milk Revenue	R$ 9.759,90	R$ 9.024,40	R$ 9.060,40	R$ 9.996,40	R$ 9.460,28	R$ 113.523,30
Revenue Cut	R$ 5.270,00	R$ 5.270,00	R$ 5.270,00	R$ 5.270,00	R$ 5.270,00	R$ 63.240,00
Balance	*R$ 15.029,90*	*R$ 14.294,40*	*R$ 14.330,40*	*R$ 15.266,40*	*R$ 14.730,28*	*R$ 176.763,30*
With depreciation	Month 1	Month 2	Month 3	Month 4	Average monthly value	Average annual value
Milk Revenue	R$ 7.276,32	R$ 6.590,62	R$ 6.626,62	R$ 7.496,22	R$ 6.997,45	R$ 83.969,34
Revenue Cut	R$ 5.270,00	R$ 5.270,00	R$ 5.270,00	R$ 5.270,00	R$ 5.270,00	R$ 63.240,00
Balance	*R$ 12.546,32*	*R$ 11.860,62*	*R$ 11.896,62*	*R$ 12.766,22*	*R$ 12.267,45*	*R$ 147.209,34*

4.2 *Simulation of Eucalyptus Plantation*

The survey includes soil preparation, planting and harvesting in two scenarios: with 67.5 ranch. Based on the forecast growth generated by SisEucalipto, Table 9 shows the growth trend as volume and curves Current Annual Increment (ICA) and Mean Annual Increment (MAI).By ICA has growth volume occurred within one year, and IMA to the result of dividing the volume by the age of the forest. Note that the curve ICA reaches a maximum value before the curve IMA, the 8th year of planting, and that the two curves intersect at the point of maximum IMA.

Table 9. Prognosis of forest yield per hectare planted eucalyptus

Age	Dominant height (m)	trees/ Ha	Average diameter (cm)	Average height (m)	Basal Area (cm^2)	Total Volume (m^3)	I.M.A.	I.C.A.
1	4,3	1650	4	3,6	2	3	3	3
2	9,7	1650	10,3	8,1	13,7	45,9	22,9	42,9
3	13,8	1650	14	11,6	25,4	122	40,7	76,1
4	17,1	1648	16,4	14,4	34,7	206,1	51,5	84,1
5	19,7	1646	18	16,6	42	288	57,6	81,9
6	22	1642	19,2	18,4	47,8	364,3	60,7	76,3
7	23,9	1638	20,2	20	52,4	434,2	62	69,9
8	25,5	1633	20,9	21,4	56,3	497,6	62,2	63,4
9	26,9	1628	21,6	22,6	59,4	555,1	61,7	57,5

Source: Research data generated by SisEucalipto.

Forest cycle is driving the growth of the buds of trees newly planted or freshly cut. To consider the horizon of 21 years and cuts every 7 years, the analysis is based on three cycles forest.

The information presented by the work of EMBRAPA are based on surveys conducted with farmers where its properties include planting pine and eucalyptus as a production system.

For operations of planting and cultivation of eucalyptus should be considered:

- Preparation area;
- Planting;
- Weed control;
- Age cut-off;
- Costs;
- Productivity;

In the simulation hypothesis of this project was performed to analyze the planting of *Eucalyptus urograndis* in Conceição de Macabu property that was taken per sample, 67.5.Estimates were calculated from the values suggested in the EMBRAPA research (2005) and CEPEA-Center for Advanced Studies in Applied Economics from USP.

It has a *stereo* equivalent to 1 m^3 of wood, commonly used measurement unit in the market forester. The CEPEA published in his newsletter in the month of Forestry September/2012 that the average value of the stereo stand for the energy market is priced at $ 50, 00.However, in considering whether a more conservative estimate, taking as reference values proposed by the Forestry Program of Arcelor Mittal (2008), the value of the stereo is $ 20.00. These values composes Table 10.

Table 10. Forecast of revenue cycle for forest planting in 65 hectares

Production and income	Volume produced by 7-year forest cycle (m^3/ha)	Stere Value (Value provided by CEPEA)	Estimated revenue per hectare	Revenue generated by forest cycle (65ha of forest)
Stere Value (supplied by CEPEA)	434,2	R$ 50,00	R$ 21.710,00	R$ 1.411.150,00
Stere Value (suggested by ArcelorMittal)	434,2	R$ 20,00	R$ 8.684,00	R$ 564.460,00

4.3 Execution of the Calculations According to the Methodology AR-AMS0001

With respect to this research project, the selected methodology is the AR-AMS0001, "Exhibit A", which is to be applied in reforestation projects carried out in the pastures.

4.3.1 Baseline GHG Removal at the Sinkhole

The calculation methodology provides equations to calculate:

- The above ground biomass estimates;
- Underground biomass estimates;
- Baseline the baseline removal of Greenhouse gases (GHG) in the vicinity of the project;
- Estimated leakage calculation;
- RCE.

Some of the equations are presented in this summary. The equation 3.1 brings as a result the stock of carbon in living biomass in the project region.

$$B(t) = \sum_{i=1}^{I} (B(t) + B_{iB(t)i}) * A_i \tag{3.1}$$

Where:

$B_{(t)}$ = the carbon stock in living biomass in the project region at time (t) the absence of project activity expressed in tC (tonne of carbon);

$_{B\,(t)} =\, _i$ is the biomass carbon stock above ground so in time *(t)* of the stratum *(i)* in the absence of the project activity expressed in t C/ha (tons of carbon per hectare);

$B_{B\,(t)} =\, _i$ is the carbon stock in below-ground biomass at time (t) of the stratum

(I) in the absence of the project activity expressed in t C/ha (tons of carbon per hectare);

$A_{(i)}$ = Area of stratum design expressed in ha (hectares);

i = stratum i (i = total number of strata).

4.3.2 Calculation of Above Ground Biomass

To calculate the above-ground biomass, uses the following equation:

$$B_{a(t)} = M_{(t)} * 0.5 \tag{3.2}$$

Where:

$B_{(t)}$ = carbon stock in aboveground biomass in the year (time) t when the lack of implementation of the project, expressed in t C/ha (tons of carbon per hectare);

$M_{(t)}$ = above-ground biomass in the year (time) t when the lack of implementation of the project, expressed in t dm / ha (tons of dry matter per hectare);

0.5 = carbon fraction of dry matter expressed in t C/t dm (tonne of carbon per tonne of dry matter).

Table 11 is the simulation of the carbon stock of above-ground biomass, year-on-year.

Table 11. Biomass carbon stocks above ground every year

M	M_{woody}	$M_{woody\,(t=n-1)} + g * Dt$	M_{woody_max}	$M_{(t=n)}$	$B_{a(t)} = M_{(t)} * 0.5$
{0}6.2{/0}{1}{/1}	0	0	0	{0}6.2{/0}{1}{/1}	3.1

4.3.3 Calculation of Biomass Below Ground

The methodology guides to be performed to calculate the below-ground biomass, as it also does capture carbon from the atmosphere by being an organism alive. This calculation considers the roots of grass and evergreen trees.

Table 12 is the simulation of the carbon stock in below-ground biomass, year-on-year.

Table 12. Biomass carbon stocks below the ground each year

Mgrass	Rgrass	Mwoody	Rwoody	Mwoody(t=n-1) + g * Δt)* Rwoody	Mwoody_max	$B_{B\,(t=n)} = 0,5 * [M_{grass} * R_{grass} + (M_{woody\,(t=n-1)} + g * \Delta t) * R_{woody}]$
6,2	1,58	0	1	0	0	4,898

4.3.4 Calculation of Baseline Actual Removal of Greenhouse Gases

To then determine the baseline actual removal of greenhouse gases in the analyzed region follows the equation:

$$\Delta_{BSL\,C,\,t} = (B_{(t)} - B_{(t-1)}) * ((44/12)) \tag{3.10}$$

Hence:

$\Delta\, CBSL,\, t$ = Baseline actual removal of greenhouse gases by sinks

$B_{(t)}$ = carbon stock in living biomass in the year (time) t would exist when the lack of implementation of the project, expressed as C t (tonne of carbon);

To be considered for Tables 12 and 13, which inform the biomass carbon in the study area, the change from baseline removal Greenhouse Gases is zero throughout the project:

$$B_{(t)} = B_{(t-1)} \Rightarrow$$

$$\Delta_{BSL\,C,\,t} = (B_{(t)} - B_{(t)}) * ((44/12))$$

$$\Delta C_{BSL,\,t} = 0 * (44/12)$$

$$\Delta C_{BSL,\,t} = 0$$

Thus, the resulting equation for calculating the baseline carbon stock in the absence of the project activity during the period of the hypothesis:

$$B(t) = \sum_{i=1}^{I} (B(t) + B_{iB(t)i})^* A_i \tag{3.11}$$

Adding Tables 11 and 12 then have the result in Table 13, with the baseline of the reference scenario.

Table 13. Baseline carbon stock of the reference scenario every year

$B_{a(t)I}$	$_B B_{(t)i}$	A	$B_{(t)}$
3.1	4898	1,000	7998

To consider that the methodology is only biomass above and below ground, not treating emissions from food, in the case of this research project, cattle, this issue was not considered in the calculations.

4.3.5 Designed Calculation Baseline in Project Implementation (ex ante)

According to the methodology, the carbon stock in the scenario that will occur in the project implementation start date of activity $(t = 0)$ must be the same as the baseline carbon stock in the same period $(t = 0)$.

4.3.6 Calculation of Carbon Stock in Aboveground Biomass

Analogous to the calculation of the baseline reference scenario, to calculate the carbon stock in living biomass above ground, must be considered stratum *i,* which comes to every geographic area of the project implementation. In the case of this research project is regarded only 1 stratum of 1000 hectares.

$$N_{A(t)} = T_{i(t)i} * 0.5 \tag{3.12}$$

Where:

$_{The} N_{(t)i}$ = carbon stocks in above-ground biomass in the year (time) t when the deployed project, expressed in t C/ha (tons of carbon per hectare);

$T_{(t)i}$ = above-ground biomass in the year (time) t when the deployed project, expressed in t dm / ha (tons of dry matter per hectare);

0.5 = carbon fraction of dry matter expressed in t C/t dm (percentage).

EMBRAPA (2002) in its methodology for estimating carbon stocks in trees, considers various land uses. In their study, Renner (2004) was used a methodology for calculating living biomass above and below ground, set by the Federal University of Paraná (UFPR) for tree species *Pinus taeda,* which resembles much as the eucalyptus Table 14.

Table 14. Biomass equations adjusted for *Pinus taeda* UFPR

		Ratios	
Compartment	Equation	a	b
PFV	$a(DAP^2H)^b$	0,0595	0,9279
PFVI		0,0012	1,0480
PFGv		0,0001	1,3922
PVR		0,4484	0,5619

Source: UFPR; ECOPLAN (2003, apud RENNER, 2004, p. 59).

Where: PVF: Weight Verde de Fuste; PVFl: Weight Green Leaves; PVGv: Weight Green Living Branches; PVR: Weight Green Roots; dbh: diameter, H: height

In continuation, the author presents the results of calculating the amount of carbon individually, here represented by Table 15. The result of these equations are expressed in t C/ha (tons of carbon per hectare).

Table 15. Equations for obtaining the amount of carbon in the compartments for individual trees

Compartment	Equation
Shaft	PCF = 0,1737 * PVF
Leaves	PCFl = 0,1422 * PVFl
Living branches	PCGv = 0,1595 * PVGv
Root	PCR = 0,1676 * PVR

Source: UFPR; ECOPLAN (2003, apud RENNER, 2004, p 60).

Where: PCF: Weight in Carbon Fuste; PCFl: Weight of Carbon in Leaves; PCGv: Weight of Carbon in Living Branches; PVF: Weight Verde de Fuste; PVFl: Weight Green Leaves; PVGv: Weight Green Living Branches; PVR: Weight Green Roots, PCR Weight Carbon in Root.

Note: Values are expressed in t C/ha (tons of carbon per hectare).

These equations allow us to estimate the carbon fixed up below ground by the implementation of the project.

The *software* SisPinus in his prognosis, presents the results of production expressed in cubic meters. The author reports that in a reference to his research, industry Klabin, believes that wood 1m equals 1 tonne mass.

She continues with the statement ROCK (2003, apud RENNER, 2004, p. 60), where a ton of carbon, the unit used by the carbon market, equivalent to 3.67 tons of CO_2, which means that in reverse situation, a ton of CO_2 equivalent to 0.27 tonnes of carbon. This information will be used for *the subsequent* calculation of REC.

Thus, in Table 16 is presents the amount of carbon estimated by summing compartments trunk, branches, foliage and roots for planting eucalyptus in the hypothesis of the research in years 7, 14 and 21, and these dates immediately before cuts.

Table 16. Fixed amount of carbon per hectare, prognosis based on lifelong project

	period	Total volume of wood produced (m^3)	Amount of Carbon (t/ha)	Qty. Carbon in 65 ha (t)	Qty. Carbon in 1000 ha (t)
First Cutting	7th year	434,2	43,70	2840,5	43700
Second Cutting	14th year	434,2	43,70	2840,5	43700
Final Cutting	21st year	434,2	43,70	2840,5	43700

The prognosis in SisPinus calculated immediately after the first cut the amount of carbon fixed remaining 15% is in reference to the volume produced, this carbon associated with the existence of roots of trees.

Table 17 shows the evolution of tree growth and the amount of carbon fixed compartments corresponding to the stem, branches, leaves and roots, which is the baseline carbon fixed for the duration of the project.

In the year of implementation of the reforestation project, as directed by the methodology, the amount of carbon fixed equals before the implementation of the project.

Table 17. Evolution of growth of trees and baseline amount of carbon fixed

First Cutting			Second Cutting			Final Cutting		
Year	Timber volume in m3/ha	Fixed carbon in tC/há	Year	Timber volume in m3/ha	Fixed carbon in tC/há	Year	Timber volume in m3/ha	Fixed carbon in tC/há
1	3,00	0,30	8	65,13	6,55	15	65,13	6,55
2	45,90	4,62	9	87,99	8,85	16	87,99	8,85
3	122,00	12,27	10	145,00	14,59	17	145,00	14,59
4	206,10	20,73	11	206,10	20,73	18	206,10	20,73
5	288,00	28,97	12	288,00	28,97	19	288,00	28,97
6	364,30	36,65	13	364,30	36,65	20	364,30	36,65
7	434,20	43,68	14	434,20	43,68	21	434,20	43,68

4.3.7 Calculating the Carbon Leakage

According to the methodology if project participants demonstrate that the activity of small-scale reforestation under the CDM concept does not result in displacement of activities or people, or not yet generates other activities that did not exist before incurring increased emission of greenhouse gases, the calculation to estimate leakage is not required.

Table 18 shows the calculation of leakage estimated for the project based on the methodology and predictions generated from timber volume.

Table 18. Estimated trail regarding baseline amount of carbon fixed

Year	Timber volume in m^3/ha	ΔC_{ACTUAL}	ΔC_{ACTUAL} * 0.15
0	0,00	0,00	0,00
1	3,00	1,11	0,61
2	45,90	16,93	9,31
3	122,00	44,99	24,74
4	206,10	76,01	41,80
5	288,00	106,21	58,41
6	364,30	134,35	73,88
7	434,20	160,13	88,06
8	65,13	24,02	13,21
9	87,99	32,45	17,84
10	145,00	53,48	29,41
11	206,10	76,01	41,80
12	288,00	106,21	58,41
13	364,30	134,35	73,88
14	434,20	160,13	88,06
15	65,13	24,02	13,21
16	87,99	32,45	17,84
17	145,00	53,48	29,41
18	206,10	76,01	41,80
19	288,00	106,21	58,41
20	364,30	134,35	73,88
21	434,20	160,13	88,06

4.3.8 Anthropogenic GHG Removals by Sinks

The net GHG removals by sinks for each year of the first period carbon credit is calculated according to the equation below:

$$ER_{AR\ CDM,\ t} = \Delta C_{PROJ,\ t} - \Delta C_{BSL,\ t} - GHG_{PROJ,\ t} - L_t \quad (3.13)$$

Where:

$ER_{AR\ CDM,\ t}$ = anthropogenic GHG removal by sinks, expressed in t CO_2 -e/year (ton of CO_2 per year);

$_{PROJ}\Delta_{C,\ t}$ = projected GHG removal by sinks at time *t*, expressed in t CO_2 -e/year (ton of CO_2 per year);

$_{BSL}\Delta_{C,\ t}$ = Baseline GHG removal by sinks, expressed in t CO_2 -e/year (ton of CO_2 per year);

$GHG_{PROJ,\ t}$ = GHG emissions of the project, expressed in CO_2 -e/year t (tonne of CO_2 per year);

L_t - Escape assigned to the project activity at time *t*, expressed in t CO_2 -e/year (ton of CO_2 per year).

Given the chance to replace a pasture area, which emits a certain amount of greenhouse gases, methane in this case, as item 4. 3.1.3 of this research, the reforested area there will be no creation of food. Thus, the emission of greenhouse gases is not considered in this project.

The result of the Certified Emission Reduction (REC) in year t_v was assumed that the check is given by the following equation:

$$tCER_{(tv)} = \sum_{t=0}^{TV} ER_{AR\ CDM}\ \Delta_t \ * \ t \quad (3.14)$$

Where:

$tCER_{(tv)}$ - = Temporary Certified Emission Reductions in the year that the scan was taken;

$ER_{AR\ CDM\ t}$ = Net GHG removal by *sinks* expressed in t CO_2 -e/year (ton of CO_2 per year);

t_v - Year that was assumed to verification;

Δt - Increment of time (1 year).

Thus, Table 19 represents the result year-on-year reduction of emissions, the amount of carbon captured at the end of each cycle forest per hectare and the total amount of carbon captured at the end of the 21 years of the project, also per hectare. Table 20 is shown the amount of carbon sequestered in a forest and 65ha of forest in a 1000 HA.

Table 19. Results of calculation for certified emission reduction project

Year	Timber volume in m^3/ha	ΔC_{ACTUAL}	ΔCACTUAL * 0,15 (escape)	$ER_{AR\ CDM}$	
0	0,00	0,00	0,00	0,00	
1	3,00	1,11	0,17	0,94	
2	45,90	16,93	2,54	14,39	
3	122,00	44,99	6,75	38,24	
4	206,10	76,01	11,40	64,61	
5	288,00	106,21	15,93	90,28	
6	364,30	134,35	20,15	114,20	
7	434,20	160,13	24,02	136,11	458.77 t
				Subtotal	CO_2/ha
				-->	end of the
8	65,13	24,02	3,60	20,42	forest cycle

9	87,99	32,45	4,87	27,58	
10	145,00	53,48	8,02	45,45	
11	206,10	76,01	11,40	64,61	
12	288,00	106,21	15,93	90,28	
13	364,30	134,35	20,15	114,20	
14	434,20	160,13	24,02	136,11	
				Subtotal -->	498.65 t CO_2/ha end of the forest cycle
15	65,13	24,02	3,60	20,42	
16	87,99	32,45	4,87	27,58	
17	145,00	53,48	8,02	45,45	
18	206,10	76,01	11,40	64,61	
19	288,00	106,21	15,93	90,28	
20	364,30	134,35	20,15	114,20	
21	434,20	160,13	24,02	136,11	
				Subtotal -->	498.65 t CO_2/ha end of the forest cycle
			Total:	1456,10	t CO_2/ha

Source: Adapted from RENNER, 2004 and survey data.

Table 20. Estimated carbon cycle captured by the two simulations forest area to be reforested

	Area (ha)	Amount of carbon captured per hectare for each forest cycle (t CO_2/ha)	Total amount of carbon captured by forest cycle on each farm (t CO_2)
1st Forest cycle	65	458,77	29820,05
	1000	458,77	458770,00
2nd Forest cycle	65	498,65	32412,25
	1000	498,65	498650,00
3rd Forest cycle	65	498,65	32412,25
	1000	498,65	498650,00

4.3.9 Monitoring Plan (ex post)

In this part of research that are treated the details of the monitoring plan of the CDM project. This monitoring plan basically performs checks throughout the duration of the project based on what was planned and considered for making the baselines.

4.3.10 Estimate (ex post) Baseline GHG Removal by Sinks

The method orients it is not necessary to monitor the baseline. The Baseline GHG Removal by Sinks will be the same as defined in section 4.3.13 of this research.

4.3.11 Estimate (ex post) of GHG Removal by Sink

The equation below shows the calculation of carbon stock to be used in monitoring.

$$P_{(t)} = \sum_{i=1}^{I} (P_{A(t)\,i} + P_{B(t)\,i}) * A_i * (44 / 12) \tag{3.15}$$

Where:

$P_{(t)}$ = the carbon stock in living biomass in the project region at time (t) in the project activity, expressed in $_{tCO2e}$ (tonnes of carbon dioxide);

$_{The}$ $P_{(t)} =_i$ is the carbon stock of above-ground biomass at time *(t)* of the stratum *(i)* the project activity, expressed in t C/ha (tons of carbon per hectare);

$P_{B(t)} =_i$ is the stock of biomass carbon underground in time (t) of the stratum;

(I) the project activity, expressed in t C/ha (tons of carbon per hectare);

$A_{(i)}$ = Area of stratum design expressed in ha (hectares);

i = stratum i (i = total amount of strata).

4.3.12 Measuring Above Ground Biomass

$$P_{A(t)} = E_{i(t)i} * 0.5 \tag{3.16}$$

Where:

$_{The}$ $P_{(t)i}$ = carbon stocks in above-ground biomass in the year (time) *t* when the project activity, expressed in t C / ha (tons of carbon per hectare);

$E_{(t)i}$ = aboveground biomass estimated in the year (time) *t* when the project activity, expressed in t dm / ha (tons of dry matter per hectare);

0.5 = carbon fraction of dry matter expressed in t C/t dm (tonne of carbon per tonne of dry matter).

To perform the estimation of aboveground biomass the following steps should be followed:

- Step 1 - Establish permanent plots and document analysis of their locations in the first monitoring report;
- Step 2 - Measure the diameter at breast height (DBH) and tree height and documented in the report;
- Step 3 - To estimate the aboveground biomass using equations developed or adapted locally or nationally.

4.3.13 Measurement of Biomass Below Ground

$$P_{B(t)} = E_{i(t)} * R_i * 0.5 \tag{3.17}$$

Where:

$P_{B(t)i}$ = Carbon stock in living biomass below ground in the year (time) *t* in the project activity, expressed in t C ha (tons of carbon per hectare);

$E_{(t)i}$ = estimate of underground biomass year (time) *t* of the project activity, expressed in t dm / ha (tons of dry matter per hectare);

R = ratio of mass of the tree and root mass (t dm/t dm);

0.5 = carbon fraction of dry matter.

It is driven methodology that values locally or regionally adapted to be used for this estimate.

4.3.14 Estimate (ex post) of Carbon Leakage

To estimate carbon leakage, you should monitor each of the following indicators during the first crediting period:

- Area in crops within the project boundary displaced due to there project activity;
- Number of animals that utilize pasture within the project boundary that have been displaced due to the project activity;
- Number of animals and volume of forage intake per animal per hectare that were brought to the project area.

As this activity of small-scale reforestation under the CDM concept does not result in displacement of activities or people, or not yet generates other activities that did not exist before incurring increased emission of greenhouse gases, the calculation to estimate the leakage is not required.

$$_{Tv}L = 0 \tag{3.18}$$

Where:

L_{tv} = Leakage attributable to the project activity at time t, expressed in CO_2 -e/ano t (tonne of CO_2 per year)

4.3.15 Estimate (ex post) Removal of Anthropogenic GHG at Sinkhole

Removal of anthropogenic greenhouse gas sink occurs at the designed result of removing less the baseline carbon fixed, less leakage. The resulting value of REC in the year the check is given as follows:

$$tCER_{(tv)} = \sum_{t=0}^{TV} P_{(t)} - \sum (GHG_{PROJ,(t)} - \Delta C_{BSL,t}) - L_{tv} \tag{3.19}$$

Where:

$tCER_{(tv)}$ = Certified Emission Reduction in year t the duration of the project;

$P(t)$ = Inventory of carbon calculated for the proximity of the project in year t the duration of the *project,* expressed in t CO_2 -e/year (ton of CO_2 per year);

$GHG_{PROJ,(t)}$ = GHG emissions of the project, expressed in CO_2 -e/ year t (tonne of CO_2 per year);

$_{BSL}\Delta_{C,t}$ = Baseline GHG removal by sinks, expressed in t CO_2 -e/ year (ton of CO_2 per year);

L_{tv} = Leakage attributable to the project activity at time t scan, expressed as CO_2 -e/ year t (tonne of CO_2 per year);

t_v - Year of the verification.

All these equations are applied in the monitoring stage, part of the CDM project upon its implementation, therefore not applicable in this research project. However, spreadsheets monitoring should be made considering these calculations and parameters.

5. Analysis and Evaluation of Results

In the previous chapter, were made in the amounts raised beef cattle and milk for the 65ha property in Conceição de Macabu. The data analysis was based on information from four months of production and operation of the farm. From those data, Table 21 shows the forecast annual cash flow of the farm without considering depreciation, so it can be compared with the results of the NPV of culture eucalyptus, conducted year-on-year.

By calculation presented in Table 8, the average annual income derived from the operation of livestock in the region studied without considering the depreciation of R \$ 176,763.30 and then the monthly income of R \$ 14,730, 28.For the same scenario, considering depreciation, the average annual income is \$ 1 47.209.34 and monthly income of R \$ 12,267.45.

Timber production of 434.2 m^3/ha at the end of each cycle forestry, informed the prognosis of SisEucalipto, served as input to calculate the revenue from the sale of the stereo to produce coking coal, whose value is shown in Table 21 to below.

Table 21. Forecast values obtained in stereo commercialization of eucalyptus for metallurgical coal production

Production and income	Volume produced by 7-year forest cycle (m^3/ha)	Stere Value (Value provided by CEPEA)	Estimated revenue per hectare	Revenue generated by forest cycle (65ha of forest)	Revenue generated by forest cycle (1000ha forest)
Stere value (supplied by CEPEA)	434,2	R$ 50,00	R$ 21.710,00	R$ 1.411.150,00	R$ 21.710.000,00
Stere value (suggested by ArcellorMittal)	434,2	R$ 20,00	R$ 8.684,00	R$ 564.460,00	R$ 8.684.000,00

To conduct a feasibility analysis of the project, mathematical equations were used-financial, which started from a baseline projections and statistical basis, with simulations.

5.1 Net Present Value (NPV) without Carbon Credits

Through the analysis of results provided by VPL, you can identify whether the project is economically viable or not. If NPV is zero, the cash flows from the investment are all consumed to pay the cost of the capital invested. If NPV is positive, the result from the project covers the debt acquired to invest in the project and the rate of return determined by investors. Therefore, search is always a project that has positive NPV (NETO, 2006).

5.2 Calculation for 65ha Area and Value-Based CEPEA

Tables 22, 23 and 24 are the cash flows and NPV calculated for planting and felling eucalyptus charcoal to market the area to 65ha, the three cycles forest, considering the amount of stereo $ 50.00, suggested by CEPEA, and annual discount rate of 10%.

Table 22. Calculated NPV of the first cycle of eucalyptus forest for culture 65ha in area, with a value of $ 50.00 stereo

	First Forest Cycle						
	Year 1	Year 2	Year 3	Year 4	Year 5	Year 6	Year 7
Eucalyptus Planting							
Total Cost	R$ 77.949,00	R$ 20.250,00	R$ 18.900,00	R$ 17.550,00	R$ -	R$ -	R$ -
Total Revenue	R$ -	R$ -	R$ -	R$ -	R$ -	R$ -	R$ 1.411.150,00
Cash Flow	-R$ 77.949,00	-R$ 20.250,00	-R$ 18.900,00	-R$ 17.550,00	R$ -	R$ -	R$ 1.411.150,00
						Eucalyptus 1st Cycle NAV	R$ 610.358,08
						Annual NAV	R$ 87.194,01
						Monthly NAV	R$ 7.266,17

Table 23. Calculated NPV of the second cycle of eucalyptus forest for culture 65ha in area, with a value of $ 50.00 stereo

	Second Forest Cycle						
	Year 8	Year 9	Year 10	Year 11	Year 12	Year 13	Year 14
Eucalyptus Planting							
Total Cost	R$18.090,00	R$14.850,00	R$14.850,00	R$14.850,00	R$ -	R$ -	R$ -
Total Revenue	R$ -	R$ -	R$ -	R$ -	R$ -	R$ -	R$ 1.411.150,00
Cash Flow	-R$18.090,00	-R$ 14.850,00	-R$14.850,00	-R$14.850,00	R$ -	R$ -	R$1.411.150,00

Eucalyptus 2nd Cycle NAV	R$ 674.125,12
Annual NAV	R$ 96.303,59
Monthly NAV	R$ 8.025,30

Table 24. Calculated NPV of the third cycle forestry culture eucalyptus 65ha in area, with a value of $ 50.00 stereo

	Third Forest Cycle						
	Year 15	Year 16	Year 17	Year 18	Year 19	Year 20	Year 21
Eucalyptus Planting							
Total Cost	R$18.090,00	R$14.850,00	R$14.850,00	R$14.850,00	R$ -	R$ -	R$ -
Total Revenue	R$ -	R$ -	R$ -	R$ -	R$ -	R$ -	R$1.411.150,00
Cash Flow	-R$18.090,00	-R$14.850,00	-R$14.850,00	-R$14.850,00	R$ -	R$ -	R$1.411.150,00

Eucalyptus 3rd Cycle NAV	R$ 674.125,12
Annual NAV	R$ 96.303,59
Monthly NAV	R$ 8.025,30

5.3 Net Present Value (NPV) with Carbon Credits

In this subchapter are the calculations considering the values of marketing ton of carbon captured in the two markets REC: Market regulated by the Kyoto Protocol, in the manner pre-2012 and average values applied, and the Voluntary Market. It is here that all revenue from the sale of each REC planned for 7 years is inputted the beginning of each forest cycle. In Table 25 are simulations of revenue earned through the sale of CERs estimated in Table 20 of the previous chapter, considering an average value of $ 20.00 for the value and market price adjusted $ 3.10 for the voluntary markets, as referenced in chapter 2. The conversion rate used was U.S. $ 1.00 = R $ 2.09, extracted on 11/28/2012 Central Bank (Central Bank, 2012).

Table 25. Projected revenue obtained through marketing REC

		Value per ton of CO_2 (US$/Ton CO_2)	Area (ha)	Amount of carbon captured per hectare for each forest cycle (t CO_2/ha)	Expected value per forest cycle (US$)	Expected value per forest cycle (R$)
Regulated Market	1st Forest cycle	20	65	458,77	$ 596.401,00	R$ 1.246.478,09
		20	1000	458,77	$ 9.175.400,00	R$ 19.176.586,00
	2nd Forest cycle	20	65	498,65	$ 648.245,00	R$ 1.354.832,05
		20	1000	498,65	$ 9.973.000,00	R$ 20.843.570,00
	3rd Forest cycle	20	65	498,65	$ 648.245,00	R$ 1.354.832,05
		20	1000	498,65	$ 9.973.000,00	R$ 20.843.570,00
Voluntary Market	1st Forest cycle	3,1	65	458,77	$ 92.442,16	R$ 193.204,10
		3,1	1000	458,77	$ 1.422.187,00	R$ 2.972.370,83
	2nd Forest cycle	3,1	65	498,65	$ 100.477,98	R$ 209.998,97
		3,1	1000	498,65	$ 1.545.815,00	R$ 3.230.753,35
	3rd Forest cycle	3,1	65	498,65	$ 100.477,98	R$ 209.998,97
		3,1	1000	498,65	$ 1.545.815,00	R$ 3.230.753,35

5.4 Calculation for 65ha Area, Value-Based CEPEA

Table 26 are the cash flows and NPV calculated for planting and felling eucalyptus market charcoal to the area of 65ha in the three cycles forest, considering the amount of stereo $ 50, 00, suggested by CEPEA, the annual discount rate of 10% and revenue from REC traded on Regulated Markets. This is the most optimistic scenario, since here applies the highest price per ton of carbon and also the highest value for the sale of stereo for the production of coking coal.

Table 26. - NPV calculated in three cycles forest for culture eucalyptus 65ha in area, with stereo value of $ 50.00, and revenue from REC in the regulated market

	First Forest Cycle						
	Year1	Year2	Year3	Year4	Year5	Year6	Year7
Eucalyptus Planting							
Total cost	R$ 77.949,00	R$ 20.250,00	R$ 18.900,00	R$ 17.550,00	R$ -	R$ -	R$ -
Revenue - Eucalyptus	R$ -	R$ -	R$ -	R$ -	R$ -	R$ -	R$ 1.411.150,00
Revenue CER-Regulated Market	R$ 1.246.478,09						
Cash Flow	R$ 1.168.529,09	-R$ 20.250,00	-R$ 18.900,00	-R$ 17.550,00	R$ -	R$ -	R$ 1.411.150,00

	Second Forest Cycle						
	Year8	Year9	Year10	Year11	Year12	Year13	Year14
Eucalyptus Planting							
Total cost	R$ 18.090,00	R$ 14.850,00	R$ 14.850,00	R$ 14.850,00	R$ -	R$ -	R$ -
Revenue - Eucalyptus	R$ -	R$ -	R$ -	R$ -	R$ -	R$ -	R$ 1.411.150,00
Revenue CER-Regulated Market	R$ 1.354.832,05						
Cash Flow	R$ 1.336.742,05	-R$ 14.850,00	-R$ 14.850,00	-R$ 14.850,00	R$ -	R$ -	R$ 1.411.150,00

	Third Forest Cycle						
	Year15	Year16	Year17	Year18	Year19	Year20	Year21
Eucalyptus Planting							
Total cost	R$ 18.090,00	R$ 14.850,00	R$ 14.850,00	R$ 14.850,00	R$ -	R$ -	R$ -
Revenue - Eucalyptus	R$ -	R$ -	R$ -	R$ -	R$ -	R$ -	R$ 1.411.150,00
Revenue CER-Regulated Market	R$ 1.354.832,05						
Cash Flow	R$ 1.336.742,05	-R$ 14.850,00	-R$ 14.850,00	-R$ 14.850,00	R$ -	R$ -	R$ 1.411.150,00

Regulated Market	NAVEucalyptus 1stforest cycle	***R$ 1.743.519,98***	NAV Eucalyptus 2ndforest cycle	***R$ 1.905.790,62***	NAV Eucalyptus 3rdforest cycle	***R$ 1.905.790,62***
	Annual NAV	***R$ 249.074,28***	Annual NAV	***R$ 272.255,80***	Annual NAV	***R$ 272.255,80***
	Monthly NAV	***R$ 20.756,19***	Monthly NAV	***R$ 22.687,98***	Monthly NAV	***R$ 22.687,98***

Table 27 are the cash flows and NPV calculated for planting and felling eucalyptus charcoal to market the area to 65ha, the three cycles forest, considering the amount of stereo $ 50.00, suggested by CEPEA, the annual discount rate of 10% and revenue from REC traded on voluntary markets.

Table 27. NPV calculated for the three cycles forest for culture eucalyptus 65ha in area, with stereo value of $ 50.00, and revenue in the voluntary market REC

	First Forest Cycle						
	Year 1	Year2	Year3	Year4	Year5	Year6	Year7
Eucalyptus Planting							
Total Cost	R$ 77.949,00	R$ 20.250,00	R$ 18.900,00	R$ 17.550,00	R$ -	R$ -	R$ -
Revenue - Eucalyptus	R$ -	R$ -	R$ -	R$ -	R$ -	R$ -	R$ 1.411.150,00
Revenue CER-Voluntary	R$ 193.204,10						
Cash Flow	R$ 115.255,10	-R$ 20.250,00	-R$ 18.900,00	-R$ 17.550,00	R$ -	R$ -	R$ 1.411.150,00

	Second Forest Cycle						
	Year 8	Year 9	Year 10	Year 11	Year 12	Year 13	Year 14
Eucalyptus Planting							
Total Cost	R$ 18.090,00	R$ 14.850,00	R$ 14.850,00	R$ 14.850,00	R$ -	R$ -	R$ -
Eucalyptus Revenue	R$ -	R$ -	R$ -	R$ -	R$ -	R$ -	R$ 1.411.150,00
Revenue CER-Voluntary	R$ 209.998,97						
Cash Flow	R$ 191.908,97	-R$ 14.850,00	-R$ 14.850,00	-R$ 14.850,00	R$ -	R$ -	R$ 1.411.150,00

	Third Forest Cycle						
	Year 15	Year 16	Year 17	Year 18	Year 19	Year 20	Year 21
Eucalyptus Planting							
Total Cost	R$ 18.090,00	R$ 14.850,00	R$ 14.850,00	R$ 14.850,00	R$ -	R$ -	R$ -
Eucalyptus Revenue	R$ -	R$ -	R$ -	R$ -	R$ -	R$ -	R$ 1.411.150,00
Revenue CER-Voluntary	R$ 209.998,97						
Cash Flow	R$ 191.908,97	-R$ 14.850,00	-R$ 14.850,00	-R$ 14.850,00	R$ -	R$ -	R$ 1.411.150,00

Voluntary Market	NAVEucalyptus 1st Forest Cycle	***R$ 785.998,17***	NAVEucalyptus2ndForest Cycle	***R$ 865.033,27***	NAVEucalyptus3rdForest Cycle	***R$ 865.033,27***
	Annual NAV	***R$ 112.285,45***	Annual NAV	***R$ 123.576,18***	Annual NAV	***R$ 123.576,18***
	Monthly NAV	***R$ 9.357,12***	Monthly NAV	***R$ 10.298,02***	Monthly NAV	***R$ 10.298,02***

5.5 Calculation for 65 ha Area and Value-Based Arcelor Mittal

Table 28 are the cash flows and NPV calculated for planting and felling eucalyptus charcoal to market the area to

65ha, the three cycles forest, considering the amount of stereo $ 20.00, suggested by Arcelor Mittal, the annual discount rate of 10% and revenue from REC traded on Regulated Markets.

Table 28. NPV calculated in three cycles forest for culture eucalyptus 65ha in area, with stereo value of $ 20.00, and revenue from REC in the regulated market

	FirstForest Cycle						
	Year 1	Year 2	Year 3	Year 4	Year 5	Year 6	Year 7
Eucalyptus Planting							
Total Cost	R$ 77.949,00	R$ 20.250,00	R$ 18.900,00	R$ 17.550,00	R$ -	R$ -	R$ -
Eucalyptus Revenue	R$ -	R$ -	R$ -	R$ -	R$ -	R$ -	R$ 564.460,00
Revenue CER-Regulated Market	R$ 1.246.478,09						
Cash Flow	R$ 1.168.529,09	-R$ 20.250,00	-R$ 18.900,00	-R$ 17.550,00	R$ -	R$ -	R$ 564.460,00

	SecondForest Cycle						
	Year 8	Year 9	Year 10	Year 11	Year 12	Year 13	Year 14
Eucalyptus Planting							
Total Cost	R$ 18.090,00	R$ 14.850,00	R$ 14.850,00	R$ 14.850,00	R$ -	R$ -	R$ -
Eucalyptus Revenue	R$ -	R$ -	R$ -	R$ -	R$ -	R$ -	R$ 564.460,00
Revenue CER-Regulated Market	R$ 1.354.832,05						
Cash Flow	R$ 1.336.742,05	-R$ 14.850,00	-R$ 14.850,00	-R$ 14.850,00	R$ -	R$ -	R$ 564.460,00

	ThirdForest Cycle						
	Year 15	Year 16	Year 17	Year 18	Year 19	Year 20	Year 21
Eucalyptus Planting							
Total Cost	R$ 18.090,00	R$ 14.850,00	R$ 14.850,00	R$ 14.850,00	R$ -	R$ -	R$ -
Eucalyptus Revenue	R$ -	R$ -	R$ -	R$ -	R$ -	R$ -	R$ 564.460,00
Revenue CER-Regulated Market	R$ 1.354.832,05						
Cash Flow	R$ 1.336.742,05	-R$ 14.850,00	-R$ 14.850,00	-R$ 14.850,00	R$ -	R$ -	R$ 564.460,00

Regulated Market	NAVEucalyptus1stForest Cycle	***R$ 1.309.034,13***	NAVEucalyptus2ndForest Cycle	***R$ 1.471.304,77***	NAVEucalyptus3rdForest Cycle	***R$ 1.471.304,77***
	Annual NAV	***R$ 187.004,88***	Annual NAV	***R$ 210.186,40***	Annual NAV	***R$ 210.186,40***
	Monthly NAV	***R$ 15.583,74***	Monthly NAV	***R$ 17.515,53***	Monthly NAV	***R$ 17.515,53***

Table 29 are the cash flows and NPV calculated for planting and felling eucalyptus charcoal to market the area to 65ha, the three cycles forest, considering the amount of stereo $ 20.00, suggested by Arcelor Mittal, the annual discount rate of 10% and revenue from REC marketed in Voluntary Markets.

Table 29. NPV calculated in three cycles forest for culture eucalyptus 65ha in area, with stereo value of $ 20.00, and revenue in the Voluntary Market REC

	Ano 8	Ano 9	Ano 10	Ano 11	Ano 12	Ano 13	Ano 14
Plantio de Eucalipto							
Total de Custos	R$ 18,090.00	R$ 14,850.00	R$ 14,850.00	R$ 14,850.00	R$ -	R$ -	R$ -
Receita - eucalipto	R$ -	R$ -	R$ -	R$ -	R$ -	R$ -	R$ 564,460.00
Receita REC Mercado Voluntário	R$ 209,998.97						
Fluxo de Caixa	R$ 191,908.97	-R$ 14,850.00	-R$ 14,850.00	-R$ 14,850.00	R$ -	R$ -	R$ 564,460.00

	Terceiro Ciclo Florestal						
	Ano 15	Ano 16	Ano 17	Ano 18	Ano 19	Ano 20	Ano 21
Plantio de Eucalipto							
Total de Custos	R$ 18,090.00	R$ 14,850.00	R$ 14,850.00	R$ 14,850.00	R$ -	R$ -	R$ -
Receita - eucalipto	R$ -	R$ -	R$ -	R$ -	R$ -	R$ -	R$ 564,460.00
Receita REC Mercado Voluntário	R$ 209,998.97						
Fluxo de Caixa	R$ 191,908.97	-R$ 14,850.00	-R$ 14,850.00	-R$ 14,850.00	R$ -	R$ -	R$ 564,460.00

Mercado Voluntário	VPL Eucalipto 1o ciclo florestal	***R$ 351,512.33***	VPL Eucalipto 2o ciclo florestal	***R$ 430,547.43***	VPL Eucalipto 3o ciclo florestal	***R$ 430,547.43***
	VPL Anual	***R$ 50,216.05***	VPL Anual	***R$ 61,506.78***	VPL Anual	***R$ 61,506.78***
	VPL mensal	***R$ 4,184.67***	VPL mensal	***R$ 5,125.56***	VPL mensal	***R$ 5,125.56***

tabela 29

	FirstForest Cycle						
	Year 1	Year 2	Year 3	Year 4	Year 5	Year 6	Year 7
Eucalyptus Planting							
Total Cost	R$ 77,949.00	R$ 20,250.00	R$ 18,900.00	R$ 17,550.00	R$ -	R$ -	R$ -
Eucalyptus Revenue	R$ -	R$ -	R$ -	R$ -	R$ -	R$ -	R$ 564,460.00
Revenue CER-Voluntary Market	R$ 193,204.10						
Cash Flow	R$ 115,255.10	-R$ 20,250.00	-R$ 18,900.00	-R$ 17,550.00	R$ -	R$ -	R$ 564,460.00

	SecondForest Cycle						
	Year 8	Year 9	Year 10	Year 11	Year 12	Year 13	Year 14
Eucalyptus Planting							
Total Cost	R$ 18,090.00	R$ 14,850.00	R$ 14,850.00	R$ 14,850.00	R$ -	R$ -	R$ -
Eucalyptus Revenue	R$ -	R$ -	R$ -	R$ -	R$ -	R$ -	R$ 564,460.00
Revenue CER-Voluntary Market	R$ 209,998.97						
Cash Flow	R$ 191,908.97	-R$ 14,850.00	-R$ 14,850.00	-R$ 14,850.00	R$ -	R$ -	R$ 564,460.00

This is the best scenario, since here applies the lower the price per ton of carbon and also the lowest value for the sale of stereo for the production of coking coal.

5.6 Evaluating the Results Properly Said

The calculations presented in the previous subchapter are considering 3 variables:

- Area for the planting of 65 ha or 1000 HA;
- Value stereo eucalyptus wood stand, with data from the CEPEA and Arcelor Mittal;
- Value applied to a ton of carbon captured in Regulated Markets and Voluntary.

Table 30 summarizes the analysis performed in this section.

Table 30. NPV calculated in three cycles forest for culture eucalyptus associate of the REC market planted on a 65ha farm

	Suggested Stere value	Monthly NAV 65ha without REC	Monthly NAV 65ha withCER–Regulated Market	Monthly NAV 65ha withCER–Voluntary Market
CEPEA	R$ 50,00	R$ 8.025,30	R$ 22.687,98	R$ 10.298,02
Arcelor Mittal	R$ 20,00	R$ 2.852,85	R$ 17.515,53	R$ 5.125,56

Thus, the estimate is more economically aggressive to plant eucalyptus in order to market the production of coking coal, considering the price suggested by CEPEA stereo and sale of REC in the Regulated Market.

6. Conclusion and Suggestion for New Research

All calculations made show that all options resulting from combinations of the variables showed positive values of NPV. Eucalyptus plantations to meet demand for coking coal generation without considering the sale of REC, in itself shows a profitable operation.

Taking up by analyzing the different NPV results for each combination made for planting in the area of 65 ha, Table 30, compared with the average revenue of Table 21, divided into two studies:

- Average revenue livestock without considering the depreciation of R$14,730.28;
- Average revenue of ranching considering depreciation of R$ 12,267.45.

For both cases, only the planting of eucalyptus shows not bring economic advantage to that get with livestock, comparing cash flows. However, in adding up the revenues obtained by the sale of REC in the regulated markets of U.S. $ 22,678.98 to U.S. $ 17,515.53, which provides a considerably higher value in the sale of carbon credits compared to the voluntary markets, R$10,298.02 to R$5,515.56, brings an economic-financial advantage when adopting this strategy in increasing the income of small producers. In bringing up the revenue from the sale of REC start of each forest cycle, income is sufficient for the cultivation of livestock can be stopped without causing harm to the producer.

In considering the specific nature of this research, the following are some suggestions of related work that could increase knowledge about the subject studied here, such as expanding the model studied here researching up on creating a cooperative of producers of eucalyptus; study creation of an association of coal for generation of steel near the region studied; analyze the logistics network in the region and suggestions for new adaptations in various modes to meet the production flow of wood to the charcoal and charcoal production undertaken by steelmakers and finally find similar solutions outside Brazil so that we can assess whether the cultures of other countries use different criteria.

References

Açobrasil. (2009). *Sustainability Report.* São Paulo. Retrieved April 11, 2013, from http://www.acobrasil.org.br/site/portugues/sustentabilidade/downloads/Relatorio_2009.pdf

Aleixo. (2007). *Change of Beneficiary Form and reoccupation of the Batch Settlement Chapel, Conception Macabu, RJ* (Master thesis). Retrieved from 211F. Dissertation. Institute of Human and Social Sciences, Federal Rural University of Rio de Janeiro, Seropédica.

AMS-Mining Association of Forestry. (2005). Perspectives and Trends Wood Supply for the Forest-Based

Industry in Brazil. *Minas Gerais.* Retrieved September 30, 2010, from http://www.silviminas.com.br/Publicacao/Arquivos/publicacao_131.pdf

Arcelor, M. (2008). Forestry Partners Program. *Minas Gerais.* Retrieved July 12, 2010, from http://www.arcelormittalbioenergia.com.br/index.asp?Grupo=2&SubGrupo=9

Arnt, R. (2000). Sequestration legal. *Exame, São Paulo: Editora Abril,* 716, 99-106.

BACEN-Banco Central do Brasil. (2013). Currency Conversion. Retrieved from http://www4.bcb.gov.br/pec/conversao/conversao.asp

Bermann, C. (2002). *Energy in Brazil: To what? For what? Crises and alternatives for a sustainable country.* Rio de Janeiro: Physics Bookstore.

BNDES-National Bank for Economic and Social Development. (2002). *The Clean Development Mechanism - Guidance.* Rio de Janeiro.

Booth, W. C., Colomb, G. G., & Williams, J. M. (2008). *The Craft of Research.* USA: Chicago. http://dx.doi.org/10.7208/chicago/9780226062648.001.0001

Brito, O. J. (2013). *Report on the Production Chain Charcoal in São Paulo.* Retrieved April 11, 2013, from http://www.ipef.br/tecprodutos/procarvao/procarvao.pdf

CEPEA-Center for Advanced Studies in Applied Economics from USP. (2012). *Forestry Bulletin.* Retrieved November 26, 2012, from http://www.cepea.esalq.usp.br/florestal/files/2012/09set.pdf

CIDE-Foundation Centre for Information and Data in Rio de Janeiro. (2012). *Data in Rio de Janeiro.* Rio de Janeiro. Retrieved September 18, 2012, from http://www.cide.rj.gov.br

Colombo, S., Pepper, A. S., & Hatakeyama, K. (2013). *Charcoal Production in Cylindrical Vertical Furnaces: A Sustainable Model.* Retrieved April 11, 2013, from http://www.simpep.feb.unesp.br/anais/anais_13/artigos/1208.pdf

Costa, F. P. (2009). *Production Costs in Cattle.* Retrieved November 17, 2009, from http://www.cnpgc.embrapa.br/publicacoes/cot/cot104/COT104.pdf

Demo, P. (1994). *Research and Knowledge Building: Methods in Scientific Way of Habermas.* Rio de Janeiro: Brazilian Time.

EMBRAPA-Brazilian Company of Support for Livestock and Agriculture. (2002). *Macroprogram 1: Strategic Research Embrapa in Brazil.* Brasília. Retrieved April 14, 2010, from http://www.macroprograma1.cnptia.embrapa.br/

Green, A. G., & Unruh, J. D. (2010). Clean Development Mechanism Afforestation and Reforestation projects: implications for local agriculture. *CAB Reviews: Perspectives in Agriculture, Veterinary Science, Nutrition and Natural Resources, 5.*

Gunther, H. (2006). Qualitative research versus quantitative research: that is the question. *Brasília, 22*(2). Retrieved 2010, from http://www.scielo.br/scielo.php?script=sci_arttext&pid=S0102-37722006000200010&lng=en&nrm=iso

Hasenclever, L. (2005). *Local development in the state of Rio de Janeiro: advanced studies in local realities.* Rio de Janeiro: Editora e-papers.

Hasenclever, L., Faure, Y., & Neto, R. S. (2003). *Local economic development in the state of Rio de Janeiro: Four exploratory studies: Fields, Itaguaí, Macae and New Freiburg.* Rio de Janeiro: Editora e-papers.

Indicators of Costs, Productivity, Income Credits and Carbon Plantations of Eucalyptus and Pinus in Small Rural Properties. (2005). Paraná. Retrieved March 3, 2010, http://www.cnpf.embrapa.br/publica/comuntec/edicoes/com_tec136.pdf

Jara, C. J. (1998). *Sustainability of Local Development-Challenges in Building a Process.* Brasilia: the Inter-American Institute for Cooperation on Agriculture (IICA).

Kyoto, Protocol. (1997). *Convention-United Nations Framework on Climate Change.* Retrieved October 17, 2009, from http://www.mct.gov.br/index.php/content/view/28739.html

LIVESTOCK. (2012). *Cotação da Arroba de Boi Para Abate.* Retrieved November 29, 2012, from http://www.pecuaria.com.br/cotacoes.php

M Ethodology of Scientific Knowledge. (2000). São Paulo: Atlas.

Manual of Forestry. (2008). *Minas Gerais*. Retrieved July 12, 2010, from http://www.arcelormittalbioenergia.com.br/arquivos/38_manual.pdf

Marafon, G. J., & Brook, M. A. (2004). *Family Farming and Rural Tourism in the State of Rio de Janeiro*. Rio de Janeiro. Retrieved October 6, 2009, from http://www.sober.org.br/palestra/5/244.pdf

Ministry of the Environment-Mma. (2002). *Eligibility Criteria and Indicators for Sustainability Assessment Project that Contribual for Climate Change Mitigation and the Promotion of Sustainable Development* (p. 42). Brasília.

Neto, A. A. (2006). *Financial Mathematics and Its Applications*. Sao Paulo: Atlas.

New Directions for Fluminense Economy. (2008). *Opportunities and Challenges of Growth of the Interior.* Rio de Janeiro: Editora e-papers.

Nozoe, N. (2009). *Sesmarias Possession and Land in Colonial Brazil*. Department of Economics, Faculty of Economics, Management and Accounting. University of São Paulo: FEA-USP, São Paulo. Retrieved October 5, 2009, from http://www.anpec.org.br/revista/vol7/vol7n3p587_605.pdf

Plant Carbon Project. (2204). *Minas Gerais*. Retrieved July 17, 2010, from http://www.plantar.com.br/portal/page/portal/plantar/projeto_carbono/introducao

Puentes, Renzo Sebastián Eduardo Solari. (2010). *Expansion Of Planted Forests in Brazil With Energy Purposes And Its Influence In Emissions Of Greenhouse Effect Gases-GHG: Analysis Of Two Future Scenarios*. (Master thesis). Retrieved from Combra Luiz Alberto Institute of Postgraduate Studies and Research (COPPE), Federal University of Rio de Janeiro, Rio de Janeiro.

Renner, Rosana Maria. (2004). *And Carbon Sequestration Feasibility of New Reforestation in Brazil*. Dissertation (Master thesis), Department of Agricultural Sciences, Universidade Federal do Paraná, Curitiba.

Rodigheri, H. R. (2010). *Forestry as an Alternative to Increased Employment and Income in Rural Land*. Retrieved May 9, 2010, from http://www.cnpf.embrapa.br/publica/circtec/edicoes/circ-tec42.pdf

Schier. (2005). *Controller CUC-As a Management Tool*. São Paulo: Editora Juruá.

Segala, C. Z. S., & Silva, I. T. (2010). *Counting the cost of milk production on a farm in the municipality of Irani-SC*. Retrieved January 29, 2010, from http://www.custoseagronegocioonline.com.br/numero1v3/Custos% 20the% 20leite.pdf

Sepúlveda, S., Wilkinson, J., Tiburcio, B., & Herrera, S. (2008). *Agroeneria and Development of Isolated Rural Communities*. Brasilia: IICA.

Shell, J. Y., Alegre, J. C., & Pocomucha, V. (2007). *Reservations Determinación of them in her Biomass Carbon Air Systems Theobroma Cacao L*. Agroflorestales of in el Department of San Martín, Peru. Costa Rica. Retrieved November 28, 2009, from http://revistas.concytec.gob.pe/pdf/ecol/v3n1-2/a25v3n1-2.pdf

Sources, A. A., Da Silva, M. L., Valverde, S. R., & Souza, A. L. (2003). Analysis of Forest Activities in the municipality of Viçosa-*MG Minas Gerais: Society of Forest Research - Tree Magazine, 27*(4), 517-525.

Spode, R. D. (2007). *Profitability Economic Culture of Eucalyptus Benefited with Carbon Credits*. 78F. Working End of Course (Specialization in Capital Markets) Graduate Program in Management, Federal University of Rio Grande do Sul, Porto Alegre.

Supply Chain Wood. (2007). volume 6, Brasília.

Tasso, D. C., & Birth, E. Q. (2005). Kyoto Protocol-Analysis of Financial Aspects of Carbon Market. In: Scientific Symposium Fucape Production, 3, 2005. *Victory. Annals electronic ... Win: Fucape, 2005*. Retrieved October 22, 2009, from http://www.fucape.br/simposio/3/artigos/carla20tasso%%%20e% 20edson 20queiroz.pdf

Thyssenkrupp Steel. (2010). *Rio de Janeiro*. Retrieved July 2, 2010, from http://www.thyssenkrupp-steel-europe.com/csa/pt/index.jsp

Toledo, P. (2001). The Clean Development Mechanism and Restoration Forestry. *Economic Information, SP, 31*(3).

UFPR; ECOPLAN. (2003). *Feasibility study for establishment of forests fixing carbon: a case study in the southern state of Paraná* (p. 93). Curitiba.

UNFCCC. (2010). *Frequently Asked Questions.* Retrieved July 2, 2010, from http://unfccc.int/ghg_data/online_help/frequently_asked_questions/items/3826.php

Vale. (2010). *What We Do Steel.* Rio de Janeiro. Retrieved July 2, 2010, from http://www.vale.com/pt-br/o-que-fazemos/siderurgia/paginas/default.aspx

Vieira, A. M. (2010). *Energy Forests and Charcoal Production in Brazil.*

Walnut, M. P. (2004). *Fastening Technology and Human Field.* São Paulo. Retrieved September 20, 2009, from Article in Portuguese.

10

Drying Steps under Superheated Steam: A Review and Modeling

Ihsan Hamawand[1]

[1] National Centre for Engineering in Agriculture (NCEA), University of Southern Queensland, Toowoomba 4350 QLD, Australia

Correspondence: Ihsan Hamawand, National Centre for Engineering in Agriculture (NCEA), University of Southern Queensland, Queensland, Australia. E-mail: Ihsan.hamawand@usq.edu.au

Abstract

Superheated steam drying has been a subject of research for many decades. Many researchers have claimed that superheated steam drying is saving energy and produces better product quality compared to air drying. Superheated steam can be distinguished from air drying by the diffusion step in the drying process. Steam drying does not include moisture diffusion from the interior of the particle to its surface. However, in air drying, diffusion is the main and the limiting step. This paper is presenting a review to the models have been used to describe drying under superheated steam. Also, modelling was carried out to all the steps that take place in the drying process. These steps are: initial condensation on the surface of the particle, diffusion of the initially condensed vapour to the interior of the particle, evaporation of the initially condensed vapour and finally drying of the whole particle due to receding core model. Understanding these steps is very important for controlling the quality of the dried products and preventing difficulties such as sticking. These steps were modelled using MATLAB software. Numerical solutions with finite difference Euler's method were used to solve the equations describe the drying steps. The theoretical results were then formulated in figures to show the behaviour of the whole drying process. Comparison of the theoretical results from this study with experimental data was left for the upcoming article.

Keywords: mathematical model, superheated steam, drying, initial condensation, receding core model

1. Introduction

Drying can be defined as removing a specific amount of water from the materials due to evaporation by applying of heat under controlled condition. The main aim of the process is to increase the storage life of the materials and decrease its transportation cost. The reduction in the amount of water content of the materials reduces the water activity. This will inhibit microbial growth at the period of storing. Goals of research in improving current drying technologies are to improve the economics of operation such as reduce energy consumption, increase capacity, reduce size of equipment, increase ease of control and environmental considerations (minimize energy consumption, reduce emissions, increase safety). In addition, improve the product quality in regards to: uniform drying, minimize chemical, physical and nutritional degradation (Pronyk et al., 2004). There are two drying methods in respect of the drying medium, drying by hot gases like air and the other method is drying by hot solvents vapour like water steam. However hot air has been used for decades as a cheap drying medium, steam emerged as a drying gas that suitable for explosive materials like coals.

1.1 Steam Drying and Air Drying

Research in steam and superheated steam drying has revealed great advantages of steam drying over air drying. Environmental pollution or odour emission to the atmosphere can be eliminated since drying occurs in closed chamber without air (Soponronnarit et al., 2006). It has been reported that higher quality products can be achieved with superheated steam drying specially for starch-based products. For example, higher head rice yield compare with hot air drying due to faster development of starch gelatinization, resulting in strengthening of intermolecular binding forces among starch granules (Soponronnarit et al., 2006). Absence of oxygen prevents oxidative reaction which is promoted by native enzymes leading to coloured compounds which undermines the product quality (Elustondo et al., 2001). Superheated steam drying of biomass gives the possibility of softening the wood tissue before palletisation. Furthermore, in regards to anaerobic environment, the oxidation is avoided (Pakowiski et al., 2007).

Hausburand in 1908 disused the concept of drying by superheated steam and the ability of recycling the steam through the process (Beeby, 1984). The recirculating drying of super heated steam and eliminating air through the dryer are energy saving, the only energy leaving the system is the extra vapour that have been evaporated from the materials under drying. Super heated steams are suitable for combustible solids such as coals where it reduces risk of explosion and fire hazards. Also, inert gases may be used for start up and shut down the system to avoid combustion or fire in the case of drying combustible solids. The control on the quality of the product is easier by monitoring the steam temperature, because the drying rates and final moisture content are function of the drying medium temperature. Dust collection in superheated steam drying is much simpler by passing the excess vapour through condensers (Trommelen & Crosby, 1969). Deventer and Heijmans (2001) gave some examples in regards to steam drying benefits and restrictions of superheated steam drying. Pronyk et al. (2004) illustrated that the use of superheated steam as a drying medium for non-temperature sensitive products has many potential benefits to the consumer and the industry. Use of superheated steam can lead to energy savings as high as 50–80% over use of hot air or flue gases. These savings can be achieved due to higher heat transfer coefficients and the increased drying rates in the constant and falling periods if the steam temperature is above the inversion temperature. The drying at the constant rate period is longer in superheated steam drying, thus providing high drying rates for longer periods of time. Some valuable volatile organic compounds generated from the drying material can be recovered and separated by a condenser (Tang & Cenkowski, 2003; Wimmerstedt & Hager, 1996). Some applications involving steam drying are: drying of fuels and biofuels with high moisture contents prior to combustion in a boiler, Cattle feed exemplified by sugar beet pulp, lumber, paper pulp, paper and sludge (Wimmerstedt & Hager, 1996).

Superheated steam cannot be used for temperature sensitive materials as it is operated at high temperature, while operation under vacuum to reduce boiling point will add more complexity to the process and equipments costs. To reduce the temperature of a drop of water from 100 °C to 86 °C, this would require operating either at 0.6 atm or with 40% air present (Trommelen & Crosby, 1969). Also better and higher insulation are required for the drying surfaces to avoid condensation. It has been recommended that before using steam for drying, hot air should be used for worming up the system until the temperature in the drying compartment reaches the desired temperature (Soponronnarit et al., 2006). Berghel and Renstrom (2002) built a drying system following the requirement of industry. Tests with sawdust carried out in a fluidized bed and willow wood chips were done in a fixed bed. They showed that even with very small steam leaks, there is a significant decline in the energy efficiency of the dryer. Also, the material input and output to/from the dryer are very difficult to seal. However there are some limitations for using superheated steam as a drying medium, proof of principles for steam drying of paper and textile have been accomplished (Deventer & Heijmans, 2001).

1.2 Drying Models

There are two different mechanisms presented in literature in regards to modelling drying under superheated steam. The first is receding core model where evaporation occurs only at a drying front, the front moves from the surface of the particle to the core as drying progresses. And the second is the thin-layer model which assumes that the loss of moisture is at an exponential rate due to internal water diffusion to the grain surface at which the vaporization occurs. These two mechanisms are describing only the falling rate period. Most of researchers assumed, for the constant rate period, a simple polynomial equation that takes in consideration two factors: temperature and flow rate of the drying medium.

1.2.1 Receding Core Model

A number of studies used the receding core model to simulate drying under superheated steam. The model presented by Chen et al. (2000) which is a modification for the model presented by Agarwal et al. (1986), assumes the removal of moisture from a porous particle is due to formation of dry-wet interface and the recession of this interface with time. Although, it takes in consideration the initial condensation phenomena, this model did not count for the increase in the moisture content of the particle during the heating period. Also, it did not count for the constant rate period as well. In another study by Looi et al. (2001), they used the receding core model studied by both Shibata et al. (1988) and Khan et al. (1991). The model assumes that evaporation occurs only at a drying front, the front moves from the surface of the particle to the core as drying progresses. Shibata et al. (1988) studied the drying mechanism of sintered spheres of coarse glass beads at atmospheric pressure and superheated steam as drying medium. They found that the drying mechanism is controlled by surface evaporation at the constant rate period where evaporation takes place only on the surface of the particles. And, it is controlled by the receding front mechanism, where the evaporation takes place in the inner surface of the particle, at the falling rate period. The study emphasised on the estimation of the critical moisture content in superheated steam compared to hot air drying. The study also found that during drying in superheated steam, evaporation takes place not only on the surface of the particles during the constant rate period but also in the thin layer below the surface. Khan et al. (1991) studied both single porous particle and a backed bed drying under superheated steam. The single particle model was incorporated into an overall model for drying in a backed bed. Interface receding theory was used to model the evaporation from the single particle. Hager et al. (1997 and 2000) developed a model for superheated steam of a single porous spherical ceramic material. In their research, different mechanisms were used to present superheated steam drying. Mass transfer due to capillary forces is assumed to be dominating initially then this mechanism replaced by pressure-driven gaseous flow as the internal pressure increase. These two mechanisms assumed to follow Darcy's law. The porous material was assumed to have cylindrical pores, and tortuosity factor was the only adjustable parameter. This model not only ignores the initial condensation and the increase in the moisture content during the heating period but it is also very complicated in practice. Elustondo et al. (2001 and 2002) described the drying process by dividing it to four period; (1) short initial unsteady warming-up period, (2) constant rate period where evaporation takes place at the external surface of the sample due to external heat transfer, (3) the falling rate period starts when diffusion is not fast enough to compensate the evaporated water from the surface, in this situation a dry layer starts to grow increasing the resistance to heat conduction and making the rate to full, (4) finally when the free water is totally removed and the final water content approach the hygroscopic equilibrium value, the temperature of the sample will start to rise. Those models discussed above did not present the all steps of the drying process under steam because they did not count for the initial condensation neither for the change in moisture content during the heating period. These two neglected phenomena are very important in any drying systems. Soponronnarit et al. (2006) used a model that count for the initial condensation and the increase in the moisture content of the materials. The model was formulated for drying in a backed bed rather than single particle. This model did not use the receding core model.

1.2.2 Thin-Layer Model

Another number of studies carried out by Prachayawarakorn et al. (2002), Babalis et al. (2006), Soponronnarit et al. (2006), Chemkhi and Zagrouba (2005), Suvarnakuta et al. (2007), Tarasov and Basargin (1975) and Pronyk et al. (2004) have used Fick's unsteady state equation to describe the drying rate of spherical geometry. The effective diffusion coefficient was estimated from the experiments to show the difference between the drying in superheated steam and hot air. Crank in 1975 presented an analytical solution of the Fick's equation for various geometries. This model was formulated to describe the drying in hot air rather than steam.

1.3 Study Aim

The aim of this study is to combine all the three major steps that take place during drying under superheated steam. First step is initial condensation which happens at the early stage of drying. This then leads to the second step which is diffusion of the condensed water to the interior of the particle. This step is responsible of increasing the particle's initial moisture content. The last step starts when the particle reaches the saturation temperature. This step includes; evaporation of the initially condensed water and then evaporation of the materials' free water content at a constant rate. At the constant rate period evaporation only takes place at the surface or the first layer of the material. The evaporated moisture from the surface will be replaced by migration from interior. Finally falling rate period begins, at this step heat conduction through the particle is assumed the limiting step. In this article, these steps were modelled separately using finite difference method and MATLAB software. As there is no study presents all these steps in a single article, for this reason this article was emerged.

2. Mathematical Modelling

Mathematical models should enable good process design and minimize costs while keeping quality constraints. A common modelling approach is to assume that one transfer mechanism is a rate controlling among the three mechanisms that control drying under superheated steam. These three mechanisms are: firstly, external heat transfer which is function of the heat transfer rate by convection and radiation from the drying medium to the surface of the material; secondly, internal heat transfer which is function of the heat transfer by conduction through the particle, and finally, internal mass transfer which is function of the permeability of the particle. Mathematical models that show the changes in temperature and moisture content as function of time for a single porous spherical particle were developed for drying under superheated steam. The following are the steps:

2.1 Initial Condensation, Heating Period

The initial condensation of vapour on the surface of particles at the beginning of the contact between the cold solid feed and the steam may lead these particles to stick to the hot surfaces of the dryer. It has been shown by Hamawand (2011) that liquid bridge which connects the particles to the surface of the dryer at the beginning may go through glass transition phase after a period. This is due to the high temperature of the superheated steam. Iyota et al. (2001) focused in their model on phenomena which occur during the initial stage of drying, condensation of superheated steam on the material's surface. This phenomenon greatly influences the material quality after drying. It is also responsible for quickly heating even the interior of the material. Furthermore, it increases the moisture content of the region slightly inward from the surface. They also showed that drying under superheated steam is independent of moisture diffusion coefficient (mass transfer) and is more dominants by heat transfer (conduction). Kozanoglu et al. (2006) did not register initial condensation as a result of additional heat supplied by the resistance installed on the wall of the dryer. They showed that there was a decrease in the amount of condensation in the initial period of the process by increasing the dryer's wall temperature. In case of reduced operating pressure, a given operating temperature correspond to a higher degree of superheating:

$$\left[\Delta T = T_o - T_{sat}\right]$$

This will participate in achieving a process without initial condensation. The decrease in the operating pressure produces slightly higher final moisture content. The impact of the vacuum conditions is decrease the saturation temperature and as a result it increases the degree of superheating.

Initially as a particle with a low temperature get in to contact with the drying medium, super heated steam (SHS) in this case, this naturally lead to steam condensation on its surface and increase in the particle moisture content. This period is called heating period where the condensation continues and the particle's temperature increases until it reaches the boiling point of the solvent. Assumptions were made that the whole particle is covered by a layer of condensed water distributed uniformly over the surface of the particle. Also assumed, there is no temperature gradient through the particle during the heating period. The temperature of the particle in this period represents the whole particle. After the temperature of the particle reaches the boiling point of the solvent, which is in case of water 100 °C, this is true under atmospheric operating pressure. Condensation will stop and evaporation will take over while diffusion of the moisture through the particle continues during the entire period of both initial condensation and evaporation.

The instantaneous location of the thin layer of the condensed water is referred to as r, note $r \geq R$, where R is the particle radius. Assuming the temperature of the particle is especially uniform at this period. Heat transfer by convection to a spherical particle, neglecting the radiation (Chen & Agarwal, 2000), can be present by Equation (1). Equation (1) is a result of energy balance over isothermal spherical particle covered by a layer of condensed water, with radius r and temperature T_p;

$$\frac{d}{dt}\cdot\left[\left(\rho_s C_{p.s}+\rho_s M_{in}.C_{p.w}+(\Phi^3-1)\rho_w C_{p.w}\right).T_p\right] = \frac{3.h_{(g-p)}}{R}\Phi^2.\left(T_g - T_p\right) - \frac{3.\Phi^2\lambda_{vap}}{R}\left(-\rho_w\frac{dr}{dt}\right) \quad r \geq R \tag{1}$$

Where;

$h_{(g-p)}$ heat transfer coefficient between the SHS medium and the surface of the particle;

T_g temperature of the drying medium (SHS);

T_p temperature of the particle (the whole particle);

$\rho_w, C_{p.w}$ density and specific heat of water;

$\rho_s, C_{p.s}$ density and specific heat of solid material;

λ_{vap} latent heat of vaporization of water;

r variation in the particle radius due to condensation and evaporation;

R initial radius of the particle;

M_{in} initial moisture content of the particle;

$\Phi = \frac{r}{R}$ dimensionless radius.

The first term represents the change in temperature of the particle with time considering the particle's material, moisture content and the initial layer of condensed water. The first term in the RHS of the equation account for the heat transfer between the particle and the heating medium. The second term represents the latent heat of condensation of the superheated steam. In a steady state conditions, superheated steam temperature varies very slowly with time so any change can be neglected and adiabatic behaviour can be assumed (drying zone is perfectly isolated) (Elustondo et al., 2002).

The following initial conditions were assumed to start the modelling:

$r_{(i)} = R$ initial particle radius;

$T_p^{(j)} = T_p^{(0)}$ initial particle temperature ($t = 0$);

Δt time interval;

i, j represents position and time, respectively.

The temperature of the particle after a time interval of Δt depends on the numbers of water layers condense over its surface. Using numerical method, finite deference Euler's method, Equation (1) can be solved for the particle's temperature as shown in Equation (2):

$$T_p^{(j+1)} = \frac{\left[\left[\rho_s C_{p.s} + \rho_s M_{in}.C_{p.w} + \left(\left(\frac{r_{(i+1)}}{R}\right)^3 - 1\right)\rho_w C_{p.w}\right].T_p^{(j)} + \Delta t.\left(\frac{3.h_{(g-p)}}{R}\left(\frac{r_{(i)}}{R}\right)^2.\left(T_g - T_p^{(j)}\right) + \frac{3\lambda_{vap}\rho_w}{R}\left(\frac{r_{(i)}}{R}\right)^2\left(\frac{r_{(i+1)} - r_{(i)}}{\Delta t}\right)\right)\right]}{\left[\rho_s C_{p.s} + \rho_s M_{in}.C_{p.w} + \left(\left(\frac{r_{(i)}}{R}\right)^3 - 1\right)\rho_w C_{p.w}\right]} \quad r \geq R \quad (2)$$

In order to solve equation (2) for the particle temperature $T_p^{(j+1)}$ as function of time intervals, it needs to consider the radius of the particle $r_{(i+1)}$, actually the radius of the particle adding to it the thickness of the condensed water, the actual particle radius was assumed constant, at each layer of water condensed on its surface.

When free water is in contact with pure vapour, the equilibrium rate of molecule transfer from vapour to the liquid phase and from the liquid to the vapour phase is given by the kinetic theory of gases. When the system is not at equilibrium, the net rate of evaporation or condensation per unit droplet (particle covered completely by water layers) surface area is governed by modified Hertz-Kundsen equation (Chen & Agarwal, 2000). This equation was first proposed by Alty 1931 (Maa, 1967). Maa (1967) studied the evaporation rate of different solvents including water. The study shows that there is little or no resistance to molecules crossing the vapour-liquid interface. This means that the evaporation coefficient is unity. Equation (3) shows the evaporation-condensation rate of water as presented by Maa (1967):

$$-\rho_w.\frac{dr}{dt} = \sqrt{\frac{M_w}{2.\pi.R_g}}.\left[\frac{P_{sat(T_p)}}{\sqrt{T_p}} - \frac{P_{atm}}{\sqrt{T_g}}\right] \quad (3)$$

Where,

P_{atm} operation Pressure of the drying medium;

$P_{sat(T_p)}$ saturation vapour pressure of condensed water at particle "surface" temperature;

M_w molecular weight of water;

R_g universal gas constant.

The saturation vapour's pressure of water at different particle's surface temperature, actually the water shell around the particle, can be given by Equation (4) (Antoine's equation);

$$LnP_{sat(T_p)} = 16.387 - 3885.7 / (T_p + 230.17) \quad (4)$$

Using numerical method, finite deference Euler's method, Equations (3) and (4) can be rewritten as following;

$$r_{(i+1)} = r_{(i)} - \frac{\Delta t}{\rho_w}\sqrt{\frac{M_w}{2.\pi.R_g}}.\left[\frac{P_{sat(T_p^{(j)})}}{\sqrt{T_p^{(j)}}} - \frac{P_{atm}}{\sqrt{T_g}}\right] \quad (5)$$

$$LnP_{sat(T_p^{(j)})} = 16.387 - 3885.7 / (T_p^{(j)} + 230.17) \tag{6}$$

Equation (5) and (6) can be used to predict the condensation-evaporation time interval and the amount of water condensed over the particle.

The evaporation will start when the temperature of the particle reaches approximately 100 °C. At this temperature the thickness of the condensed water on the surface of the particle is known from Equations (5) and (6). The evaporation will occur at constant particle temperature. The evaporation will be dominated by heat transfer between the condensed water and the drying medium. At this period an assumption was made that the whole particle is a droplet of water at 100 °C until the entire condensed water evaporates. Energy balance over a droplet of water can be present by Equation (7);

$$\frac{\rho_w C_{p.w}.R}{3} \cdot \frac{dT_p}{dt} = h_{(g-p)}.(T_g - T_p) + \rho_w \lambda_{vap} \frac{dr}{dt} \tag{7}$$

For constant particle temperature at 100 °C, $\frac{dT_p}{dt} = 0$

$$-\frac{dr}{dt} = \frac{h_{(g-p)}}{\rho_w \lambda_{vap}}.(T_g - T_p) \tag{8}$$

Where $T_p = T_w = 100$ °C

The radius of the condensed water depend on the rate of evaporation from the droplet surface, by using numerical method, finite deference Euler's method, Equation (8) can be rewritten as below;

$$r_{(i+1)} = r_{(i)} - dt.\left(\frac{h_{(g-p)}}{\rho_w \lambda_{vap}}.(T_g - T_p)\right) \tag{9}$$

From the above Equations (2, 5, 6 and 9) final temperature of the particle can be found at the end of the heating period. The condensation–evaporation period (heating period) is very important for the calculation of the amount of moisture absorbed. Also, the thickness of the condensed water with time can be tracked.

Trommelen and Crosby (1969) demonstrated the use of correlations of heat transfer coefficient which is applicable for evaporation of small pure water drops in air is also valid for evaporation in superheated steam. Forced convective heat transfer can be assumed to take place between the bulk superheated steam and the particle surface either covered with water layer or not, proposed by Hager et al. (1997), Equation (10);

$$\frac{h_{(g-p)}.d_p}{k_g} = 2 + 0.61.\,Re^{0.52}.\,Pr^{0.33} \tag{10}$$

$h_{(g-p)}$ Average convective heat transfer coefficient, J/m^2.k.s;

k_g Thermal conductivity of steam, J/m.k.s.

$$Re = \frac{\rho_g.u_g.d_p}{\mu_g}$$

$$Pr = \frac{\mu_g.C_{p.g}}{k_g}$$

ρ_g density of the superheated steam;

u_g superficial velocity of the superheated steam;

μ_g viscosity of the superheated steam;

$C_{p.g}$ heat capacity of the superheated steam.

2.2 Water Diffusion from the Surface Through the Particle

The moisture content of the particle will increase during the heating period due to diffusion of the condensed water at the surface to the interior of the particle. This phenomenon is important in dryers such as rotary drum dryer where the particles have much time for rest (contact with the surface) (Hamawand & Yusaf, 2013). Unlike fluidized bed dryers the water condensed on the particle surface evaporates in a shorter time due to continuous contact between the particles and the drying medium. The first layer of the particle's surface will be saturated with moisture during the heating period. This can be described by the diffusion equation for a spherical particle (Soponronnarit et al., 2006; Chemkhi & Zagrouba, 2005), as shown in Equation (11);

$$\frac{\partial M}{\partial t} = D_{eff}\left[\frac{\partial^2 M}{\partial r^2} + \frac{2}{r}\frac{\partial M}{\partial r}\right] \qquad r \leq R \tag{11}$$

M moisture content kg water / kg dry basis;

D_{eff} effective diffusivity coefficient m^2/s.

The effective diffusivity coefficient can be determined from the drying experiment itself as shown by Chemkhi and Zagrouba (2005) and Soponronnarit et al. (2006). Chemkhi and Zagrouba (2005), investigated the moisture diffusivity through a porous clay materials in air drying at different humidity. The moisture diffusivity can be estimated by: analysis of the drying data, sorption kinetics and permeability measurements. They estimated the effective diffusivity from the drying data, from the slope of the drying curve and/or from comparison of the experimental data with that predicted by Fick's second law. Also, the equilibrium moisture content was estimated from the sorption isotherm of the particle. The effective diffusivity can be predicted by Equation (12);

$$D_{eff} = D_o \exp\left(-\frac{E_a}{R.T}\right) \tag{12}$$

Where;

E_a activation energy;

D_o equation constant;

R_g Universal gas constant (J/mol.K);

T_p particle temperature (K);

D_{eff} effective diffusivity coefficient m^2/s.

$\frac{\Delta r^2}{D_{eff}.\Delta t} \geq 2$ in order to have the numerical solution stable.

Initial and boundary condition at this period are as following (Soponronnarit et al., 2006);

$$M_{(r,t)} = M_{in} \qquad t = 0$$

$$M_{(R,t)} = M_{eq} \qquad t > 0, \quad r=R$$

$$\left.\frac{\partial M}{\partial r}\right|_{r=0} = 0 \qquad t > 0$$

Where M_{eq} correspond to the maximum moisture content for which the particle can hold, this can be predicted by soaking the particle in water over night (Soponronnarit et al., 2006). The boundary conditions propose constant moisture content at the centre and maximum at the surface. The surface of the particle will be in equilibrium with the condensed water. Using numerical method, finite deference Euler's method, Equation (11) can be rewritten as Equation (13);

$$M_{(i)}^{(j+1)} = M_{(i)}^{(j)} + \Delta t.D_{eff}\left[\frac{M_{(i+1)}^{(j)} - 2M_{(i)}^{(j)} + M_{(i-1)}^{(j)}}{\Delta r^2} + \frac{2}{r_{(i)}}\frac{M_{(i+1)}^{(j)} - M_{(i)}^{(j)}}{\Delta r}\right] \quad (13)$$

$$t = 0 \qquad M_{(r,t)} = M_{in} = M_{(0)}^{(0)} = M_{(1)}^{(0)} = M_{(2)}^{(0)} = \ldots\ldots\ldots$$

$$r = R \qquad M_{(R,t)} = M_{eq} = M_{(0)}^{(0)} = M_{(0)}^{(1)} = M_{(0)}^{(2)} = \ldots\ldots \qquad t > 0,$$

$$r = 0 \qquad M_{(n)}^{(j)} = M_{(n-1)} \qquad t > 0$$

Where i and j represent position and time, respectively.

Equation (13) can be used to find the moisture content in each layer of the particle along the entire heating period where the diffusion occurs. During the heating period the particle absorb water through its surface. This absorbed water will distribute through the particle radially. The average moisture content at the end of heating period cab be estimated using Equation (14) (Soponronnarit et al., 2006);

$$\bar{M}_{(t)} = \frac{4\pi}{V_p}\int_0^R r^2.M_{(r,t)}dr \quad (14)$$

Using numerical integration, Simpson rule;

$$\bar{M}_{(t)} = \frac{4\pi}{V_p}\cdot\frac{\Delta r}{3}\left[R^2.M_{(R,t)} + 4.r_1^2.M_{(r_1,t)} + r_2^2.M_{(r_2,t)}\right] \quad (15)$$

Where V_p is the particle volume (m^3).

The amount of water absorbed by the particle after an interval of $t_{heatingperiod} = t_{diffusion}$;

$$m_{absorbed(t)} = m_p.\left[\bar{M}_{(t)} - \bar{M}_{in}\right] \quad (16)$$

$m_{absorbed}$ the amount of water absorbed during the heating period (kg);

m_p solid weight of the dry particle (kg);

At the end of the heating period the average temperature and moisture content of the particle will be 100 °C, and $\overline{M}_{(t)}$ respectively.

2.3 Constant Drying Rate Period

At the initial stage of drying, the drying rate is constant due to the vaporization of water at particle's grain surface. This period is indicated by a line presenting moisture content versus drying time. The assumption is made that evaporation will start when the temperature of the materials reaches approximately 100 °C. The evaporation will be dominated by the heat transfer between the particle surface and the drying medium. The amount of water evaporates from the particle's surface will be replaced by liquid diffusion from the inner region. At this period, another assumption was made that the whole particle is at 100 °C until the entire free water evaporates. Evaporation from a particle's surface can be represents by Equation (17);

$$\frac{\rho_w C_{p.w}.R}{3}\cdot\frac{dT_p}{dt} = h_{(g-p)}.(T_g - T_p) + \rho_w\lambda_{vap}\frac{dr}{dt} \quad (17)$$

For constant particle temperature at 100 °C, $\frac{dT_p}{dt} = 0$

$$-\frac{dr}{dt} = \frac{h_{(g-p)}}{\rho_w \lambda_{vap}}.(T_g - T_p) \qquad (18)$$

Where $T_p = T_w = 100$ °C.

At this period, the evaporation will occur at the surface of the particle, so there is no change in the wet zone radius. Equation (18) can be rewritten in form of moisture content as shown in Equation (19);

$$-\frac{dM}{dt} = \frac{h_{(g-p)}.A_s}{\lambda_{vap}}.(T_g - T_p) \qquad (19)$$

The amount of moisture can be calculated by Equation (20);

$$M_{(t)} = M_{(o)} - h_{(g-p)}.\left[\frac{A_s}{\lambda_{vap}}.(T_g - T_p).dt\right] \qquad (20)$$

Where $M_{(o)}$ is the initial amount of moisture in the sample (kg water), and $h_{(g-p)}$ is the heat transfer coefficient at the boundary of the sample as shown in Figure (1). By dividing Equation (20) by the initial weight of the sample, moisture content of the particle can be estimated. The initial weight of the sample can be formulated as in Equation (21);

$$Sample\ Weight = \left[\rho_s.\varepsilon_s + \rho_w.(1-\varepsilon_s)\right].V_s \qquad (21)$$

Then, Equation (20) can be rewritten as moisture content on wet basis as in Equation (22);

$$MC_{(t)} = MC_{(o)} - h_{(g-p)}.\left[\frac{3.(T_g - T_p).dt}{\lambda_{eva}.[\rho_s.\varepsilon_s + \rho_w(1-\varepsilon_s)].r}\right] \qquad (22)$$

Where; $MC_{(t)}$ and $MC_{(o)}$ is the time dependent and initial moisture content on wet basis (kg water/ kg sample), ρ_s is the density of the solid, ρ_w is the density of water, ε_s is the fraction of solid in the sample and $(1-\varepsilon_s)$ is the fraction of water in the sample, A_s is the cross sectional surface area, and r is the radius of the sample.

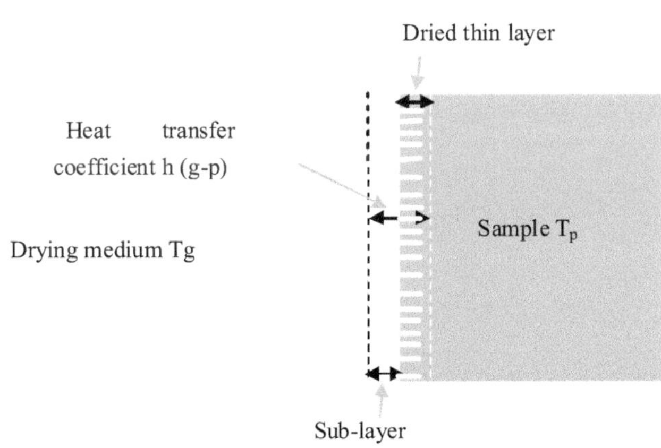

Figure 1. A schematic diagram for the heat transfer boundary

2.4 Falling Rate Period

The temperature distribution in a spherical dry particle can be presented by the differential equation below. The heat transfers from the drying medium to the particle surface by convection and then through the particle by conduction;

$$\frac{\partial T_p}{\partial t} = \alpha . \left[\frac{\partial^2 T_p}{\partial r^2} + \frac{2}{r} \frac{\partial T_p}{\partial r} \right] \qquad r \leq R \tag{23}$$

Using numerical method, finite deference Euler's method, with an explicit formulation, Equation (23) can be rewritten as;

$$T_{p(i)}^{(j+1)} = T_{p(i)}^{(j)} + \Delta t . \alpha . \left[\frac{T_{p(i+1)}^{(j)} - 2T_{p(i)}^{(j)} + T_{p(i-1)}^{(j)}}{\Delta r^2} + \frac{2}{r_{(i)}} \frac{T_{p(i+1)}^{(j)} - T_{p(i)}^{(j)}}{\Delta r} \right] \tag{24}$$

Where, $\alpha = \frac{k_p}{\rho_p . C_{p.p}}$, and $\frac{\Delta r^2}{\alpha . \Delta t} \geq 2$ in order to have the numerical solution stable;

Where;

k_p thermal resistance of the particle materials, it is between $k_{p.dry}$ and $k_{p.wet}$;

ρ_p density of the particle, it is between $\rho_{p.dry}$ and $\rho_{p.wet}$;

$C_{p.p}$ specific heat capacity, it is between $C_{p.dry}$ and $C_{p.wet}$.

The initial and boundary condition at this period are;

$$T_p = T_{p(i)}^{(0)} = 100 \ C \qquad t = 0$$

$$\frac{\partial T_p}{\partial r} = \frac{h_{(g-p)}}{k_p} \left[T_g - T_p \right] \qquad r = R$$

$$\frac{\partial T_p}{\partial r} = 0 \qquad r = 0$$

Equation (24) can be used to predict the change in temperature of each layer dr with time directly after the recession of its moisture. The Assumption was made that the temperature of each layer is constant during the recession of its moisture and kept at 100 °C. After this period, the temperature of the particle's layer went through recession will start to rise due to equation (24). In order to solve equation (24) for temperature gradient through the particle, the time interval for recession of moisture from each layer need to be predicted. Energy balance over the first layer can be presents by Equation (25);

$$h_{(g-p)} (T_g - T_p) = \lambda_{vap} (M_{in} - M_c) . \rho_p . \frac{dr}{dt} \tag{25}$$

Equation (25) can be rewritten in numerical formula as;

$$dr_1 = \frac{dt . h_{(g-p)} (T_g - T_{p(i)}^{(j)})}{\lambda_{vap} (M_{in} - M_c) . \rho_p}$$

$$T_{p(1)}^{(0)} = 373 K \tag{26}$$

Solve for the temperature of the first layer, this can be done by equating the heat transferred to the first layer by convection to the heat transferred by conduction through the particle. The temperature of the first layer can be predicted via Equation (27);

$$T_{p(i)} = \frac{T_{p(i+1)} + dr.\frac{h_{(g-p)}}{k_p}.T_g}{1 + dr\frac{h_{(g-p)}}{k_p}} \qquad r = R \tag{27}$$

By assuming the value of dt in Equation (26), the location of the interface after moisture recession from the first layer can be predicted, during this period the temperature of each layer is assumed constant at 100 °C. After elapsed time of dt, the first layer become dry. At the next time interval dt, the first layer will start to heat up over 100 °C, its temperature can be calculated using Equation (27). For the second and the following layers, Equation (28) below can be used to estimate the thickness (or location) of each layer goes under recession;

$$k_p \frac{dT_p}{dr} = \lambda_{vap}(M_{in} - M_c).\rho_p.\frac{dr}{dt} \tag{28}$$

Equation (28) can be written numerically as;

$$r_{(i+1)} = r_{(i)} - \sqrt{\frac{k_p dt}{(M_{in} - M_c)}.\frac{\left(T_{p(i+1)} - T_{p(i)}\right)}{\lambda_{vap}.\rho_p}}. \tag{29}$$

Where M_c is the critical moisture content of the particle.

The moisture content can be calculated as function of time by conducting material balance over each layer, as shown in Equation (30);

$$M = M^0 . \left[\frac{4/3\pi.(r_{(i)}^3)}{V_t}\right] \tag{30}$$

The equation above gives the accumulated moisture content as function of location.

3. Materials Properties

A porous particle is consisting of the solid structure of the particle and the moisture filling the vacancies. The properties of the particle will changes after evaporation of the entire water content. When the water begin to evaporate from the particles pores, the particle will be consists of the solid structure and vapour in the pores. For this reason the properties of the particle such as; density, specific heat, and thermal conductivity $[\rho_p, C_p, \rho_p, k_p]$ most be predicted at different period and positions, the effect of temperature can be neglected (Iyota et al., 2001).

Shibata et al. (1988) presented a simplified quantitative presence of each component to the relationship. At the wet zone, the particle is composed of solid structure with water filling the pores; therefore the density of the particle at this zone is;

$$\rho_{p.wet} = (1 - \varepsilon_w)(1 + M_{in}).\rho_s$$

As the water in the pores evaporate due to conductive heat transfer, the water in the pore will replaced by the steam. Then the density of the particle at the dry zone is;

$$\rho_{p.dry} = (1 - \varepsilon_v)(1 + M_c).\rho_s + \varepsilon_v \rho_v$$

The same treatment for the density will be used for predicting the specific heat and the thermal conductivity, the specific heat capacity in the wet zone becomes;

$$C_{p.wet} = \frac{\rho_s}{\rho_{p.w}}(1 - \varepsilon_w)(C_{p.s} + C_{p.w}M_{in})$$

The specific heat capacity at the dry zone is;

$$C_{p.dry} = \frac{\rho_s}{\rho_{p.d}}(1-\varepsilon_v)(C_{p.s} + C_{p.w}M_c) + \frac{\rho_v}{\rho_{p.d}}.\varepsilon_v.C_{p.v}$$

The thermal conductivity at the wet zone will be;

$$k_{p.wet} = k_s(1-\varepsilon_w) + k_w\varepsilon_w$$

The thermal conductivity at the dry zone can be represented as;

$$k_{p.dry} = k_s + (k_w - k_s).\left(\frac{1-\varepsilon_v}{(\rho_w/\rho_s.M_c)+1}\right) + (k_v - k_s)\varepsilon_v$$

Where subscripts w, s, v represent the liquid, solid, and vapour components, $\varepsilon_v, \varepsilon_w$ are the mass fraction of vapour and water in the particle. As the vapour will replace the water during the recession, then $\varepsilon_v = \varepsilon_w$ and the mass fraction of the solid can be represented by;

$$\varepsilon_s = (1-\varepsilon_w)$$

Iyota et al. (2001) assumed that the physical properties are constant in regards to temperature, and the apparent physical values of compound can be driven from the specific solid and water values with regards to moisture content. Tang and Cenkowski, (2000, 2003), gave a formula for the density of super heated steam as;

$$\rho_s = 0.7401 - 0.001505.T_s$$

C_{ps} for steam change between 100 to 200 °C as 2024 to 1975 J/kg.C

C_w specific heat of moisture , 4193 between 0 to 100 °C , and 4313 between 100 to 200 °C.

4. Modelling Results

4.1 Initial Condensation

This step estimates both the temperature of the particle which is calculated using Equation (5) and the amount of liquid condensed over the surface of the particle as an indication of the particle radius using both Equations (6) and (7). Equation (5) is an energy balance, it shows that the temperature of the particle increases as steam condensed over its surface. The condensation of steam will stop whenever the temperature of the whole particle reach water boiling point. As shown in Figure (2), the temperature of the particle raises sharply starting from initial particle temperature. This temperature then flattens at 100 °C, the boiling point temperature of drying solvent (in this case water). At this temperature steam condensation will stop and the evaporation of the condensed steam will begins. The radius of the particle (the condensed layer of water) starts to decrease from this point forward. Figure (2) shows a sharp rise in temperature of the particle from its initial temperature to approximately 100 °C, this is due to the high specific heat of the condensed water, the speed of this rise in temperature is also function of the particle size.

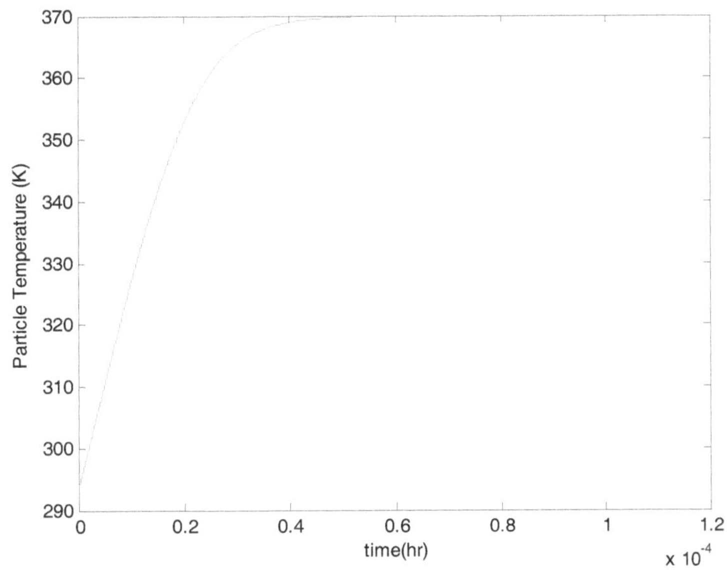

Figure 2. Particle temperature during initial condensation

4.2 Initial Evaporation

This step estimates the evaporation rate of the initial liquid condensed during the heating period. Equation (8) estimates this by considering the radius of the particle as indication of the amount of the evaporated water. By dividing the changes in the radius of the particle over the time interval, the rate of evaporation can be predicted. In the first part of Figure (3), the rising part of the curve presents the data predicted from both Equations (6) and (7). These two equations estimate the amount of condensed vapour using the radius of the particle as indication. Whenever the temperature of the condensed vapour is below water boiling temperature or its partial pressure is below the saturation, the condensation will continue. In the second part of Figure (3), the declined line is presenting the evaporation period. As the figure shows, the evaporation happens at a constant rate (constant slope), this is because the evaporation is happening from a surface covered with free water. Although, the surface area of evaporation is changing because of the reduction in the radius of the condensed layer; it was neglected due to its small thickness.

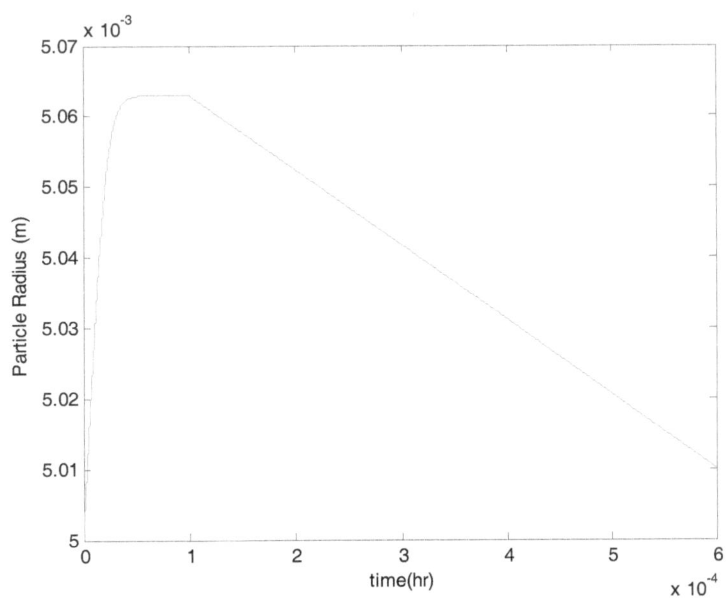

Figure 3. Particle radius during initial condensation and later evaporation

4.3 Diffusion of Moisture from the Surface to the Interior of the Particle

This step estimates the moisture distribution through the interior of the particle during the heating period using Equation (13). Also, it estimates the change in the amount of the particle's moisture content (average moisture content) using Equation (14). As shown in Figure (4), the average moisture content of the particle was raised from 0.15 (initial moisture content) to 0.1683 during the heating period. The curve in Figure (4) shows the distribution of moisture in the particle starting from the surface to the centre. The first layer of the particle is at saturation because the surface is covered totally by a layer of condensed water. The moisture diffuses from the surface to the interior in a decline sequence due to the resistance to diffusion. The shape of the curve in Figure (4) depends on the time interval of this period (heating period). As the interval of the heating period increases, the average moisture content of the particle increases as well.

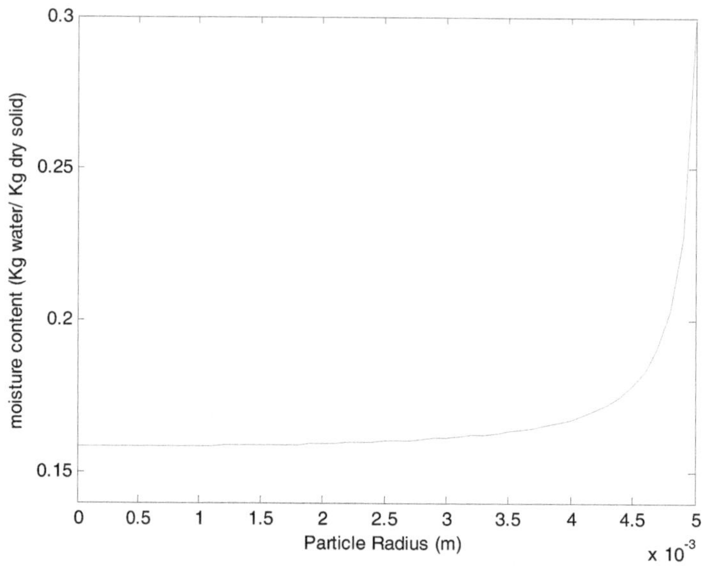

Figure 4. Distribution of moisture through the particle during heating period, the Average Moisture content of the particle rose from 0.15 to 0.1683

4.4 Drying

This period was divided to two steps; the constant rate period and the falling rate period.

4.4.1 Constant Rate Period

This step estimates the moisture content of the particle as function of heat transfer coefficient. This period assumes that the evaporation of the moisture happen only at the surface of the particle because migration of moisture from the interior of the particle to the surface is higher than the evaporation rate at the surface. Equation (22) estimates the amount of moisture loses as function of time as shown in Figure (5). Figure (5) shows a linear decline in the moisture content because at the constant rate period it was assumed that evaporation is happening from a layer of free water at the surface. In this period the surface of the particle will be covered by a layer of migrated moisture until no more moisture can be supported by diffusion from inside of the particle.

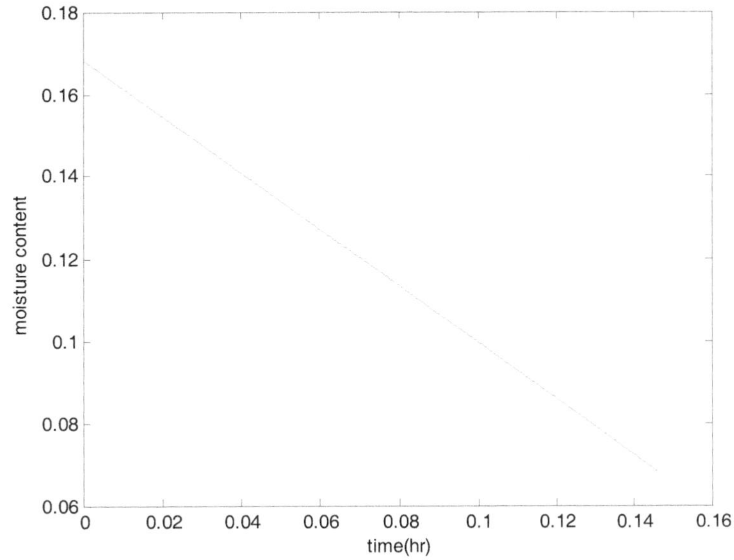

Figure 5. Moisture content during constant rate period

4.4.2 Falling Rate Period, Recession of Moisture Content Controlled by Heat Conduction

This step estimates the temperature distribution along the radius of the particle using Equation (24). Also, Equation (27) was used to predict the temperature of the first layer of the particle. To estimate the recession of moisture from each layer of the particle starting from the surface, Equation (29) was used to locate the dried layers movements through the particle. The result from this equation then can be used to calculate the moisture content of the particle via Equation (30) when the geometry of the particle is known.

Figure (6) shows the temperature at two points in the particle, the surface and the centre of the particle. The curve at the top is representing the particle surface temperature. The temperature of the first layer at the surface of the particle rises to water boiling temperature in a very short period. Recalling Figure (3), it takes less than one second for the temperature at the surface of the particle to rise to 100 °C (depend on the particle size). This temperature then remains constant at 100 °C until the particle's moisture content reaches a critical. After this point the curve rises sharply because the surface of the particle is dry and it is no more been cooled via evaporation. Similar result is shown in Figure (6) for the particle's centre temperature. The lower curve shows that the temperature at the centre remains at its initial temperature, after rising to 100 °C, for the whole heating period. This is because the heat transferred from the drying medium is consumed by the evaporation at the surface. For this reason its temperature will only start to rise when the surface is dry and the heat intrude the interior of the particle.

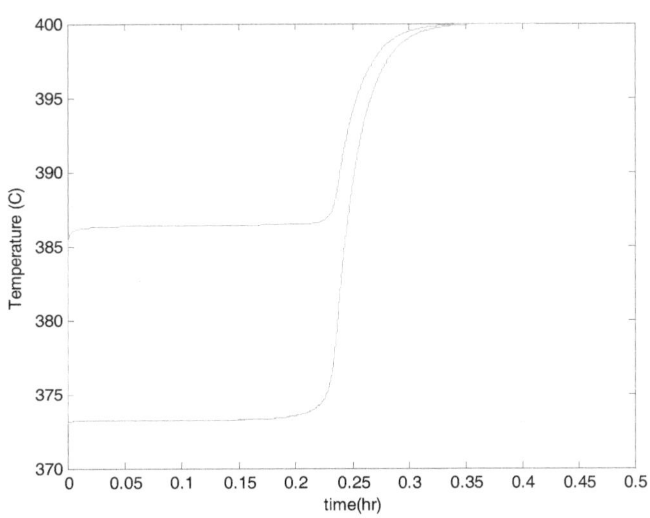

Figure 6. Temperature at the surface and the centre of the particle

The moisture content of the particle during this period decreases smoothly from the critical moisture content to the equilibrium moisture content. In this period, the heat will transfer through the interior of the particle via conduction and recession of moisture will take place in each layer starting from the surface. As shown in Figure (7), this period takes longer time by many magnitudes compared to the constant rate period. This is due to high resistance to heat conduction and the difficulties associated with the receding moisture from a critical value to equilibrium moisture content. Critical moisture content represents the moisture that is available in the cell of the materials. As the recession go deeper through the particle, the resistance to the heat conduction increases. This explains the semi-flatten curve after long time of drying. The model also shows the position of the moisture as receding through the particle. Figure (8) illustrates the time required for each layer of the particle to dry to their equilibrium moisture content. The linearity in this relation is due to regular increase in the time of recession of each layer. When all the layers of the particle reach the equilibrium moisture content, the line flatten because there is no more moisture is available, except that chemically bonded.

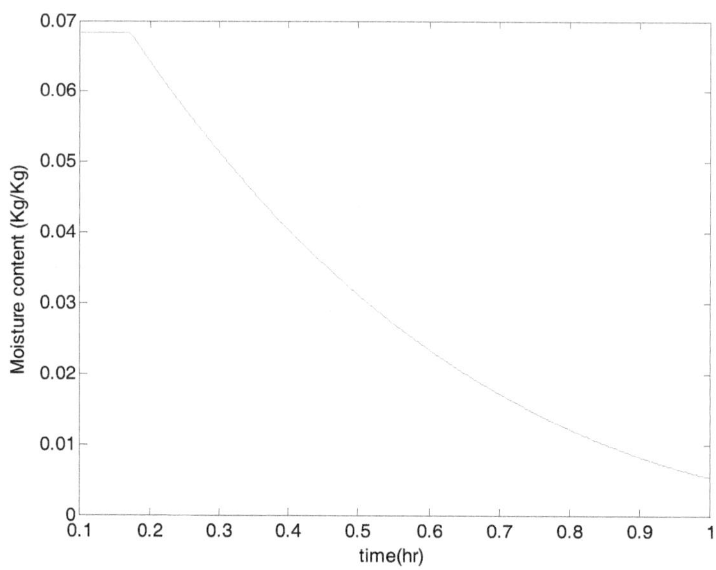

Figure 7. Moisture content during the falling rate period

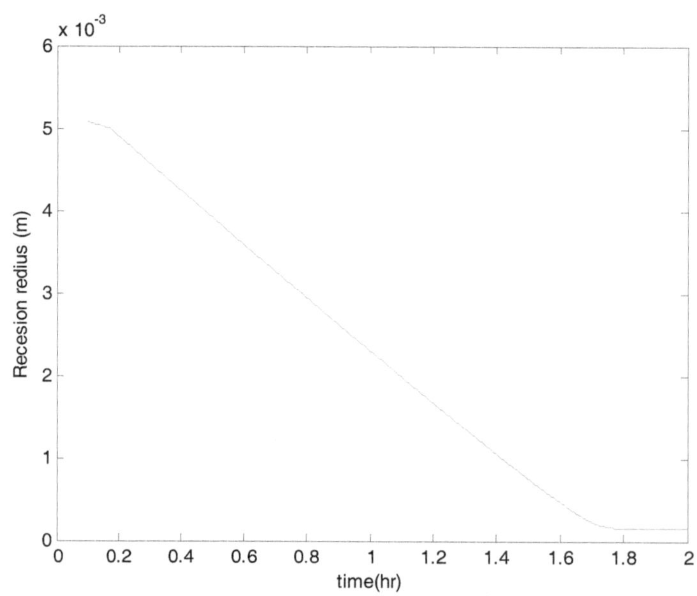

Figure 8. Recession of moisture through the particle

5. Conclusion

The models have been applied in this research are results of literature review conducted in the field of drying under superheated steam. Also, energy balance was conducted in some other occasions. This paper is combining different models presented by many researchers in this filed to come up with a complete modelling sets for the entire drying steps under superheated steam. Finite difference with the aid of MATLAB software was used in order to apply the models. The theoretical results predicted from the models illustrate the drying behaviour under superheated steam. However, experiments were not conducted, the trends of the results are agreeing with the actual behaviour of drying under superheated steam. This article successfully presented a literature review focused on the modelling of drying under superheated steam. Also, full details in regards to modelling the all steps that take place under superheated steam drying were demonstrated.

Nomenclature

Symbol	Description	Unit	Values assumed
$h_{(g-p)}$	heat transfer coefficient between the SHS medium and the surface of the particle.	J /m^2.k.s	110-135
T_g	temperature of the SHS	K	398-423-448
$T_p^{(0)}$	initial temperature of the particle	K	298
ρ_w	density of water	Kg/ m^3	1000
$C_{p.w}$	specific heat of water	KJ/ Kg.K	4.27
ρ_s	density of particle material (ceramic)	Kg/ m^3	1600
$C_{p.s}$	specific heat of particle material	KJ/ Kg.K	0.64
ρ_v	density of the steam	Kg/ m^3	0.565
$C_{p.v}$	specific heat of the steam	KJ/ Kg.K	2.007
λ_{vap}	latent heat of vaporization of water	KJ / Kg	2598
R	radius of the particle	m	5 E-3
M_{in}	initial moisture content of the particle	Kg water/ Kg dry solid	0.15
M_c	critical moisture content of the particle	Kg water/ Kg dry solid	0.01
Δt	time interval	s	3 E-8
P_{atm}	vapour pressure of the super heated steam	KN/ m^2	100
M_w	molecular weight of water	Kg / mol	18
R_g	universal gas constant	J / mol.K	8.314
k_g	thermal conductivity of steam	J / m.k.s	2.48E-02
u_g	superficial velocity of the steam	m/ s	2.7
μ_g	viscosity of the steam	Pa.s	1.30211E-05
D_{eff}	effective diffusivity coefficient m^2/s	m^2/ s	

E_a	activation energy	J / mol	
D_o	equation constant	m^2/ s	
M_{eq}	moisture content for which the particle can hold	Kg water/ Kg dry solid	
m_p	solid weight of the dry particle	Kg	
Δr	thickness of the element	mm	10 – 100 E-8
k_s	thermal conductivity of the solid	J / m.k.s	2.5
k_w	thermal conductivity of the water	J / m.k.s	0.688
K	permeability	m^2	2 E-15
m_g	superheated steam flux	Kg / m^2 .s	0.15, 0.25, 0.35
ε_w	mass fraction of water	-------	0.41
ε_v	mass fraction of vapour	--------	0.41
ε_s	$(1-\varepsilon_w)$ mass fraction of solid	-------	0.59

References

Agarwal, P. K., Genetti, W. E., & Lee, Y. Y. (1986). Coupled drying and devolatilization of coal in fluidized bed. *Chem. Eng. Science, 41*(9), 2373-2383. http://dx.doi.org/10.1016/0009-2509(86)85087-4

Babalis, S. J., Papanicolaou, E., Kyriakis, N., & Belessitis, V. G. (2006). Evaluation of thin-layer models for describing drying kinetics of figs. *Journal of Food Engineering, 75*, 205-214. http://dx.doi.org/10.1016/j.jfoodeng.2005.04.008

Beeby, C. (1984). *Drying in Superheated steam-fluidized bed* (PhD thesis). University of Monash, Chem. Eng. Dept, Australia.

Berghel, J., & Renstrom, R. (2002). Basic design criteria and corresponding results performance of a pilot-scale fluidized superheated atmospheric condition steam dryer. *Biomass and Bioenergy, 23*, 103-112. http://dx.doi.org/10.1016/S0961-9534(02)00040-5

Chemkhi, S., & Zagrouba, F. (2005). Water diffusion coefficient in clay material from drying data. *Desalination, 185*(1-3), 491-98. http://dx.doi.org/10.1016/j.desal.2005.04.052

Chen, Z., Wu, W., & Agarwal, P. K. (2000). Steam drying of coal part 1. Modeling the behavior of a single particle. *Fuel, 79*(8), 961-74. http://dx.doi.org/10.1016/S0016-2361(99)00217-3

Deventer, H. C., & Heijmans R. M. H. (2001). Drying with superheated steam. *Drying Technology, 19*(8), 2033-2045. http://dx.doi.org/10.1081/DRT-100107287

Elustondo, D. M., Mujumdar, A. S., & Urbicain, M. J. (2002). Optimum operating conditions in drying foodstuffs with superheated steam. *Drying Technology, 20*(2), 381-402. http://dx.doi.org/10.1081/DRT-120002548

Elustondo, D., Elustondo, M. P., & Urbicain, M. J. (2001). Mathematical modelling of moisture evaporation from foodstuffs exposed to subatmospheric pressure superheated steam. *Journal of Food Engineering, 49*, 15-24. http://dx.doi.org/10.1016/S0260-8774(00)00180-1

Hager, J., Hermansson, M., & Wimmerstedt, R. (1997). Modelling steam drying of single porous ceramic sphere: Experiments and simulation. *Che. Eng. Science, 52*(8), 1253-64. http://dx.doi.org/10.1016/S0009-2509(96)00493-9

Hager, J., Wimmerstedt, R., & Whitaker, S. (2000). Steam drying a bed of porous spheres: Theory and experiment. *Che. Eng. Science, 55*(9), 1675-1698. http://dx.doi.org/10.1016/S0009-2509(99)00398-X

Hamawand, I. (2011). Effect of Colloidal Particles associated with the Liquid Bridge in Sticking during Drying in Superheated Steam. *International Journal of Engineering, 24*(2), 119-126.

Hamawand, I., & Yusaf, T. (2013). Particles Motion in a Cascading Rotary Drum Dryer. *Canadian Journal of Chemical Engineering*. In Press. http://dx.doi.org/10.1002/cjce.21845

Iyota, H., Nishimura, N., Yoshida, M., & Nomura. (2001). Simulation of superheated steam drying considering initial steam condensation. *Drying Technology, 19*(7), 1425-1440. http://dx.doi.org/10.1081/DRT-100105298

Khan, J. A., Beasley, D. E., & Alatas, B. (1991). Evaporation from packed bed of porous particles into superheated vapour. *Int. J. Heat Mass Transfer, 34*(1), 267-280. http://dx.doi.org/10.1016/0017-9310(91)90193-I

Kozanoglu, B., Vasquez, A., Chanes, J., & Patino, J. (2006). Drying of seed in superheated steam vacuum fluidized bed. *Journal of Food Engineering, 75*, 383-387. http://dx.doi.org/10.1016/j.jfoodeng.2005.03.063

Looi, A. Y., Golonka, K., & Rhodes, M. (2002). Drying kinetics of single porous particles in superheated steam under pressure. *Chemical Engineering Journal, 87*, 329-338. http://dx.doi.org/10.1016/S1385-8947(01)00244-3

Maa, J. (1967). Evaporation coefficient of liquid. *Ind. Eng. Chem. Fundam, 6*(4), 504-18. http://dx.doi.org/10.1021/i160024a005

Pakowiski, Z., Krupinka, B., & Adamski, R. (2007). Prediction of sorption equilibrium both in air and superheated steam drying of energetic variety of willow salix viminalis in wide temperature range. *Fuel, 86*, 1749-1757. http://dx.doi.org/10.1016/j.fuel.2007.01.016

Prachayawarakorn, S., Soponronnarit, S., Wetchacama, S., & Jaisut, D. (2002). Desorption isotherms and drying characteristics of shrimp in superheated steam and hot air. *Drying Technology, 20*(3), 669-684. http://dx.doi.org/10.1081/DRT-120002823

Pronyk, C., Cenkowski, S., & Muir, W. E. (2004). Drying foodstuff with superheated steam. *Drying Technology, 22*(5), 899-916. http://dx.doi.org/10.1081/DRT-120038571

Shibata, H., Mada, J., & Shinohara, H. (1988). Drying mechanism of sintered spheres of glass beads in superheated steam. *Industrial & engineering chemistry research, 27*(12), 2353-2362. http://dx.doi.org/10.1021/ie00084a020

Soponronnarit, S., Prachayawarakorn, S., Rordprat, W., Nathakaranakule, A., & Tia, W. (2006). A superheated steam fluidized bed dryer for Parboiled Rice: testing of pilot-scale and mathematical Model Development. *Drying technology, 24*(11), 1457-67. http://dx.doi.org/10.1080/07373930600952800

Suvarnakuta, P., Devahastin, S., & Mujumdar, A. S. (2007). A mathematical model for low pressure superheated steam drying of biomaterial. *Chemical Engineering and Processing, 46*(7), 675-683. http://dx.doi.org/10.1016/j.cep.2006.09.002

Tang, Z., & Cenkowski, S. (2000). Dehydration dynamics of potatoes in superheated steam and hot air. *Canadian Agricultural Engineering, 42*(1), 1-13.

Tang, Z., & Cenkowski, S. (2003). Modelling the superheated steam drying of a fixed bed of Brewers' spent grain. *Biosystems Engineering, 87*(1), 67-77.

Tarasov, V. M., & Basargin, B. N. (1975). Kinetics of drying of moist materials in superheated steam. *Translated from Inzhenerno-Fizicheskii Zhurnal, 28*(1), 34-39.

Trommelen, A. M., & Crosby, E. J. (1969). Evaporation and drying of drops in superheated vapours. *AICHE Journal, 16*(5), 857-67. http://dx.doi.org/10.1002/aic.690160527

Wimmerstedt, R., & Hager, J. (1996). Steam drying–modelling and applications. *Drying Technology, 14*(5), 1099-1119. http://dx.doi.org/10.1080/07373939608917141

A Literature Review on Analyzing the Architectural Design Process of Post-Disaster Houses

Behnam Ghasemzadeh[1], Atefeh Fathebaghali[2], Kasra Rahbaripour[3] & Hamed Pouyafar[4]

[1,3] Young Researchers and Elite Club, Tabriz Branch, Islamic Azad University, Tabriz, Iran

[2] M.A. in Architectural Engineering, Islamic Azad University of Tabriz, Tabriz, Iran

[4] Department of Art and Architecture, Tabriz Branch, Islamic Azad University, Tabriz, Iran

Correspondence: Behnam Ghasemzadeh, Young Researchers and Elite Club, Tabriz Branch, Islamic Azad University, Tabriz, Iran. E-mail: Behnam.Ghasemzadeh@yahoo.com

Abstract

This study focused on presenting a secondary source academic analysis on the nature of the architectural design processes used in the creation of post-disaster housing. The specific focus was on sustainability and the extent to which architects have managed to develop sustainable designs, both in terms of economic factors and longevity (including possible experience of future disasters). This study was necessary due to the rise in natural disasters globally over the past two decades as well as the realization that those individuals most affected by the disasters tended to be those poorest in society.

The secondary data analysis found that the existing data supported the hypotheses. The current sustainable architectural designs in disaster-hit areas are linked to high economic expenditure. It is clear though that no current cheap and affordable method of sustainable housing exists apart from one example in Sri Lanka and this needs to be highlighted in the literature. It should also be noted that there are specific political and economic factors that prevent the sustainable development of post-disaster housing. Recommendations are made as to how future studies can use primary data to help discover methods of design that could be used by the wider population in regions hit frequently by natural disaster. By drawing attention to the inequality in housing in these regions through this current study, it is felt that the work has been a success and should lead to further debate and action in the near future.

Keywords: sustainability, natural disaster, economically-viable, longevity

1. Introduction

The recent impact of Hurricane Isaac in the United States has highlighted once again that natural disasters have the ability to impact on every individual in society, regardless of the community that they live in, the developed nature of their education or their location geographically. This study focused on the analysis of the architectural design process of post-disaster houses. The study concerns itself with this topic for a number of reasons but mainly due to the fact that there have been great improvements in architectural design over the past half century and this, coupled with the rise in climate change which has led to a growth in the number and severity of natural disasters on the global scale (Porfiriev, 2010). These two reasons suggest that it was necessary to provide a detailed analysis of how architectural design is adapted for the purpose of post-disaster houses, aiming to strengthen and prevent a similar level of damage and need for reconstruction in the future. Oliver (2006) indicates that there are a number of considerations that need to be met in terms of post-disaster housing in any specific region and that these are cultural and social, as well as political and economic.

The occurrence of natural disasters in a variety of locations around the world have led to the development and increase of knowledge concerning sustainable housing and how architectural designs can be implemented to help prevent future damage, in both the short, medium and long-term period. This research is vital in terms of furthering the knowledge of how certain architectural developments and designs can help to repair the damage caused by natural disasters, as well as helping to prevent future damage through the creation of sustainable housing. It is underlined that 'natural disasters have extensive and violent effects, cause loss of life and property, substantial affect on the communal life. The huge number of damaged and collapsed buildings after the earthquakes creates a housing problem needing urgent attention' (Limoncu & Celebioglu, 2006, p.2). Moreover, it is assessed that it is those poorest in society that are most likely to feel the highest level of negative impact after a natural disaster.

A work that assessed the impact of an earthquake on the Peruvian coastline stated that 'the poorest households are least prepared when the earthquake hits and most likely to suffer from a lack of funds, rights and access to recover their lives in the aftermath, often falling into a vicious cycle whereby a poor effort to rebuild their homes leaves them even more exposed to the next natural disaster' (Blanco-Lion, Pelsmakers, & Taylor, 2011, p. 435). Through these assessments, it is clear that it is necessary to analyze how effective the architectural design process is for post-disaster housing and this is the main objective of this research study. This study used both qualitative and quantitative research methodologies in a mixed method approach to critically evaluate the current architectural designs for post-disaster housing in terms of its sustainability and cost.

2. Research Objectives

To ensure that the main aim of this study was achieved successfully, it was necessary to present tangible research objectives and clear focus issues that enabled the researcher to remain on task during the longevity of the study. The central research objectives of this study were created to highlight the pressing need to study architectural design in the form of post-disaster housing. Therefore, the objectives were:

1) Conduct a thorough review of the available empirical literature on the subject of architectural design and the creation of sustainable post-disaster housing, allowing the researcher to gain an enhanced knowledge of the current state of the literature.
2) Design a study using both qualitative and quantitative research methodologies to help present recent case studies in the field, providing a secondary source academic analysis of the topic.
3) Ensure that all information was reliable and helped to further the knowledge on the subject so that a conclusive summary could be achieved and that further recommendations of study could be made at the end of the research work.

3. Research Questions

From the research objectives and the focus issues of this study, it was possible to create three main research questions that were answered through the completion of the study. The main research questions were:

1) Have architects been able to ensure sustainability of post-disaster houses?
2) Has the sustainability developed over the past decade?
3) To what extent can this sustainability be maintained in the face of another potential disaster and can it be achieved through limited economic means?

Through these research questions, it was felt that the empirical literature could be studied effectively, allowing a detailed understanding of the architectural design process of post-disaster houses. Moreover, the focus on sustainability within certain economic and cultural limits provided the research with a working context within which to place this current study.

4. Research Hypothesis

Following the creation of the research questions, it was prudent to offer certain research hypotheses. The research hypotheses are two-tailed hypotheses. These are those hypotheses that 'make a prediction that the effect of an independent variable may go in either direction. For a two-tailed hypothesis, there is double the probability in either direction for a positive outcome' (Greene & D'Oliveira, 2005, p.19). The main hypothesis for this work is that successful architectural designs achieving sustainability and protection from disaster are positively related to the economic expenditure used during its creation. Therefore, it is assumed that those houses with greater access to financial support, will be more effective at creating sustainable design in a post-disaster region.

The second hypothesis focuses on the longevity of the sustainability and suggests that the longer the time period that the house exists, the less likely it is to be successful in preventing damage from possible disasters. The conclusions of this study could have a dramatic impact in terms of government support (financial) for regions hit by natural disasters as well as for architectural techniques and designs used in the process.

Table 1. Hypothesis testing in research

Variable	Type of relationship		
	Negative relationship	Positive relationship	Lack of relationship
Sustainable Housing and Economic Expenditure		This is assumed to be a positive relationship, with the higher the expenditure, the more sustainable and protected a house will be against disaster	-
Architectural Design and Time Length	This is viewed as having a negative relationship between these two variables. The longer the time period, the less likely it is that the architectural design would be sufficient in preventing or negating the impact of a disaster.		-

5. Literature Review

The nature of architectural design for post-disaster housing is one that is a 'considerable challenge' (Hayles, 2010, p. 103) for all of those involved in the rebuilding planning process. However, it is evident that the design of post-disaster housing is a central part of the architectural world's agenda, given the increase in natural disasters over the past two decades. An analysis of a brief period of the last decade presents a clear indication of the challenge that is imparted on architects with the view that 'the recent Indonesian Earthquake and Indian Ocean Tsunami (2004), Bam Earthquake (2003) and Gujarat Earthquake (2001)' (Boen & Jigyasu, 2005, p. 1) key examples of how developing countries in particular are impacted by such events. The literature highlights that a large part of the challenge in these particular areas is that cultural and social factors need to be taken into account as well as structural, safety and economic concerns. A 1995 study by Ladinski found that following a 1963 earthquake in Skopje, the 'decision to protect buildings with technology solely based on earthquake engineering principles led to damage of integrity, identity and the originality of the built heritage' (Awotana, 1997, p. 221). To place this viewpoint in context, it is apparent that 'spatial patterns and architectural features of housing are reflection of social order and cultural practices in many societies around the globe' (Vatsa, 2001, p. 6). This highlights that the architectural design process is a vital part of the post-disaster rehabilitation effort and it is necessary for the design to represent both a respect towards the local culture as well as the need to utilize new technologies so that sustainability can be achieved in the design. This is perhaps a difficult balance to achieve and this forms the main research aim of the entire study.

Finally, it has also been found that due to a lack of economic and technological resources, it has often been the case that architectural design processes on post-disaster housing has been unsafe and unsuitable. It was noted by Boen and Jigyasu (2005) that there have been numerous cases, especially in the developing world, where the improper use of technology has led to poorer housing designs and the failure to ensure sustainability in certain disaster-hit areas, with the study noting Guatemala as an extreme case in point. This point clearly indicates the importance of economic assistance, allowing architectural designs to use the needed technology and design to help ensure that safety is guaranteed in the post-disaster housing.

6. Research Structure

This research study relies heavily on existing studies within the architectural literature. Due to the limitations placed on the study, the researcher was not able to conduct any primary research. Therefore, the study relies on existing studies in the field to help analyze sustainability in architectural design and the extent to which various architects and architectural companies around the world have developed certain designs that allow cheap and sustainable housing to be constructed in areas that experience natural disasters. The study is a critical exploration of these designs, providing a discussion of certain innovative designs in the field and assessing them on how far they can sustain future natural disasters.

In this manner, the work aims to assess the followings:

- The techniques employed by architects and;

- How far this has changed in the last decade through the establishment of innovative techniques in industry;

- The ability to use technology to predict the extent of sustainable housing and;

- Whether these designs can be implemented in areas where there are low or limited economic funds, such as in South East Asia and Central America;

- Which are two of the poorest regions on the globe in modern society?

7. Methodology

The research methodology follows a mixed method approach immersed within a secondary source analysis research paradigm. The employment of the mixed methods approach is appropriate for this study because the combination of both qualitative and quantitative methods in one research study is thought to reduce the number of limitations normally associated with single methodology research (Creswell & Clark, 2010). The combination of statistical data and qualitative insights from studies already conducted in this particular research area will therefore be able to add strength to the study and make the findings much more reliable than they would be if a single research methodology was applied (Hesse-Bieber, 2010).

As well as the mixed methods approach, the secondary source analysis form of research is thought to be effective because it allows a researcher to draw together a range of findings from a particular field of study (Gravetter & Forzano, 2011). The definition of the approach is highlighted as a 'form of research in which the data collected and processed by one researcher (or more) is reanalyzed, often for a different purpose, by another' (Rubin & Babbie, 2010, p. 250). In this case, it is possible to collate information relating to architectural design in the world's poorest areas and critically assess the findings to come to conclusions concerning the main research questions in this current study. This is important because of the lack of resources currently held by this researcher and the fact that a primary, on-location study concerning sustainable housing over a long period of time would simply not be possible. However using the secondary approach, results from past studies can be used to help the researcher find the necessary data and to conclude successfully on the subject.

8. Research Tools

The research hypotheses assumed that sustainability was linked to economic expenditure and that longevity of a house life reduced the effectiveness of the sustainability against natural disasters. Therefore, it was deemed necessary to find a research methodology and the research tools that would be able to measure the economic relationship with sustainability as well as the impact of time on certain designs. The research used a search of the architectural literature to find studies relating to sustainability projects, cross-referencing them with the length of the project. Ideally studies were used that had been conducted in the last ten years and had analyzed the effectiveness of certain architectural designs over a period of time, rather than just assuming results in the future. This research method focusing on the key word of sustainability is a common method in recent studies in the architectural field. It is highlighted that 'over the last decade, increased research interest in sustainability has reintroduced many issues such as energy conservation, funding and protection against disaster, but framed within a new conceptual model' (Groat & Wang, 2002, p. 6). Using this search method, appropriate studies were found that could be applied to this current research study. These are highlighted in the results and discussion chapters of this project.

9. Limits and Limitations

At the beginning of this section it is pertinent to point out that limitations of a research study should not be viewed negatively. Kumar (2002) states that by placing limitations on a study, the aims become narrow and realistic. Having said that, there are certain limits placed on this study through a lack of resources that prevented a primary study of the data. The study therefore relies on secondary analysis data and this has limitations. The research study is limited to the data that exists currently, as well as having possible issues with validity (Cargan, 2007). Moreover, the issue of reliability and validity is heightened due to the reliance on other studies that may have used different variables or parameters to that set by this study. Therefore, it is vital that each study used for the analysis is thoroughly checked and critically assessed in terms of its validity before its use is sanctioned for the purposes of the current study.

10. Results and Discussion

This section presents an overview of the findings of the study using both forms of methodology through a case study of regional areas in Central America and South-East Asia. It should be highlighted during this that the architectural world acknowledges that 'safe, healthy and well-built houses serves a basic human right…unfortunately millions of citizens of marginalized communities around the world are denied this right and struggle to meet the basic needs of daily life through a lack of money and poor government investment' (UN, 2012). Therefore, during this assessment of sustainable housing, it is clear that the knowledge exists that regions of natural disasters need help to strengthen their architectural designs but that this is not necessarily occurring through low investment. Moreover, it is noted by Boano and Hunter (2012) that in each situation when a disaster occurs, 'Spatially speaking, there are massive needs for physical reconstruction of damaged structures, provision of new housing, and the conservation of heritage buildings that challenge faster responses and open up opportunities for incorporating preventive measures in relation to the occurrence of future disasters' (2012, p. 1) but that these decisions also rest on the surrounding political and economic arena that often dictates that preventive methods of architecture cannot be used for the rebuilding effort, due to restrictions placed on the community such as a lack of funding. The study by Boano and Hunter is also critical on the role of architects, stating that to be successful, post-disaster housing efforts requires 'a necessary change in the traditional identity and approach of architects, meaning that they need to understand the complexity of the situation and the community at risk' (2012, p. 10).

The study by Berke (1995) indicated that sustainable development could be achieved through dual planning with local natural disaster commissions and international aid, but that the low economic expenditure on housing, coupled with the desire to rebuild quickly meant that sustainability often was overlooked in the rush to give the local population its housing back. He identified areas of flooding in Bangladesh as a key example of this. The low government spending meant that housing was simply rebuilt in the style prior to the natural disaster, meaning that future sustainability was not guaranteed. This finding was supported by a study by Dikmen (2011) who analyzed post-disaster housing in Turkey. He found that due to economic limitations, the residents rated the traditional housing (pre-disaster) as better than the post-disaster housing in two independent regions in the country and that the residents felt more secure in their old housing, highlighting that through restrictions economically, the post-housing was not viewed as sustainable by the residents.

One study that highlights a successful situation in which sustainability has been increased in post-disaster housing was that of Hidellage and Usoof (2009) who found that in post-tsunami Sri Lanka, the local government and architectural firms worked together to produce a low economic and highly sustainable method of housing for those affected by the tsunami. Although the study highlights that this sustainability has not yet been tested over a long period of time, the fact that economic limitations were not necessarily an issue and that the housing was created relatively cheaply, should be a cause for celebration.

11. Conclusion

In conclusion, the studies analyzed present clear findings that suggest that the most effective methods of architectural design for sustainability in areas of natural disaster are those that are more expensive. It is clear from the study that architects have not been able to ensure sustainability of post-disaster houses, with the literature stating that their fundamental beliefs do not allow for this to happen and that they need to put people and communities first for this to change. As well as this, it is clear that the issue of sustainability has increased over the past decade or so but this has happened slowly, with only a few examples, such as that in Sri Lanka where sustainability has been achieved at an affordable rate. For the most part, sustainability is restricted by economic and political factors and this shows no signs of abating, to the detriment of the majority of the population in disaster-affected regions of the world. Finally, there is little evidence to suggest that longevity reduces the sustainable nature of housing in disaster-hit areas, but this is due to the fact that the issue has only recently been covered by architectural literature and a longer period of time needs to pass before this can be assessed adequately.

12. Further Research

The study concluded in support of the two hypotheses, finding that economic and time factors were important considerations in terms of the ability of housing to remain sustainable throughout its lifetime in areas hit by natural disasters. However, this study relied on secondary data. It would be necessary to conduct (with the appropriate resources) a case study primary analysis of forms of architectural design and how effective they are in terms of sustainability in the face of disasters. This type of study could have a tremendous impact on the living conditions of individuals living in poor areas of the world. It is necessary to discover a cheap and economically viable method of engineering sustainable housing in these regions and this type of future study could help uncover vital data that is needed for this progression to happen. At the current time, those living in natural disaster areas are only able to have sustainable housing if they can afford it and it puts at risk a great majority of the local populations, especially in areas such as South East Asia and Central America. It is imperative that this subject be further analyzed and through primary study, as recommended in this section.

References

Ary, D., Jacobs, L., Razavieh, A., & Sorenson, C. (2009). *Introduction to research in education.* London: Cengage Learning.

Awotana, A. (1997). *Reconstruction after disaster-issues and practices.* Bristol: Ashgate Publishing.

Berke, P. (1995). Natural-hazard reduction and sustainable development: a global assessment. *Journal of Planning Literature, 9*(4), 370-382. http://dx.doi.org/10.1177/088541229500900404

Blanco-Lion, C., Pelsmakers, S., & Taylor, M. (2011). Sustainable solution for the reconstruction of low income housing in post-disaster zone. In M. Bodart & A. Evrard, *Architecture and sustainable development. 27th International conference on passive and low energy architecture, volume 2.* Belgium: Louvain University Press.

Boano, C., & Hunter, W. (2012). Architecture at risk?: the ambivalent nature of post-disaster practice. *Architectoni, 1*, 1-13. http://dx.doi.org/10.5618/arch.2012.v1.n1.1

Bodart, M., & Evrard, A. (2011). *Architecture and sustainable development. 27th International conference on passive and low energy architecture, volume 2.* Belgium: Louvain University Press.

Boen, T., & Jigyasu, R. (2005). *Cultural considerations for post disaster reconstruction: post-tsunami challenges.* Jakarta: WSSI.

Cargan, L. (2007). *Doing social research.* London: Rowman & Littlefield.

Creswell, J., & Clark, V. (2010). *Designing and conducting mixed methods research.* London: SAGE.

Dikmen, N. (2011). Comparative analysis of permanent post-disaster houses constructed in Cankiri and Dinar. *Disasters, 35*(2), 404-416. http://dx.doi.org/10.1111/j.1467-7717.2010.01220.x

Gravetter, F., & Forzano, L. (2011). *Research methods for the behavioral sciences.* California: Cengage Learning.

Greene, J., & D'Oliveira, M. (2005). *Learning to use statistical tests in psychology.* New York: McGraw-Hill International.

Groat, L., & Wang, D. (2002). *Architectural research methods.* London: John Wiley & Sons.

Hartas, D. (2010). *Educational research and inquiry: qualitative and quantitative approaches.* California: Continuum International Publishing Group.

Hayles, C. (2010). An examination of decision-making in post-disaster housing reconstruction. *International Journal of Disaster Resilience in the Built Environment, 1*(1), 103-122. http://dx.doi.org/10.1108/17595901011026508

Hesse-Bieber, S. (2010). *Mixed methods research: merging theory with practice.* New York: Guilford Press.

Hidellage, V., & Usoof, A. (2009). *Up-scaling owner-driven reconstruction: lessons from post-Tsunami Sri Lanka.* London: South Bank University Press.

Kumar, A. (2002). *Research methodology in social science.* London: Sarup & Sons.

Limoncu, S., & Celebioglu, B. (2006). *Post-disaster sustainable housing system in Turkey.* Turkey: Yildiz Technical University Press.

Oliver, P. (2006). *Built to meet needs: cultural issues in vernacular architecture.* Abingdon: Routledge.

Porfiriev, B. (2010). *Climate change as environmental and economic hazard.* Washington D.C.: Earthscan.

Rubin, A., & Babbie, E. (2009). *Essential research methods for social work.* California: Cengage Learning.

UN. (2012). *United Nations Environment Program.* Retrieved October 1, 2012, from http://www.unep.org/environmentalgovernance/Events/HumanRightsandEnvironment/tabid/2046/language/en-US/Default.aspx

Vatsa, K. (2001). *Rhetoric and reality of post-disaster rehabilitation after the Latur Earthquake of 1993: a rejoinder.* New Delhi: Personal Communication Publishing.

Comparative Study of Price Variations of Basic Civil Engineering Construction Materials

Oginni, Felix A.[1], Ogunbiyi, Moses A.[1] & Balogun, Sodiq O.[1]

[1] Department of Civil Engineering, Osun State University, P.M.B. 9544, Oshogbo, Nigeria

Correspondence: Oginni, Felix A., Department of Civil Engineering, Osun State University, P.M.B. 9544, Oshogbo, Nigeria. E-mail: kafnog@gmail.com/oginnifa@uniosun.edu.ng

Abstract

A study of varying prices of basic civil engineering construction materials was undertaken. The basic materials considered were two varieties of cement, three types of sand, and different sizes of two types of timber. Prices of these construction materials were collected from local Contractors in Oshogbo region over a period of twenty seven months. The period considered was from January 2010 to March 2012. Variation pattern in the prices was studied on monthly basis. Results indicate that prices of all the basic construction materials follow an annual cyclic pattern. Prices of the two brands of Cement were the same except at months 12, 23 and 24. The prices are stable from April to August of each year (months 4-8, and 16-20). During these periods the price of this material increased by 13.3% from ₦1,500.00 in 2010 to ₦1,700.00 in 2011. There were two rises per annum in the prices of sand. The first rise occurred in March and the second in the following November for sharp and fine sands. Timber prices are relatively stable within 5% variability during the months of November – March and December – March respectively for hardwood and whitewood except types H4 and W1. H4 and W1 prices were the same and they followed the same trend. The trends presented by types H4 and H5, which are same size but different species, are similar, indicating that hardwood specie is not a factor in the price variation of timbers. It can generally be concluded that the prices of the basic construction materials will stabilize during the months of April to August of every year. Although construction activities are light every January, January prices keep increasing from year to year.

Keywords: prices, variation, construction, materials, management, period, cement, sand, timber

1. Introduction

In almost all construction projects, changes or variations are common occurrences during the design and construction phases, (Entrusty Group, 2008). The prices of construction materials fluctuate mainly due to supply, supply environment and the demand for the material for other uses. The prices of sawn wood and other forest products fluctuate mainly due to supply and demand.

The major construction materials in the field of civil engineering include the following:

> Cement, Sand (sharp sand, fine/plastic sand), Granite, Hardcore, Reinforcement Bars, Timbers, Structural steel frames.

Construction is a complex process that exhibits a high degree of uncertainty. The following are factors affecting prices of construction materials:

> Size of project; Location of the project; Environment of the project; Political climate; Award season; War situation; Economic climate; Exposure of the Contractor; Regional variation; Disparity in pricing – importation.

Ssegawa et al., 2002 indicated that it is hardly possible to complete a project without changes to the plans or the construction process. This makes variation to become so prevalent in construction works. Revay, 2002 stated that this may lead to changes to scope, time, cost and/or quality on most, if not all, construction projects. Various studies have identified variation orders among the causes of project cost and time overruns, which Sterman, 1992 considered as endemic problems in international construction. Tyagi and Kushwaha, 2012 proposed the revision of the price variation clause first in using the price indices to be followed by the use of the revised PVC for updating the rates. Makwana and Pitroda, 2013 used frequency analysis in studying region wise price variation of construction raw materials. Other workers in this field include Oladapo, 2007; Entrusty Group, 2008; Evija Greģe – Staltmane, 2009; Kalu et. al., 2009; Arowosoge et. al, 2011 and Long, 2013.

Generally, it has been estimated that building materials and components constitute between 50 to 60 percent of the total cost of construction. Okupe, 2000, observed that the cost could be about 75 percent in Nigeria. He outlined factors contributing to this to include inadequate infrastructural facilities like bad roads, which may increase the cost of delivery of materials and tariffs charged on the materials that also shoot up the price.

Duration of major Civil Engineering projects on the average is about three years. Within the period of project execution and delivery, major construction materials exhibit price variations. These variations may have a direct impact on the lifespan of the project.

Factors affecting the prices of construction materials vary with the type of construction material under consideration. The materials can be categorized according to the level of local contents and inputs in their production. Cost of production of cement, iron rods and steel frames have low local contents/inputs. Factors contributing to the variation in pricing of these materials in Nigeria include the following:

Cost of production   Impact of production scale   Transportation infrastructure

Regional variation in demand and supply   Disparity in pricing-importation.

Cost of production of sand, granite and timber have high local contents/inputs. Variation in prices of these materials depends on vegetation and climatic factors within the region of the country. They may also be affected by seasonal factors. Aje, 1995 revealed that finishes and substructure are the building elements that attract the highest percentage of variation on building projects. Izekor and Izekor, 2011 revealed that factors that account for variations in the prices of sawn timber in Benin Metropolis, in Nigeria include transportation cost representing 40%, seasonality (42%) and power supply (18%). Variations within an area depends on constraints to delivery within the area, general constraints to delivery within the production center/sector seasons – festival, weather related constraints. The scope of this study is limited to a South Western Nigeria vegetation zone.

2. Materials and Methods

The construction materials considered in this study are as follows:

2 brands of Cement, Elephant cement and Dangote cement

3 types of sand, laterite, fine sand and sharp sand

2 types of timber, hardwood and whitewood

The 2 types of timber are further considered on the basis of the following sizes:

25mm x 300mm x 3600mm

50mm x 75mm x 3600mm

50mm x 100mm x 3600mm

75mm x 100mm x 3600mm (50mm x 150mm x 3600mm)

The various types of hardwood are denoted as H1, H2, H3, H4, and H5and are described respectively as indicated below:

Type H1 - Harwood (Ayunre, ire, ila, ayo ire) 25mm*300mm*3600mm

Type H2 – Hardwood (Mahogany, Idigbo, Osun) 50mm*75mm*3600mm

Type H3 - Harwood (Ayunre, ire, ila, ayo ire, e.t.c) 50mm*100mm*3600mm

Type H4 - Harwood (Ayunre, ire) 75mm*100mm*3600mm(50mm*150mm*3600mm)

Type H5 - Harwood (Mahogany, etc) 75mm*100mm*3600mm(50mm*150mm*3600mm)

Types H4 and H5 are of the same size but of different species.

The various types of whitewood are denoted as W1, W2, and W5and described as indicated below.

Type W1 - Whitewood (Oro, Opon, Eapu, Ofee) 25mm*300mm*3658mm type prices

Type W2 - Whitewood (Oro, Opon, Eapu) 50mm*75mm*3658mm type prices

Type W3 - Whitewood (Oro, Opon, Eapu, Ofee,) 50mm*100mm*3658mm type price

2.1 Data Collection

Monthly prices of the construction materials were collected for the 2 brands of cement, 3 types of sand, and 2 types timbers (of various sizes), from Retail Dealers around Olaiya area in Oshogbo, Osun state. The data collected was for a period of twenty seven calendar months spanning from January 2010 to March 2012.

The prices of the two brands of cement and three types of fine aggregates, sharp sand, fine sand and lateritic soil materials considered in the study are presented in Table 1.

Table 1. Monthly prices of 2 brands of cement (Dangote and Elephant) and 3 types of sand

			Cement Price/Type (₦/bag)		**Sand Price / Type (₦/5m³Tipper load)**		
Year	**Month**	**Nth month**	**Dangote**	**Elephant**	**Sharp Sand**	**Fine Sand**	**Laterite**
2010	January	1	1625	1625	3000	3000	4500
2010	February	2	1625	1625	3000	3000	4500
2010	March	3	1575	1575	3500	3500	5000
2010	April	4	1500	1500	3500	3500	5000
2010	May	5	1500	1500	3500	3500	5000
2010	June	6	1500	1500	3500	3500	5000
2010	July	7	1500	1500	3500	3500	5000
2010	August	8	1500	1500	3500	3500	5000
2010	September	9	1625	1650	3500	3500	5500
2010	October	10	1625	1650	3500	3500	5500
2010	November	11	1700	1700	4000	4000	5500
2010	December	12	1750	1900	4000	4000	5500
2011	January	13	1900	1875	4000	3500	4000
2011	February	14	1900	1875	4000	3500	4000
2011	March	15	1800	1800	4500	4000	4500
2011	April	16	1700	1700	3500	4000	4500
2011	May	17	1700	1700	3500	4000	4500
2011	June	18	1700	1700	3500	4000	4500
2011	July	19	1700	1700	3500	4000	4500
2011	August	20	1700	1700	3500	4000	4500
2011	September	21	1900	1875	5000	4000	5000
2011	October	22	1950	1950	5000	4000	5000
2011	November	23	1950	2000	5000	4500	5000
2011	December	24	2300	2300	5000	4500	5000
2012	January	25	2250	2150	6000	3000	6000
2012	February	26	2150	2150	6000	3000	6000
2012	March	27	1725	1750	6000	3500	6000

Prices of the two types of timber, (hardwood and whitewood) for basic sizes which are used in construction works, denoted as H1, H2, H3, H4, H5, W1, W2, and W3 are shown in Table 2.

Table 2. Monthly prices of wood types

			Price/ Wood Type (₦)							
Year	**Month**	**Nth month**	**H1**	**H2**	**H3**	**H4**	**H5**	**W1**	**W2**	**W3**
2010	January	1	1300	330	430	660	600	660	200	320
2010	February	2	1300	330	430	660	600	660	200	320
2010	March	3	1300	330	430	660	600	660	200	320
2010	April	4	1350	345	435	690	650	690	230	340
2010	May	5	1350	345	435	690	650	690	230	340
2010	June	6	1350	345	435	690	650	690	230	340
2010	July	7	1350	345	435	690	650	690	230	340
2010	August	8	1350	345	435	690	650	690	230	340
2010	September	9	1300	335	430	670	650	670	230	340
2010	October	10	1300	335	430	670	630	670	200	330
2010	November	11	1350	345	430	670	630	670	200	330
2010	December	12	1350	345	430	670	650	670	230	340
2011	January	13	1350	350	440	700	650	700	230	340

2011	February	14	1350	350	440	700	650	700	230	340
2011	March	15	1350	350	440	700	650	700	230	340
2011	April	16	1450	380	445	760	700	760	250	350
2011	May	17	1450	380	445	760	700	760	250	350
2011	June	18	1450	380	445	760	700	760	250	350
2011	July	19	1450	380	445	760	700	760	250	350
2011	August	20	1450	380	445	760	700	760	250	350
2011	September	21	1350	360	440	720	700	720	250	350
2011	October	22	1350	360	440	720	680	720	230	350
2011	November	23	1450	380	440	720	680	720	230	350
2011	December	24	1450	380	440	720	700	720	250	350
2012	January	25	1450	400	450	800	700	800	250	350
2012	February	26	1450	400	450	800	700	800	250	350
2012	March	27	1450	400	460	800	700	800	250	350

3. Results and Analysis

3.1 Cement Price Variation

The variation in price of Cement not minding the brand can be seen to follow an annual pattern. The price presents an annual variation. These are shown in Fig. 1.

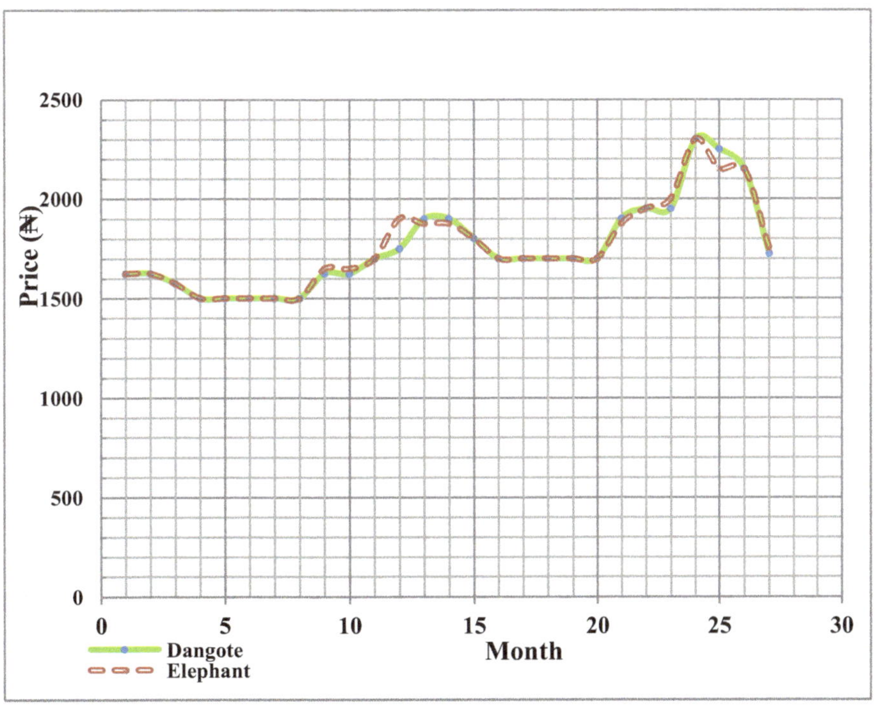

Fig. 1. Variation of Dangote and Elephant cement prices from Jan 2010 - March 2012

Price of Elephant cement has two peaks during the period of study. The first peak was at month 12 while the second was 12 months later at month 24. Price of Dangote cement also has two peaks during the period of study, though it reached its first peak, a month after that of Elephant Cement at month 13 through month 14. This implies that Dangote cement prices had to wait a month after Elephant cement prices. Generally the prices of the two brands of Cement were the same except at months 12, 23 and 24 as shown in Fig. 1. It can also be observed that there is price stability from April to August of each year (months 4-8, and 16-20). During these periods the price of this material increased by 13.3% from ₦1,500.00 in 2010 to ₦1,700.00 in 2011.

3.2 Analysis of Price Variation of Sand

The first eight months of year 2010, showed that from January, the prices of fine sand, sharp sand and laterite rose from ₦3,000, ₦3,000 and ₦4,500 respectively in February by ₦500. This implies a 16.67% increase for the fine and sharp types of sand, and just 11.11% for laterite. These prices remained stable for eight months after which it experienced another rise of ₦500, except that of laterite that was stable for 6 months before the rise of another ₦500 in the following month 9. These are shown in Fig. 2.

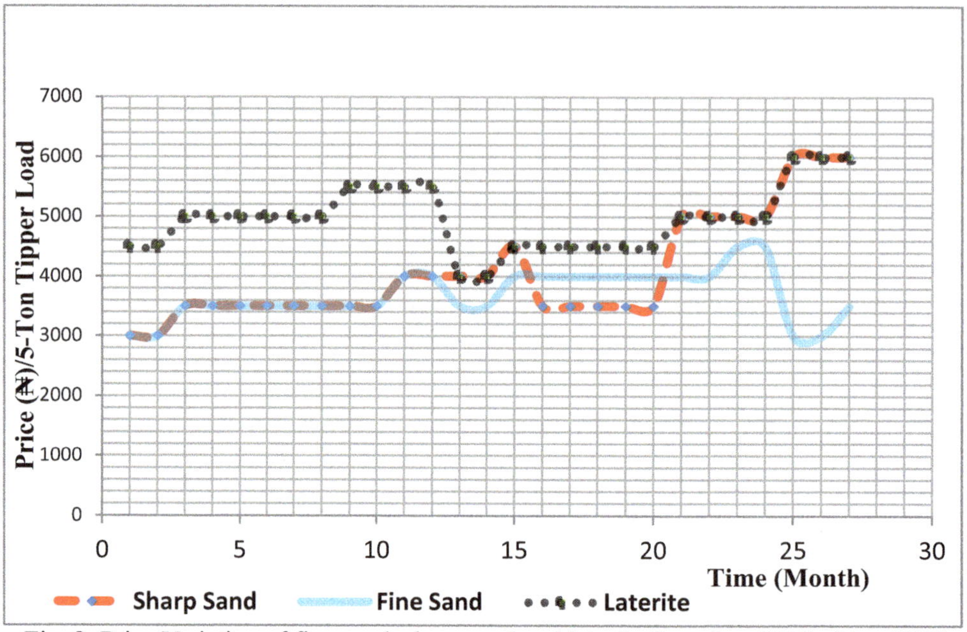

Fig. 2. Price Variation of fine sand, sharp sand and laterite from Jan. 2010 - Mar. 2012

Prices of fine sand and laterite stabilized for 2 and 4 months respectively from November and September to December of year 2010. Price of sharp sand stabilized for 4 months at ₦4,000 from month 11 to month 14 (February 2011) before another ₦500 increase in a unit of 5-ton tipper load. This is followed by a fall of ₦1,000 in month 16 which then stabilized for 5 months from month 16 to 20. This price rose by ₦1,500 in month 21 and stabilized for the next 4 months, another end of year (month 24). It then rose by ₦1,200 stabilizing there till month 27. The price of laterite followed similar pattern to that of fine sand, except in month 24 when price of laterite fell sharply by ₦1,500 from ₦4,500 to ₦3,000.

Generally all the sharp and fine sands and the laterite experienced 5 rises or 4 rises in 2 years at the rate of twice / year. The first rise was observed in March while the second was observed in November for sharp and fine sands. Laterite and sharp sand experienced only one fall in price, while fine sand had two falls during the study period. Prices of these materials can be said to follow an annual pattern of rise and fall dictated by end of year festivity periods, and the level of economic activities within the year using cement price pattern as an indicator of the level of economic activity. Increases are mostly ₦500/5-Ton tipper load considered except after the second year. It is also observed that when the price of cement begins to fall, that of sands begins to rise.

3.3 Analysis of Price Variation of Timber Types, Species and Sizes

The general trend in price variation of hardwood types and sizes except type H3 are noticed during the months of March and April, as indicated in months 3 and 4 and months 15 and 16 respectively. These trends are shown in Fig. 3. April to August experienced stable timber prices. Lower prices are experienced in the months of September and November. Price of timber Type H3 increases steadily at the rate of 0.26% from January 2010 to 7% in March 2012.

Timber prices are relatively stable within 5% variability during the months of November to the following March, as indicated in months 11-15 and 23-27 in Fig. 3, with the exception of type H4. The trends presented by types H4 and H5, which are same size but different species, are similar. This indicates that specie is not a factor in the price variation of timbers.

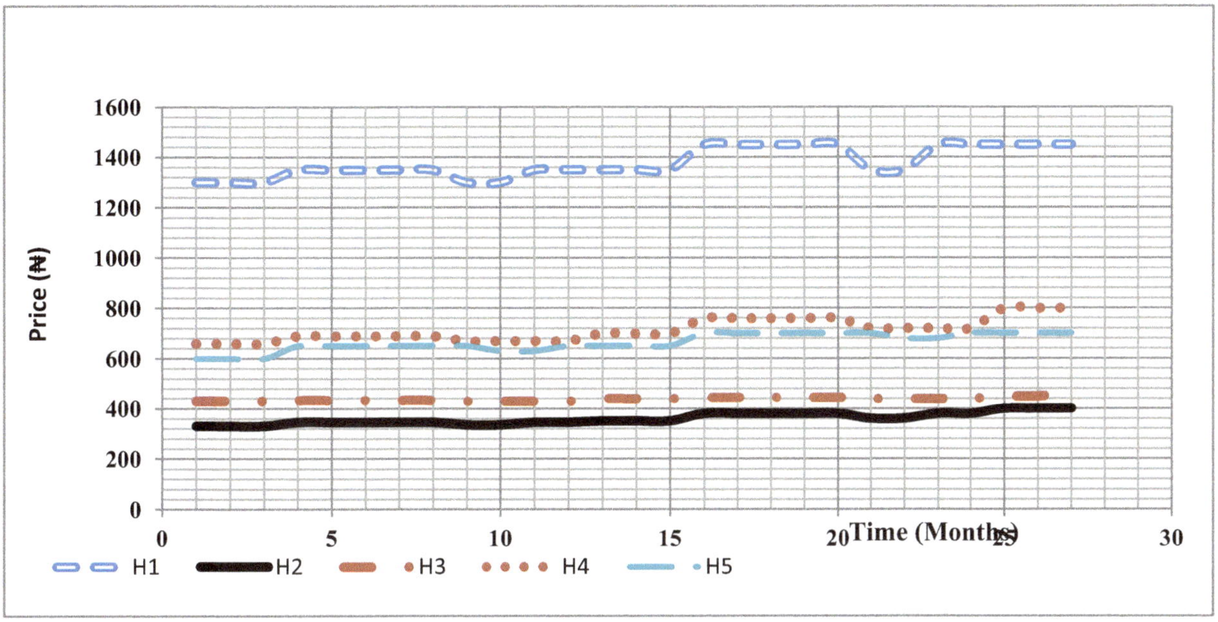

Fig. 3. Variation in prices of hardwood of different species and sizes from Jan. 2010 - Mar. 2012

A general trend of price increase for whitewood types and sizes is noticed between the months of March and April, as indicated in Fig. 4 for months 3 and 4 and months 15 and 16 respectively. The price stabilized for at least five months for type W1 and six months for Types W2 and W3. This implies that the months of April to August experience stable timber prices. Generally, relatively lower prices are experienced in the months of October and November.

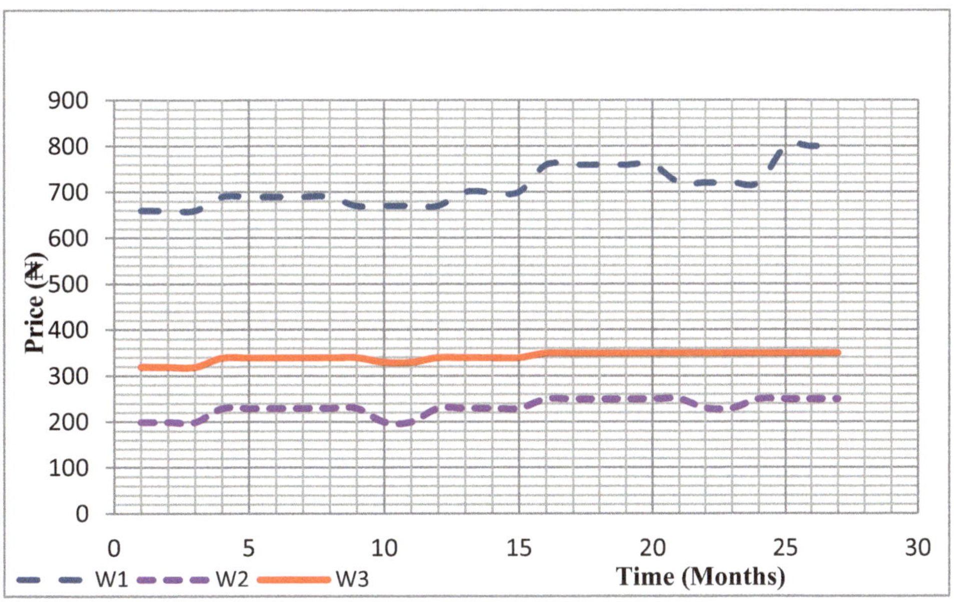

Fig. 4. Variation in prices of whitewood of different sizes from Jan. 2010 - Mar. 2012

Whitewood prices are relatively stable within 5% variability during the months of December to the following March, as indicated in months 12-15 and 24-27, with the exception of type W1.

H4 and W1 prices are the same and they follow the same trend.

Probable reasons for the cyclic trend in the variation of the prices of the basic construction materials are that more construction works take place during the annual dry season. Most people want to complete especially their building construction works so as to have something to show for the year before it runs out. This makes December the peak of civil engineering construction activities, with associated temporal peak demand for construction materials.

4. Conclusion and Recommendation

Analysis of the price variations for the basic construction materials investigated revealed as follows:

Prices of all the basic construction materials follow an annual cyclic pattern.

By February of each year, the price of cement begins to fall while that of sand begins to rise with the exception of fine sand which had a delay of rising by a month in the third year of study.

Prices of the two brands of Cement were the same except at months 12, 23 and 24 and there is price stability from April to August of each year (months 4-8, and 16-20). During these periods the price of this material increased by 13.3% from ₦1,500.00 in 2010 to ₦1,700.00 in 2011.

Increases in price of sand are mostly ₦500/5-Ton Tipper load except in the beginning of 2012. It can be concluded that there are there are 2 rises per annum in the prices of sand. The first rise occurred in March and the second in the following November for sharp and fine sands.

Prices of all types of timber except H3, increases by April and stabilizes through August. Lower prices are experienced in the months of September and November for types H1, H2, H4, H5, and W1; and October and November for types W2 and W3. Price of timber Type H3 increases steadily at the rate of 0.26% from January 2010 to 7% in March 2012.

The trends presented by types H4 and H5, which are same size but different species, are similar, indicating that hardwood specie is not a factor in the price variation of timbers.

Timber prices are relatively stable within 5% variability during the months of November – March and December – March respectively for hardwood and whitewood except types H4 and W1. H4 and W1 prices are the same and they follow the same trend.

It can generally be concluded that the prices of the basic construction materials stabilize during the months of April to August of every year. Although construction activities are light every January, January prices keep increasing from year to year.

Considering that price variation of construction materials can be a function of local contents factors, political, festival and weather related factors, it is recommended that more work can be extended to determine the levels at which each factor can affect price variation.

References

Aje Olanivi Isaac. (2000). The effect of Variation on Construction Cost and Contract Period of Building Projects. *An M. Tech. (Quantity Surveying) Thesis in The Department of Quantity Surveying*, Federal University of Technology, Akure. August, 2000.

Arowosoge, O. G. E., Ogunsanwo, O. Y., & Labode, P. (2011). Spatial And Temporal Price Variations of Sawn Wood Utilized For Furniture Making In Selected Cities in Nigeria. *Journal of Agriculture and Social Research (JASR)*, *11*(1), 2011.

Entrusty, G. (2008). What Constitute Variations And How To Evaluate Them? Your Contractual Questions Answered. *Master Builders MBAM Journal* 4th Quarter 2008 HT Ong at 22-1& 2 Jalan 2/109E, Desa Business Park, Taman Desa, 58100 Kuala Lumpur, Malaysia.

Evija, G. S. (2009). Standing timber prices and factors which affect them. *Forest Faculty, Woodworking department* Latvia University of Agriculture Dobeles street 41, Jelgava, LV 3001, 2009.

Izekor, D. N., & Izekor, O. B. (2011). Analysis Of Sawn Timber Enterprise In Benin Metropolis, Edo State, Nigeria. *Nigerian Journal of Agriculture, Food and Environment, NJAFE*, *7*(2), 19-23.

Kalu, C., Atagana, S., & Ani, P. I. (2009). Marketing of Round Logs in Benin City, Edo State, Nigeria. *Journal of Applied Sciences Research, 5*(7), 785-789. INSInet Publication.

Long, R. J. (2013). Construction Claims for Variation in Quantity. *Long International Inc. 2013*. Retrieved from www.long-intl.com

Makwana, A. H., & Jayeshkumar P. (2013). A Study on Region wise Price Variation of Construction Raw Materials using Frequency Analysis through SPSS Software. *International Journal of Engineering Trends and Technology (IJETT), 4*(7).

Okupe, L. (2000). The Role of Private Sector in Nigeria. A seminar paper on Effective Approach to Housing. *The Yaba Journal of Environmental Research.* School of Environmental Studies Yaba College of Technology.

Oladapo, A. A. (2007). A quantitative assessment of the cost and time impact of variation orders on construction projects. *Journal of Engineering, Design and Technology, 5*(1), 35-48. Emerald Group Publishing Limited. http://dx.doi.org/10.1108/17260530710746597

Revay, S. O. (2002). Coping with extras, *The Revay Report*, Vol. 21.

Ssegawa, J. K., Mfolwe, K. M., Makuke, B., & Kutua, B. (2002). Construction variations: A scourge or a necessity. *Proceedings of the 1st CIB-W107 International Conference on Creating a Sustainable Construction Industry in Developing Countries, Cape Town*, South Africa.

Sterman, J. D (1992). Systems Dynamic Modelling for Project Management, Sloan School of Management, MIT Cambridge.

Vishvas, K. T., & Kushwaha, R. C. (2012). Development of methodology for revision of Price Variation Clauses for Track Fittings. *A project undertaken as a part of the Senior Professional Course (P. Way) Course no 12207* from 18-06-2012 to 13-07-2012 for Indian Railways Institute of Civil Engineering, Pune, July-2012.

13

Desalination Behavior of Calcined Hydrotalcite From Seawater for Preparation of Agricultural Cultivation Solution Using Natural Zeolite

Takaaki Wajima[1]

[1] Graduate School of Engineering, Chiba University, Chiba, Japan

Correspondence: Takaaki Wajima, Graduate School of Engineering, Chiba University, 1-33, Yayoi-cho, Inage-ku, Chiba, Japan. E-mail: wajima@tu.chiba-u.ac.jp

Abstract

The desalination behavior of calcined hydrotalcite (CHT) from seawater used to treat seawater that was subsequently subjected to zeolite treatment prior to application of irrigation water was investigated. The results revealed that CHT removed Cl^-, SO_4^{2-}, Mg^{2+} and Ca^{2+} from seawater, but that Na^+ and K^+ were almost unchanged, and that CHT treatment made the pH of the solution alkaline. As the solution temperature increased, the solution pH increased, the rate of Mg^{2+}, Ca^{2+}, Cl^- and SO_4^{2-} removal improved, and greater Cl^- removal was observed. Following CHT treatment, the obtained solution (CHT-solution) had a high Na^+ content and high pH (approximately 12), while treatment of CHT-solution with natural zeolite neutralized the pH of the CHT-solution and reduced the Na^+ content. Although germination of radish sprouts did not observed using seawater and CHT-solution, the growth of radish sprouts could be observed using the final solution after natural zeolite treatment of CHT-solution.

Keywords: desalination, calcined hydrotalcite, seawater, natural zeolite, radish sprout

1. Introduction

In the 21st century, problems associated with energy shortages and environmental conditions have increased, and food shortage has become one of the greatest problems worldwide. Securing irrigation water for agriculture is one of the most important factors in food production; however, regions subjected to shortages of agricultural water are increasing as a result of desertification and global warming. Indeed, during the 20th century, the global human population increased fourfold, but indispensable amount of water from natural freshwater sources increased eightfold (Abdul, 2006).

Effective utilization of untapped resources is an important issue in the twenty-first century (World Water Council [WWC], 2002). One method of the technology for supplying water resources is seawater desalination. The multi-stage flash (MSF) and the reverse osmosis membrane (RO) are well-known desalination technologies (Goto, 2000) in which almost all ions contained in seawater are removed, producing high-quality fresh water for domestic or industrial use; however, 70% of global fresh water use is used for irrigation (Tilzer, 2006). Irrigation water for agriculture use should contain elements for crops rather than being highly purified, and use of MSF or RO for agriculture is not cost effective. High-purity water requires the fertilizer addition before application to the agricultural use. Seawater contains essential elements needed for plant growth, but its high concentration of NaCl causes salt damage, which prevent its direct use. Accordingly, irrigation water produced from seawater by reducing NaCl content using a simple process would be extremely advantageous.

In an earlier study, we attempted a novel two-step process to reduce the NaCl concentration in seawater using two clays, calcined hydrotalcite (CHT) and natural zeolite, which resulted in generation of a solution capable of supporting the growth of radish sprouts (Wajima et al., 2006; Wajima et al., 2008,2009, 2010). Hydrotalcite is a layered double hydroxide composed of a brucite-like positively charged layer due to the partial substitution of Mg^{2+} and Al^{3+}, which is compensated for by negatively charged interlayers containing anions and water (Constantino & Pinnavaia, 1995; Rives & Ulibarri, 1999; Pavan et al., 1999). Hydrotalcite calcined at 500–800 °C is transformed into Mg-Al oxide, which can adsorb anions from aqueous solutions through the reconstruction of their original structure (Pavlovic et al., 2005). Anion contaminants removed from the solution by CHT have been reported and shown to have high removal capacity for acids (Vreysen & Maes, 2008; Yang, Song, & Qian, 2010, 2013), heavy metal ions (Goswamee et al., 1998; Yang et al., 2005), pesticides (Inachio et al., 2001; You, Zhao, & Vance, 2002; Legrouri et al., 2005; Li et al., 2005; Cardoso & Valim, 2006), dyes (Zhu et al., 2005; Gaini et al., 2009; Asouhidou et al., 2011), halogens (Lv et al., 2006; Lv & Luo, 2007), phosphates (Shin et al., 1996; Chitrakar et al., 2005; Das et al., 2006; Cheng et al., 2009), surfactants (Pavan, Crepaldi, & Valim, 2000) and thiocyanate (Li et al., 2008), but there have been no studies of the treatment of seawater using CHT. Natural zeolite is generally associated with grassy volcanic rock in natural deposits,, and is available in large quantities at low cost (Tominaga, 2001). Due to the ion-exchange capacity, absorption characteristics, water retention and low cost, natural zeolites have been used in various fields, such as agronomy, horticulture and industry (Akgü et al., 2006). Natural zeolite contains exchangeable cations, such as Mg^{2+}, Ca^{2+}, K^+ and Na^+. When seawater is treated with natural zeolite, Na^+ in seawater is exchanged with exchangeable cations in zeolite, and Na^+ can be removed from seawater and supply Mg^{2+}, Ca^{2+} and K^+ to the solution (Wajima, 2013). This process can be applied to seawater after its use as a coolant in electric power plants and chemical plants, and to high salinity water in some areas; accordingly, we expect this process to be useful for vegetation in arid areas. Therefore, we investigated the ion exchange properties of natural zeolite in a two-step process (Shimizu et al., 2007; Wajima et al., 2013). However, little information regarding the treatment of seawater using CHT is available.

In this study, we examined the desalination properties of calcined hydrotalcite from seawater to determine the conditions for anion reduction. In addition, solutions obtained via a combined process of calcined hydrotalcite and natural zeolite were tested for the growth of radish sprouts to examine their potential for use as irrigation water.

2. Experimental

2.1 Samples

Two types of Japanese seawater were used in this study. Seawater-1, which was used for the CHT treatment experiment, was obtained from the surface of Imari Bay, Saga Prefecture, Japan. Seawater-2, which was used for the cultivation test, was collected from other points below 100 m depth in Japan. The chemical compositions and pH values of seawater used in this study are shown in Table 1. The properties of the seawater used were similar and typical of seawater found worldwide (Nakashima, 2002).

Table 1. Chemical compositions and pH of seawater used in this study

	Chemical composition (mg/L)						pH
	Na^+	K^+	Mg^{2+}	Ca^{2+}	Cl^-	SO_4^{2-}	
Seawater-1	10,753	361	1256	378	16715	2476	7.4
Seawater-2	11,283	412	1382	334	16370	2823	8.0

Commercial CHT, KW-2000, which was supllied from Kyowa Kagaku Kogyo Company Limited, Japan and Japanese natural zeolite, Iizaka-zeolite obtained from Fukushima prefecture were used. The exchangeable cations and cation exchange capacity (CEC) of natural zeolite are shown in Table 2. Natural zeolite samples had high CEC values, and most of the exchangeable cation sites were occupied by Ca^{2+} and Na^+ ions.

Table 2. Exchangeable cations and cation exchange capacity of natural zeolite used in this study

	Exchangeable cations				Cation exchange capacity
	Na^+	K^+	Mg^{2+}	Ca^{2+}	
Amount (mmol/g)	0.56	0.21	0.03	0.44	1.67

2.2 Removal Behavior of Calcined Hydrotalcite

We examined the effects of reaction temperature and reaction time on the pH of the solution and levels of Na^+, K^+, Mg^{2+}, Ca^{2+}, Cl^- and SO_4^{2-} in seawater after CHT treatment.

Briefly, 50 g of CHT were added to 500 mL of seawater and stirred with a magnetic stirrer at 5, 15, 25, 40 and 60 °C. During stirring, 3-mL aliquots of the slurry were removed at varying time intervals to determine the pH of the solution and concentrations of Na^+, K^+, Mg^{2+}, Ca^{2+}, Cl^- and SO_4^{2-} in the seawater to monitor the reaction process.

2.3 Combination of CHT Treatment and Natural Zeolite Treatment

We examined the possibility of preparing agricultural cultivation solution using a two-step combination process of CHT treatment and natural zeolite treatment. Briefly, 50 g of CHT were added to 500 mL of seawater-2 and stirred with a magnetic stirrer at 25 °C for 6 h. After stirring, the slurry was filtered to give CHT-solution. Next, 20 g of natural zeolite were added to 200 mL of CHT-solution and stirred for 1 h with a magnetic stirrer. After stirring, the slurry was filtered and fresh natural zeolite was added. This procedure was repeated ten times to obtain the final solution. The pH of the seawater-2, CHT-solution and final solution, and the contents of Na^+, K^+, Mg^{2+}, Ca^{2+}, Cl^- and SO_4^{2-} in these solutions were determined, after which they were applied as potential agricultural cultivation solutions.

2.4 Cultivation Test

The solutions obtained during the process were applied to growth test using radish sprouts (*Raphanus sativus*). During this investigation, cultivation was conducted for 10 days at room temperature to investigate the possibility of using the obtained solutions.

2.5 Characterization

The pH of the solution was measured with a pH meter (MA-130, Mettler Toledo). Concentrations of Na^+, K^+, Mg^{2+}, Ca^{2+}, Cl^- and SO_4^{2-} in the solution were determined by ion chromatography (DX-120, Japan Daionex). Removal percentages of Cl^- and SO_4^{2-}, R (%), were calculated using the following equation:

$$R = (C_0 - C)/C_0 \times 100 \quad (1)$$

where C_0 and C are the concentrations of Cl^- and SO_4^{2-} in the initial solution and the treated solution (mg/L), respectively.

Exchangeable cations and cation exchange capacities of natural zeolite were measured by the modified Schollenberger's method (Development of New Utilization of Materials [DNUM], 1994).

3. Results and Discussion

3.1 Desalination Properties of Calcined Hydrotalcite

Figure 1 shows the change in pH of the solution at 5, 25 and 60 °C after adding CHT. Before the addition of CHT, the pH of the seawater was 7.4, while after the addition of CHT, the pH of the solutions became highly alkaline. The pH of the solution gradually increased to about 12 as temperature increased from 5 °C to 60 °C.

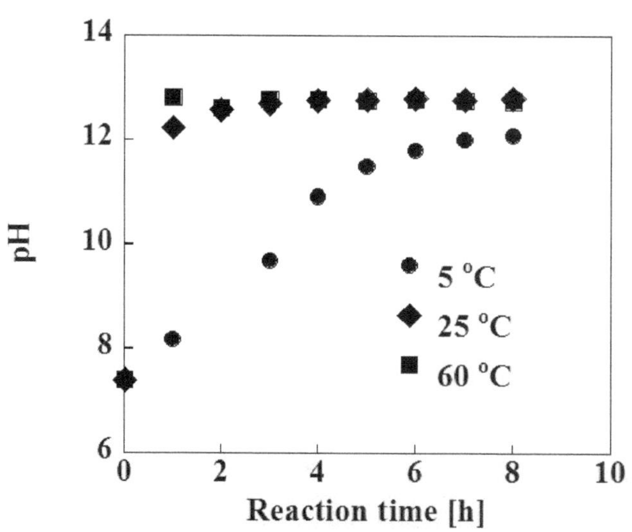

Figure 1. Change in solution pH at 5, 25 and 60 °C after adding calcined hydrotalcite

Figure 2 shows themain cations concentrations in seawater, (a) Na^+, (b) K^+, (c) Mg^{2+} and (d) Ca^{2+}, at various temperatures after adding CHT. During stirring, the concentrations of Na^+ and K^+ remained almost constant, regardless of temperature, indicating that there levels were unchanged following treatment with CHT. However, the level of Mg^{2+} and Ca^{2+} decreased slowly at 5 °C, and they reached zero after 1 h at 15 °C. These behaviors correlate well with the behavior of pH (Figure 1), which was likely because increasing pH would cause Mg^{2+} and Ca^{2+} to hydroxylate and precipitate to reduce the contents of Mg^{2+} and Ca^{2+} in seawater (Kragten, 1978).

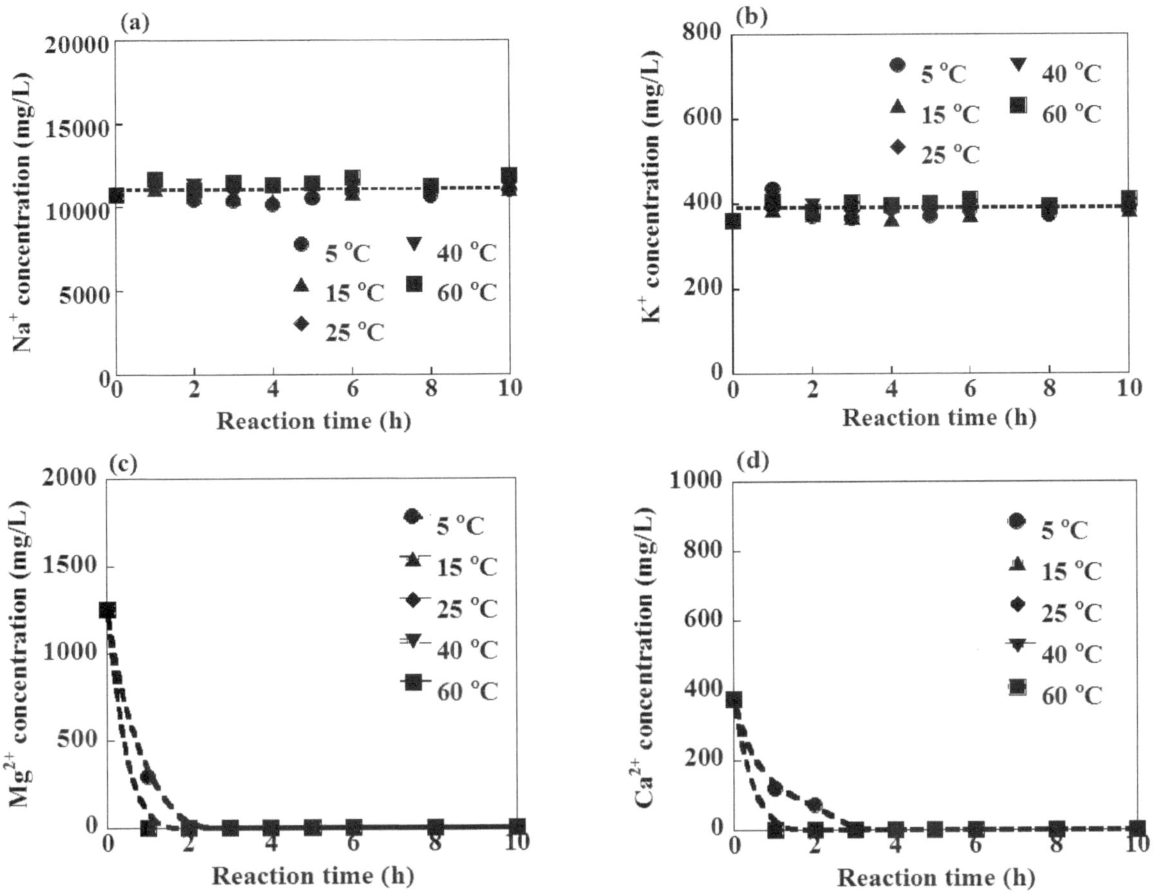

Figure 2. Change in the concentrations of main cations in seawater, (a) Na^+, (b) K^+, (c) Mg^{2+} and (d) Ca^{2+}, at various temperatures after the addition of calcined hydrotalcite

Figure 3 shows the changes in the concentrations of the main anions in seawater, (a) Cl^- and (b) SO_4^{2-} at various temperatures after adding CHT. During stirring, the concentrations of Cl^- and SO_4^{2-} decreased and became almost constant. For Cl^-, as temperature increased, the reaction time to equilibrium concentration increased, as did the reduction in Cl^-. It is noted that the removal percentages of Cl^- at 1 h treatment are 4.8% (5 °C), 5.0% (15 °C), 15.2% (25 °C), 28.1% (40 °C) and 34% (60 °C). For SO_4^{2-}, no significant difference of reduction rate was observed at all setting temperatures in this experiment. The enough time to reach equilibrium for the removal of Cl^- and SO_4^{2-} was over 2 h under all conditions.

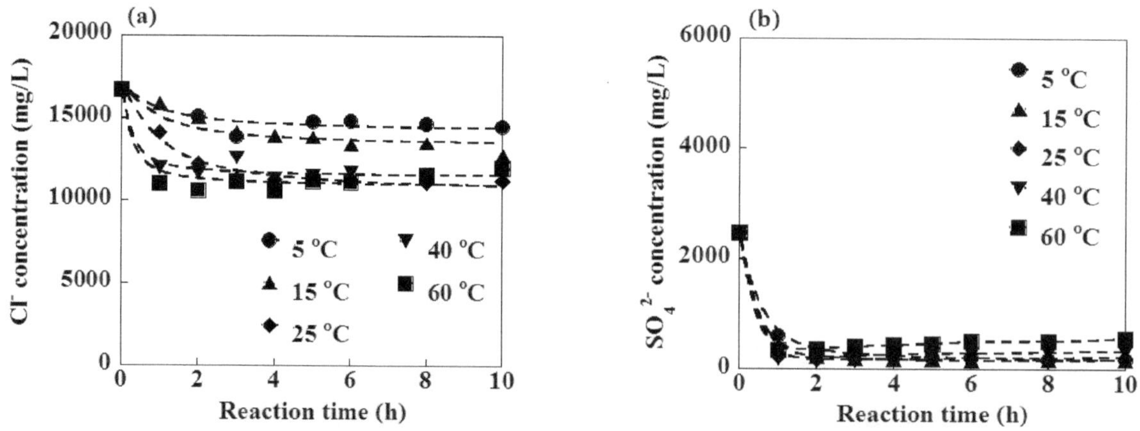

Figure 3. Change in concentrations of main anions in seawater, (a) Cl^- and (b) SO_4^{2-}, at various temperatures after adding calcined hydrotalcite

Figure 4 shows the temperature dependence of the removal of Cl^- and SO_4^{2-} from seawater, and the pH of the solution after 6 h of stirring. Regardless of the temperatures, the removal of SO_4^{2-} remained almost constant (80–90%). As the temperature increased to 25 °C, removal of Cl^- and pH of the solution increased, but above 25 °C, these values remained almost constant.

Based on these results, the condition using CHT at 25 °C over 2 h is enough to remove anions from seawater, is the optimum condition.

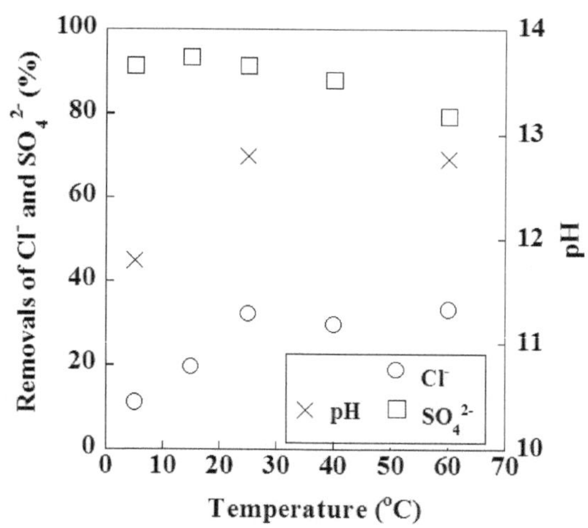

Figure 4. Temperature dependence of Cl^- and SO_4^{2-} removal from seawater and pH of the solution after 6 h of stirring

3.2 Cultivation

Radish sprouts were cultivated with seawater-2, CHT-solution and final solution.

Table 3 shows the chemical compositions and pH of seawater, CHT-solution and final solution. Seawater contained high concentrations of some ions, especially Na^+ (11,283 mg/L) and Cl^- (16,370 mg/L). The concentrations of Cl^- and SO_4^{2-} were highly reduced in CHT-solution, and the pH of solution was high (pH 12.1) because CHT fixes anions and releases OH^- (Kameda, Yoshioka, & Okuwaki, 2001). The concentrations of Mg^{2+} and Ca^{2+} also decreased. In the final solutions, the pH became neutral, Na^+ decreased, and Mg^{2+} and Ca^{2+} (which were lost in the first step) increased because of ion exchange between Mg^{2+} and Ca^{2+} in the natural zeolite and Na^+ in the CHT-solution.

Radish sprouts were grown in each solution for 10 days at room temperatuer (Figure 5). We could observe the growth of radish sprouts using the product solution through a two-step process (Figure 5 (c)), although we could not observe the germination of radish sprout using seawater and the CHT-solution (Figure 5 (a), (b)). Solutions prepared from seawater by this process were possible for use as cultivation solutions, and confirm that this process can be applied to seawater collected from various points.

Table 3. Chemical compositions and pH of seawater, Solution-1 and Solution-2

	Chemical composition (mg/L)						pH
	Na^+	K^+	Mg^{2+}	Ca^{2+}	Cl^-	SO_4^{2-}	
Seawater-2	11283	412	1382	334	16370	2823	8.0
CHT-solution	13640	504	8	36	13483	584	12.1
Final solution	1291	62	54	1347	8622	433	7.1

Figure 5. Observation of Raphanus sativus after 10-days of growth using (a) seawater, (b) CHT-solution, and (c) final solution

4. Conclusion

A novel process for preparing agricultural cultivation solution from seawater using two clays, hydrotalcite and natural zeolite was attempted. We investigated the desalination properties of water treated using calcined hydrotalcite, and those of water treated by a combination of ion exchange processes using natural zeolite, after which treated water was applied to determine if it could be used to grow Radish sprouts. Treatment with calcined hydrotalcite reduced Mg^{2+}, Ca^{2+}, Cl^- and SO_4^{2-} from seawater and increased the pH to high levels. The optimum treatment conditions using calcined hydrotalcite were 6 hours at 25 °C. Treatment with natural zeolite could neutralize highly alkali solution after calcined hydrotalcite treatment, and reduce Na^+ and supply Mg^{2+} and Ca^{2+} to the solution via ion exchange with natural zeolite. Radish sprouts could be grown using the solution obtained from treatment with a combination of ion exchange processes using natural zeolite.

From these results, it is possible to prepare agricultural cultivation water from seawater using combined treatment with calcined hydrotalcite and ion exchange using natural zeolite.

Acknowledgements

This study was performed under the Cooperative Research Program of the Institute of Ocean Energy, Saga University (13003A). Hydrotalcite, KW-2000, was kindly supplied by KYOWA KAGAKU KOGYO Company, Limited.

References

Abdul, A. K. (2006). Sustainability of water resources in Afghanistan. *Journal of Developments in Sustainable Agriculture, 1*, 53-66. http://dx.doi.org/10.11178/jdsa.1.53

Akgü, M., Karabakan, A., Acar, O., & Yürüm, Y. (2006). Removal of silver (I) from aqueous solutions with clinoptilolite. *Microporous and Mesoporous Materials, 94*, 99-104. http://dx.doi.org/10.1016/j.micromeso.2006.02.023

Asouhidou, D. D., Triantafyllidis, K. S., Lazarridis, N. K., & Matis, K. A. (2011). Adsorption of reactive dyes from aqueous solutions by layered double hydroxides. *Journal of Chemical Technology and Biotechnology, 87*, 575-582. http://dx.doi.org/10.1002/jctb.2755

Cardoso, L. P., & Valim, J. B. (2006). Study of acids herbicides removal by calcined Mg-Al-CO_3-LDH. *Journal of Physics and Chemistry of Solids, 67*, 987-993. http://dx.doi.org/10.1016/j.jpcs.2006.01.015

Cheng, X., Huang, X., Wang, X., Zhao, B., Chen, A., & Sun, D. (2009). Phosphate adsorption from sewage sludge filtrate using zinc-aluminium layered double hydroxides. *Journal of Hazardous Materials, 169*, 958-964. http://dx.doi.org/10.1016/j.jhazmat.2009.04.052

Chitrakar, R., Tezuka, S., Sonoda, A., Sakane, K., Ooi, K., & Hirotsu, T. (2005). Adsorption of phosphate from seawater on calcined MgMn-layered double hydroxides. *Journal of Colloid and Interface Science, 290*, 45-51. http://dx.doi.org/10.1016/j.jcis.2005.04.025

Constantino, V. R., & Pinnavaia, T. J. (1995). Basic properties of $Mg_{1-x}^{2+}Al_x^{3+}$ layered double hydroxides intercalated carbonate hydroxide chloride and sulfate anions. *Inorganic Chemistry, 34*, 883-892. http://dx.doi.org/10.1021/ic00108a020

Das, J., Patra, B. S., Baliarsingh, N., & Parida, K. M. (2006). Adsorption of phosphate by layered double hydroxides in aqueous solutions. *Applied Clay Science, 32*, 252-260. http://dx.doi.org/10.1016/j.clay.2006.02.005

Development of new utilization of materials [DNUM]. (1994). *Natural Zeolite and its Utilization*. Tokyo: No. 111 committee, Japan Society for the Promotion of Science.

Gaini, L. E., Lakraimi, M., Sebber, E., Meghea, A., & Bakasse, M. (2009). Removal of indigo carmine dye from water to Mg-Al-CO_3 calcined layered double hydroxides. *Journal of Hazardous Materials, 161*, 627-632. http://dx.doi.org/10.1016/j.jhazmat.2008.04.089

Goswamee, R. L., Sengupta, P., Bhattacharyya, K. G., & Dutta, D. K. (1998). Adsorption of Cr(VI) in layered double hydroxides. *Applied Clay Science, 13*, 21-34. http://dx.doi.org/10.1016/S0169-1317(98)00010-6

Goto, T. (2000). Kaisui-tansuika no hukyuudoukou to gijyutsukadai. *Suidou-Kouron, 36*(2), 43-46.

Inacio, J., Taviot-Guého, C., Forano, C., & Besse, J. P. (2001). Adsorption of MCPA pesticide by Mg/Al layered double hydroxides. *Applied Clay Science, 18*, 255-264. http://dx.doi.org/10.1016/S0169-1317(01)00029-1

Kameda, T., Yoshioka, T., & Okuwaki, A. (2001). Development of new synthetic process for hydrotalcite using seawater and calcined dolomite and its application to waste water treatment. *Chemical Industry, 52*, 810-814.

Kragten, J. (1978). *Atlas of Metal-legand Equibria in Aqueous Solution*. Chichester: Ellis Horwood Limited.

Legrouri, A., Lakraimi, M., Barroug, A., Roy, A. D., & Besse, J. P. (2005). Removal of the herbicide 2,4-dichlorophenoxyacetate from water to zinc-aluminum-chloride layered double hydroxides. *Water Research, 39*, 3441-3448. http://dx.doi.org/10.1016/j.watres.2005.03.036

Li, F., Wang, Y., Yang, Q., Evans, D. G., Forano, C., & Duan, X. (2005). Study on adsorption of glyphosate (N-phosphonomethyl glycine) pesticide on Mg/Al layered double hydroxides in aqueous solution. *Journal of Hazardous Materials, 125*, 89-95. http://dx.doi.org/10.1016/j.jhazmat.2005.04.037

Li, Y., Gao, B., Wu, T., Chen, W., Li, X., & Wang, B. (2008). Adsorption kinetics for removal of thiocyanate from aqueous solution by calcined hydrotalcite. *Colloids and Surfaces A: Physicochemical and Engineering Aspects, 325*, 38-43. http://dx.doi.org/10.1016/j.colsurfa.2008.04.046

Lv, L., & Luo, L. (2007). Adsorption behavior of calcined layered double hydroxides towards removal of iodine contaminants. *Journal of Radioanalytical and Nuclear Chemistry, 273*, 221-226. http://dx.doi.org/10.1007/s10967-007-0740-x

Lv, L., He, J., Wei, M., Evans, D. G., & Duan, X. (2006). Factors influencing the removal of fluoride from aqueous solution by calcined Mg-Al-CO_3 layered double hydroxides. *Journal of Hazardous Materials, 133*, 119-128. http://dx.doi.org/10.1016/j.jhazmat.2005.10.012

Nakashima, T. (2002). *Kaiyou-shinsousui-no-Riyou*. Tokyo: Midori-Shobou.

Pavan, P. C., Crepaldi, E. L., Gomes, G. D., & Valim, J. B. (1999). Adsorption of sodium dodecylsulfate on a hydrotalcite-like compound. Effect of temperature, pH and ionic strength. *Colloids and Surfaces A: Physicochemical and Engineering Aspects, 154*, 399-410. http://dx.doi.org/10.1016/S0927-7757(98)00847-4

Pavan, P. C., Crepaldi, E. L., & Valim, J. B. (2000). Sorption of anionic surfactants on layered double hydroxides. *Journal of Colloid and Interface Science, 229*, 346-352. http://dx.doi.org/10.1006/jcis.2000.7031

Pavlovic, L., Barriga, C., Hermosín, M. C., Cornejo, J., & Ulibarri, M. A. (2005). Adsorption of acidic pesticides 2,4-D clopyralid, and picloram on calcined hydrotalcite. *Applied Clay Science, 30*, 125-133.

Rives, V., & Ulibarri, M. A. (1999). Layered double hydroxides (LDH) intercalated with metal coordination compounds and oxometalates. *Coordination Chemistry Reviews, 181*, 61-120. http://dx.doi.org/10.1016/S0010-8545(98)00216-1

Shimizu, T., Wajima, T., & Ikegami, Y. (2007). Ion exchange properties of natural zeolite in the preparation of an agricultural cultivation solution from seawater. *Journal of Ion Exchange, 18*, 540-543. http://dx.doi.org/10.5182/jaie.18.540

Shin, H., Kim, M., Nam, S., & Moon, H. (1996). Phosphorous removal by hydrotalcite-like compounds (HTLcs). *Water Science and Technology, 34*, 161-168. http://www.iwaponline.com/wst/03401/wst034010161.htm

Tilzer, M. M. (2006). Renewable, but not inexhaustible: the fresh water supply for a growing human population. *Proceedings of the 2nd Autoanalyzer Symposium in Osaka.*

Tominaga, H. (2001). *Zeolite no kagaku to ouyou.* Tokyo: Kodansha Corp.

Vreysen, S., & Maes, A. (2008). Adsorption mechanism of humic and fulvic acid onto Mg/Al layered double hydroxides. *Applied Clay Science, 38*, 237-249. http://dx.doi.org/10.1016/j.clay.2007.02.010

Wajima, T. (2013). Ion exchange properties of Japanese natural zeolites in seawater. *Analytical Sciences, 29*, 139-141. http://dx.doi.org/10.2116/analsci.29.139

Wajima, T., Shimizu, T., & Ikegami, Y. (2006). New simple process of making agricultural cultivation solution from seawater. *Bulletin of the Society Seawater Science, Japan, 60*, 201-202.

Wajima, T., Shimizu, T., Yamato, T., & Ikegami, Y. (2008). Simple two-step process for making agricultural cultivation solution from seawater using calcined hydrotalcite and natural zeolite. *Clay Science, 14*, 7-12.

Wajima, T., Shimizu, T., Yamato, T., & Ikegami, Y. (2009). Fundamental study on simple desalination treatment of natural zeolite for the production of irrigation water from seawater. *Journal of the Society of Materials Engineering for Resources of Japan, 22*, 18-22. http://dx.doi.org/10.5188/jsmerj.22.18

Wajima, T., Shimizu, T., Yamato, T., & Ikegami, Y (2010). Removal of NaCl from seawater using natural zeolite. *Toxicological & Environmental Chemistry, 92*(1), 21-26. http://dx.doi.org/10.1080/02772240902762958

Wajima, T., Shimizu, T., Yamato, T., & Ikegami, Y. (2013). Effect of HNO_3 and H_3PO_4 on ion exchange of natural zeolite for making agricultural cultivation solution from seawater. *International Journal of the Society of Materials Engineering for Resources.*

World Water Council [WWC]. (2002). World Water Vision. Making Water Everybody's Business, Earthscan, London. Semenak, S. (1995, December 28). Feeling right at home: government residence eschews traditional rules. *Montreal Gazette*, p. A4.

Yang, D., Song, Z., & Qian, X. (2010). Study on adsorption of benzonic acid by hydrotalcites and their calcined products. *Environmental Engineering Science, 27*, 853-860. http://dx.doi.org/10.1089/ees.2010.0072

Yang, D., Song, Z., & Qian, X. (2013). Adsorption of abietic acid from colloidal suspension by calcined Mg/Al hydrotalcite compounds. *Industrial and Engineering Chemistry Research, 52*, 6956-6961. http://dx.doi.org/10.1021/ie303513n

Yang, L., Shahrivari, Z., Liu, P. K. T., Sahimi, M., & Tsotsis, T. T. (2005). Removal of trace levels of arsenic and selenium from aqueous solutions by calcined and uncalcined layered double hydroxides (LDH). *Industrial and Engineering Chemistry Research, 44*, 6804-6815. http://dx.doi.org/10.1021/ie049060u

You, Y., Zhao, H., & Vance, G. F. (2002). Adsorption of dicamba (3,6-dichloro-2-methoxy benzoic acid) in aqueous solution by calcined layered double hydroxide. *Applied Clay Science, 21*, 217-226. http://dx.doi.org/10.1016/S0169-1317(01)00102-8

Zhu, M., Li, Y., Xie, M., & Xin, H. (2005). Sorption of an anionic dye by uncalcined and calcined layered double hydroxides: a case study. *Journal of Hazardous Materials, 120*, 163-171. http://dx.doi.org/10.1016/j.jhazmat.2004.12.029

14

Spatio-Temporal Implications of Inadequate Water and Sanitation on Residents of Ijora-Badia Area of Lagos State, Nigeria

Yussuf Lukeman[1], Bako, A. I.[2], Omole, F. K.[3], Nwokoro, I. I. C.[1] & Olajuyigbe, A. E.[3]

[1] Department of Urban and Regional Planning, University of Lagos, Nigeria

[2] Department of Urban and Regional Planning, University of Ilorin, Nigeria

[3] Department of Urban and Regional Planning, the Federal University of Technology, Akure, Nigeria

Correspondence: Omole, F. K., Department of Urban and Regional Planning, the Federal University of Technology, P.M.B 704, Akure, Nigeria. E-mail: fkyomole@yahoo.co.uk

Abstract

In recent times, it has been proofed that industrial revolution, urbanization, inadequate provision of facilities, poor urban management and poverty were the determinant of environmental condition of city dwellers. In this paper, the sanitary condition of the environment of residents of Ijora Badia was investigated. Some of the sanitary conditions data involving sources of drinking water, sources of cooking, method of waste disposal, types of toilet facility, drainage system and road conditions were collected through questionnaires, personal interview, direct observation and focus group discussion. The research population was based on the total number of existing buildings from which the total numbers of household head were determined and 5% of the total household head population was taken for the interview. Findings reveal that 65.5% of the residence sources of drinking water is through water vendor which has aggravated the water borne diseases, 93.1% of residence sources of cooking is stove which cause respiratory disease among the households. The paper suggests among other things that all stakeholders in the built environment should see to the welfare and provision of environmental facilities in order to make life comfortable and functional for the slum dwellers.

Keywords: environmental condition, development, facility, poverty and slum

1. Introduction

In Nigeria like in other developing countries, poor sanitation in various cities pose population health hazard through pollution (air, water and noise). Disposal and treatment of sewage is available for less than 2% of the population as the case is in Nigeria (UNICEF, 1999). Human excreta pose serious threat to the environment as well as public health, if not properly handled. Human excreta contain pathogen such as bacteria, viruses, protozoa, and helminthes, which lead to a range of infection in humans such as dysentery, diarrhea, and cholera among others. Uncontrolled disposal of excreta in soil and water can lead to an overload of organic compounds and nutrients in the environment resulting in the loss of plant, animals, and human lives.

Brundtland, (2003) identified environmental risks to include inadequate access to safe drinking water, poor hygiene and sanitation, inadequate water resources management, air pollution, chemical hazards, and unintentional injuries. According to him, estimated 1.1 billion people lacked access to an improved water source in the year 2000. In year 2003, around 2.4 billion people lack access to any type of improved sanitation facilities, while air pollution becomes a serious risk factor for respiratory disease and major contribution to ill-health among children around the world.

Along this line, Egunjobi, (1986) posited that a lot of people die every day from various diseases caused by their inadequate shelter or lack of access to such basic services as clean water or sanitary excreta disposal. Available global evidence in the literature including the works of Smith et al. (1999) and that of Tibaijuka (2003) suggest that lack of access to clean water and sanitation and indoor air pollution are principal risk factors of illness and death, mainly affecting children and women in poor families.

The public worldwide has now become acutely aware that health and life sustainability are put at risk with the present level of poor sanitary conditions of the environment. Some of the waste dumped around proved extremely toxic and hazardous, which through their indiscriminate dumping, have brought about the death of livestock, environmental pollutions, and an increasing number of incidents that have caused ill-health and epidemic of diseases like malaria and respiratory diseases. A potentially more serious risk to human health is the contamination of source of water supply (Adeola, 1996).

In recent times, the residents of Ijora-Badia (the study area) have been faced with many environmental conditions problems such as poor water supply, inadequate toilet facilities, poor drainage and sanitation conditions. The aftermath of the shortfall in the provision of some of these facilities has resulted in epidemic and deadly diseases such as diarrhea, cholera, typhoid among others. In view of these, this paper examines the environmental conditions of the residents of Ijora-Badia in Lagos Nigeria.

2. The Study Area

Ijora- Badia is one of the blighted area identified by the UNDP which was also validated by the consulting firm SNC-Lavalin employed by the Lagos State Government in the World Bank assisted Infrastructure Upgrading programme for the metropolitan Lagos (UNDP, 1997).

The specific study area is Ijora –Badia, one of the suburbs in the Apapa Local Government of Lagos State and it is located in the Southern fringe of the Lagos Metropolitan area. It is situated at the interaction point of the geographical latitude 3° 23' and longitude 4° 22'. It is bounded in the North by the Lagos Badagry expressway link bridge into the National Theatre. A railroad to Apapa forms the Eastern boundary. The South is bounded by Ajegunle another low-income residential suburb.

3. Literature Issues

Extreme inequality and idleness lead people to anti-social behavior, which is more pronounced in the slum area.

Slums are the places where all the evils come together, where peace and security is elusive and where young people cannot be protected. Tibaijuka, (2003) is of the view that deteriorating situation which potentially threatened the security and increases pressures on immigration to rich countries should be addressed. He observed that some slums were now as large as cities. The Kibera district in Nairobi which was classed as the largest slum in the world has as many as 600,000 people. The Dharavi area of Mumbai and the Orangi district of Karachi have only slightly fewer people, while the Ashaiman slum is now larger than the city of Tema in Ghana, around which it grew.

Other cities, such as Dhaka in Bangladesh, have several hundred small slums or squatter settlements, which have no access to services and are liable to be moved on at short notice. The world is entering a significant stage, say the report's authors. Over the next 30 years, the urban population in the developing world will double to about 4 billion people, at the rate of about 70 million a year. Rural populations will barely increase and begin to decline after 2020." The authors also predicted that three quarters of the world's anticipated population growth would take place in relatively small cities with populations of between 1 million and 5million. The report found that the world's urban population had increased by 36% in the 1990s, and that city authorities had been unable or unwilling to keep up. Slums are the product of failed policies, bad governance, corruption and a lack of political will, the report says. Very few countries have recognized this critical situation and very little effort is going into providing jobs or services." But the authors roundly blamed laissez-faire globalization and "neo-liberal" economic policies imposed on poor countries by global institutions such as the International Monetary Fund and the World Trade Organisation for much of the damage caused to cities over the past 20 years.

The authors say people are encouraged to move to the cities by factors including the privatization of public services, job losses, and the removal of subsidies and tax breaks from key industries. Such effects, they say, increase inequality, and make sure that those who move to the cities remain in deep poverty. One of the few direct benefits that slum dwellers have gained from globalization is greater access to aid agencies, the report says. But the very limited advantages are outweighed by a truly formidable array of disadvantages so many, in fact, that some governments might be excused for not wishing to take part at all in globalization if they have the welfare of the urban poor at heart. In a form of colonialisation that is probably more stringent than the original, many developing countries have become suppliers of raw commodities to the world, and fall further and further behind. The authors conclude that as cities have become a dumping ground for people working in unskilled, unprotected and low-wage industries and trade the slums of the developing worlds well.

Innocent, (2009) in his study on the Millennium Development Goal on Slum Eradication in River state, stated that the urban crisis in Nigeria is not only manifested in the acute shortage of dwelling unit which has resulted in overcrowding, high rents, and slum settlements. There is also the inadequacy, both in terms of quality and quantity of basic infrastructural facilities and social services such as pipe-born water, electricity, roads, schools and health institutions. It is also portrayed by the gradual breakdown of the urban economic way of life which is reflected in rising unemployment, diminishing incomes and poverty especially among the less privileged.

4. Methodology

In order to ensure that this research work achieves its purpose, the study embraces both primary and secondary sources of information. These were obtained from the field through the use of research instrument such as administration of questionnaires, personal interviews, observations and focus group discussion. Secondary information was collected from reports, (published and unpublished sources), textbooks, journals, file of government agencies and parastatals. These include the National Population Commission, the local government, and on the Internet, among many others.

There are 539 residential buildings in Ijora-Badia, Lagos (Tenament Rate Department Apapa Local Government, 2011). It has an average of minimum 7 households in a building, being a high residential area of metropolitan Lagos, which is in accordance with the National Population Commission (NPC) information of 2006. Since, the minimum average households per building is 7 households, the targeted population household in the study area translate to 539 x 7 = three thousand seven hundred and seventy three (3,773) households. The targeted household population of this study is 3,773 and the sample size is 5% of the targeted household population which translate to one hundred and eighty nine (189) and this becomes the total number of questionnaires administered for the study. In each district in the study area, 5% sample size was considered reasonable for this study because of the homogeneous characteristics that was peculiar to the residents in the study area.

The research adopts multi stage sampling technique in the research procedure. First stage is the delineation and dividing Ijora Badia into three districts namely Ijora Badia West, Ijora Badia central, and Ijora Badia East (Onyeche, 2011). Second stage is the identification of the buildings in each of the district by systematic sampling technique. All buildings in each district were arranged serially, from which the sampled buildings were selected. The 3rd building forms the nth term while every 3rd building was sampled from each district.

A randomly sampling technique was adopted in the third stage for the selection of household head sampled. This sampling method was based on household size in selected residential building. Household with highest number of people was selected.

Finally, in each of the districts, questionnaires were administered systematically and randomly selection of household head on every 3rd buildings in each district. The questionnaires were administered according to the number of buildings in the study area.

Table 1. Distribution of questionnaires within the district in the study area (Ijora-badia)

Name of district	Buildings in each District	No. of Questionnaires Administered
Ijora-badia West	129 buildings	45
Ijora-badia Central	230 buildings	81
Ijora-badia East	180 buildings	63
Total	539 Buildings	189

Source: Tenement Rate Department Apapa Local Government (2012).

5. Research Findings and Discussion

The environmental variables considered in this study focus on the sanitary conditions in the study area which comprises of sources of drinking water, types of toilet facility, sources of cooking, sources of energy, road conditions and types of drainage facility.

Table 2. Major source of drinking water

Sources of Water	**Frequency**	**Percent**
Borehole	37	19.6
Water vendor	124	65.6
Well	3	1.6
Pipe borne water	25	13.2
Total	189	100.0

Source: Authors' Field Survey, 2012.

The sources of drinking water in the study area as shown in table 2 indicates that 19.6% of respondents sources of drinking water is borehole, 65.6% of respondents sources of drinking water is through water vendor, 1.6% of respondents sources of drinking water is through well and 13.2% of respondents sources of drinking water is from pipe-borne water. It is concluded that, majority of respondents sources of drinking water is through water vendor which very dangerous to the health of residents.

Table 3. Treatment of water before drinking

Options	Frequency	Percentages
Yes	51	27.0
No	138	73.0
Total	189	100.0

Source: Authors' Field Survey, 2012.

The responses of the residents in the study area shows that majority of the respondents do not treat their water before drinking. 27% of respondents treat their water before drinking while 73% of respondents do not treat the water before drinking. This made them to be vulnerable to water borne diseases.

Table 4. Distance of water from home

Distance	Frequency	Percentages
Less than 100m	165	87.3
200m-500m	23	12.2
501m-1km	1	.5
Total	189	100.0

Source: Authors' Field Survey, 2012.

The distance of water from various home shows that majority of respondents sources of water is not too far from their homes in the study area. Table 4 shows that 87.3% of respondents' distance of source of water from home is less than 100m, 12.2% of respondents home is between 200m-500m to the source of water and only 0.5% respondents' home is between 501m-1km to the source of water. It can be inferred that majority of respondents source of water meet up with United Nations (UN) standard of less than 100m to various homes. The water was found to be public borehole which was constructed by Local Government Chairman and the Lagos states Legislature representing Apapa consistency.

Table 5. Types of toilet facility

Types of Toilet	**Frequency**	**Percentages**
Pit latrine	33	17.5
Bush/open dump Drainage channel Water closet	58	30.7
	72	38.0
	26	13.8
Total	189	100.0

Source: Authors' Field Survey, 2012.

The types of toilet facilities in the study area as shown in table 5 show that 38.0% of the respondents used drainage canal, 17.5% made use of pit latrine, 30.7% used bush/ open dump site, while 13.8% of the respondents used water closet. This means that majority of the respondents used drainage canal as their toilet facilities. The canal was meant to collect water from the primary drainage and enter into the lagoon. This may also caused unnecessary odour which pollute the whole environment.

Table 6. Sources of energy for Cooking

Sources of energy	Frequency	Percentages
Stove	176	93.1
Gas	4	2.1
Charcoal	7	3.7
Firewood	2	1.1
Total	189	100.0

Source: Authors' Field Survey, 2012.

Table 6 shows the sources of energy for cooking in the house by the respondents in study area shows that 93.1% of respondents' source of energy for cooking is stove, 2.1% used Gas, 3.7% of respondents used charcoal and 1.1% used firewood. It can be inferred that majority of respondents' sources of energy for cooking was stove which may lead to respiratory diseases and air pollution.

Table 7. Types of drainage facility

Type of drainage facility	Frequency	Percentages
Open	176	93.1
Closed/covered	12	6.3
Tunnel	1	.5
Total	189	100.0

Source: Authors' Field Survey, 2012.

The types of drainage facility available in the study area show that 93.1% were open drainage, 6.3% of drainage facility available were closed and only 0.5% of drainage facility was tunnel types. From this, I can be inferred that the drainage facility available in the study area mainly the open type. The implications of this among others were that, the residents dropped waste inside the drainage, which resulted in the blockage of the drainage. Based on this observation, the study area may likely to be affected by the menace of flood.

Table 8. Methods of waste disposal

Methods of waste disposal	Frequency	Percentages
Burning	9	4.8
PSP/LAWMA	116	61.4
Open dump site	18	9.5
Cart pusher	46	24.3
Total	189	100.0

Source: Authors' Field Survey, 2012.

The waste generated was disposed by various means in the study area. 4.8% of respondents disposed their waste through burning system. 61.4% of respondents disposed their waste through Private Sector Participation/ Lagos State Waste Management Authority(PSP/LAWMA), 9.5% of respondents disposed waste through open dump site and 24.3% of respondents disposed waste through cart pursher. This shows the effort of the Lagos state government in waste disposal by the PSP/LAWMA which is very good services regardless of the study area being among slum environment in Lagos state.

Table 9. Frequency of waste disposal

Frequency of Waste disposal	Frequency	Percentages
Daily	15	7.9
Once in a week	110	58.2
Twice a week	54	28.6
Monthly	5	2.6
Occasionally	5	2.6
Total	189	100.0

Source: Authors' Field Survey, 2012.

The waste generated was disposed by various means and the frequent rate of disposal is that 7.9% of respondents disposed their waste daily, 58.2% of respondents disposed waste twice a week, 2.6% of respondents disposed waste monthly and occasionally respectively. It is deduced that the activities of PSP/LAWMA is not well appreciated in the study area.

Table 10. Major sources of energy

Major sources of energy(electricity)	Frequency	Percentages
PHCN	74	39.2
Generating set	114	60.3
Inverter	1	.5
Total	189	100.0

Source: Authors' Field Survey, 2012.

The major sources of energy in the study area shows that 39.2% of respondents source of energy is Power Holding Company of Nigeria (PHCN), 60.3% of respondents major source of energy is generating set and 1% of respondents source of energy is inverter. It has been deduced that the major source of energy in the study area is generating set which is hazardous to human health.

Table 11. Road conditions

Options	Frequency	Percentages
Tarred	49	25.9
Untarred	140	74.1
Total	189	100.0

Source: Authors' Field Survey, 2012.

The roads types in the area were either tarred or untarred. 0.25.9% of the roads in the study area was tarred while 74.1% of roads were not tarred (See table 11). This situation can be described as not too good, especially during the raining seasons. This exposes the residents to various dangers such as flooding. In the other hands, during the dry seasons, it generates too much of dust which is very dangerous to human health.

6. Policy Recommendations

This study has identified various environmental conditions affecting the lives of slums and shanty dwellers in Nigeria as epitomized in Ijora Badia. The following are some of the policy recommendations of this study.

- In the first instance, there should be a strong and effective method of enforcing environmental policies, particularly in the area of refuse disposal. The Local Government and the Lagos State Waste Management Authorities (LAWMA) are to be reinforced with sufficient funds and equipments for effective functioning, while sanitary inspections are put in place to monitor and arrest the culprits.
- It is equally recommended that essential infrastructure facilities like portable water supply inform of mini-water works or boreholes and public toilets should be provided in strategic places. This will reduce the rate of communicable diseases in the area as well as boosting its physical outlook.
- Provision of good drainages and the maintenance of that drainage should be saddled by the communities and the government to allow for free flow of the drainage to reduce the growing number of mosquitoes.
- The government and non- governmental organization should be able to organized seminar, conferences and the provision of chemicals for the residents in the study area for the proper treatment of their water before drinking, this will reduce the rate of water borne diseases such as diarrhea, dysentery e.t.c.
- All the stakeholders involved in the build environment should see the welfare of the slum dwellers as a collective responsibility in the provision of environmental facility that will make life comfortable for the dwellers.
- In addition, the people need proper enlightenment on the importance and benefits of a healthy environment. This is needful so as to sustain every improvement efforts put in place. For this reason, environmental education should be made compulsory in the school curriculum as well as using public media to educate the

public particularly the residents of Badia Lagos Nigeria.

References

Adeola, F. O. (1996). Environmental Hazards, Health and Racial Inequality in Hazardous Waste Distribution. *Environmental Behaviour*, *2*(26), 99.

Brundtland, H. G. (2003). *Start New Movement to Address Environmental Crisis Affecting Children's Health*.w3.whosea.Org/En/Section260/Section484/Section 484.World Health Day. A paper delivered at the launch of World Health Day; New Delhi, India.

Egunjobi, L. (1986). Human Elements in Urban Planning and Development in *Ibadan. Habitat International, 10*(44), 147-153. http://dx.doi.org/10.1016/0197-3975(86)90079-2

Expert Group Meeting. (2002). Slum environment in developing countries, meeting held in Nairobi. Retrieved April 7, 2011, from http://www.unhumansettlement.org

Inocent, M. (2009). Slum Development in Cross River, State. *Journal of Nigeria Institute of Town Planners, 22*(2), 22-24.

Lagos State Economic Summit. (2001). Economic summit on formal and informal economy of Lagos State' held at Eko Hotel July 27-29. Retrieved from http://www.lagosbureau of Statistic on 7/4/2011 at 1.25 am

Mabogunje, A. L. (2002). *2002 National Housing and Urban Development Policy: Catalyst for Mass Housing Delivery in Nigeria.* Lead Paper Presented at the Second Abuja International Housing Summit Held at the Le Meridien Hotel Abuja.

National Population Commission (NPC). (2006). Census report in Nigeria. Collected from the Lagos State NPC Office Surulere, Nigeria.

National Population Commission of Nigeria Official Web Site. Retrieved from http//www.npcnigeria.org

NDP, (1997). Human Development Report to Eradicate Poverty. Retrieved April 13, 2011, from http://www.unhabitat.org

Onyeche, E. (2011). Ijora Badia: Slum for decades. Retrieved June 17, 2011, from http//www.slumnigeria.org

Smith, K. R., Corvalàn, F. C., & Kjellström, T. (1999). How much ill health is attributable to environmental factors? *Epidemiology, 10*(5), 573-584. Retrieved July 17, 2011, from http://www.environmental_health.org http://dx.doi.org/10.1097/00001648-199909000-00027

Tenements Rate Department. (2011). Apapa Local Government, Lagos State, Nigeria.

Tibaijuka, A. (2003). *UN Report on slum development across the globe*. Retrieved from http://www.unhabital.org on 7/4/2011 at 1.30 am

UN-Habitat. (2006). State of The World's Cities 2006/2007. *The Millennium Development Goals and Urban Sustainability.* London, U.K. Retrieved April 13, 2014, from http://www.unhabitat.org

UNICEF. (1999). 'Disposal of sewage where no mains drainage is available…' website. Retrieved July 4, 2011, from http://www.unicef.org

World Bank. (2007). *Dhaka, Improving the Living Conditions for the Urban Poor.* World Bank Report No. 35824-BD, Washington, D.C. Retrieved May 15, 2011, from http//www.worldbank.org/urban

Study on Continuous Production of Biodiesel Using Fixed Bed Reactors Filled With Anion-Exchange Resins

Takuya Ito[1], Yusuke Kakuta[2], Katsumi Hirano[2] & Toshinori Kojima[1]

[1] Department of Materials and Life Science, Seikei University, Japan

[2] Department of Materials and Applied Chemistry, Nihon University, Japan

Correspondence: Takuya Ito, Department of Materials and Life Science, Seikei University, 3-3-1 Kichijoji-kitamachi, Musashino-shi, Tokyo, Japan. E-mail: takuya.ito@st.seikei.ac.jp

Abstract

The biodiesel attracts attention as a fossil alternative fuel. It has turned out to be effective to use the methoxide type anion-exchange resin as a catalyst for the biodiesel production. However, the problem of the resin deactivation in the fixed bed reactor still remains. The deactivation mechanism of the resin in the reactor for the biodiesel production was investigated. As a result, it was clarified that the contact of the resin with triglycerides of the reactant was obstructed by glycerin of the byproduct absorbed by the resin. It was suggested that increasing the linear velocity in the reactor decreased the adsorption of glycerins, so that the deactivation of the resin could be controlled. Moreover, the setting up drift plates in the reactor could improve the yield of biodiesel.

Keywords: biodiesel, anion-exchange resin, transesterification, fixed bed reactor, catalyst lifetime

1. Introduction

According to the Petroleum Association of Japan Petroleum Industry (2013), about 33% of the world primary energy demand is met by petroleum, and the world petroleum demand in 2012 was 89.7 million barrels/day. According to the estimates by the International Energy Agency, demand will reach 95.7 million barrels/day by 2017, an increase of 7% of 2012 levels or 5 million barrels/day, with the increase in demand centering on China and Asia. However, the reserves-to-production ratio is about 59 years, and there is a concern of depletion (Petroleum Association of Japan Petroleum Industry, 2013).

Since most of the petroleum is consumed in internal combustion engines, development of a light liquid fuel as an alternative to petroleum for use in existing internal-combustion engines is urgent. Biodiesel is attracting attention as one of the possible alternative fuel to petroleum. Basically, biodiesel is comprised of fatty acid methyl esters (FAMEs), which are produced by the transesterification of triacylglycerols, the main component of fats and oils, with methanol and homogeneous alkali catalysts (Ex. NaOH, KOH, ect.). Biodiesel can be used in existing diesel engines, and as it vanishingly contain sulfur, SO_X is hardly emitted. In addition, the flash point of biodiesel is high and complete combustion can be achieved, with reduced emissions of HC, CO, and black smoke.

In Europe, production of biodiesel from virgin vegetable oils exceeds 3 million tons per year in 2007, and production is increasing (Matsumura, 2007). In Japan, it is difficult to secure enough land to cultivate sufficient quantities of raw plant material. Therefore, biodiesel production from post-consumer fat and oil, such as waste cooking oil, is sought-after. However, the used cooking oil contains free fatty acids (FFAs). FFAs react with the alkali catalyst and produce soap. Soap takes BDF into the water in the washing process and decreases the product yield (Horie, Sugano, Hirano, & Mashimo, 2006). Moreover, the use of the homogeneous alkali catalysts causes the procution of the harmful alkaline waste fluids in the water washing process (Saka, 2007). Then, the authors paid attention to the two step ion exchange resin method. This method involves two steps: esterification of FFAs using cation exchange resin in the 1st step and transesterification of triacylglycerols using anion exchange resin in the 2nd step. As a result, the continuous process with a fixed-bed reactors can be constructed without the soap generation or the water washing process. The high durability of the cation exchange resin has already been reported (Kouzu, Nakagaito, & Hodaka, 2011). However, the durability of the anion exchange resin was poor, because the hydroxyl ions of the anion exchange resins caused the hydrolysis of triacylglycerol and the fatty acids were generated (SHIBASAKI & KITAKAWA, 2007). The authors reported that when the hydroxyl ion of the anion exchange resin is exchanged with the methoxide ion, the fatty acid is not generated (Hirano, Ito, Kakuta, Sugano, & Mashimo, 2010). Hence, the applicability to the fixed bed reactor of the methoxide (CH_3O) type resin is examined in this paper.

2. Methods

2.1 Catalyst Preparation

Anion exchange resin (Organo Corporation, IRA204Cl) of 400 mL was swelled in methanol and was filled the column with glass filter. The sodium methylate-methanol solution (1 mol/L) of 4 L was poured into the column at the speed of SV = 4h^{-1}. Next, methanol of 8 L was poured into it at the speed of SV = 10 h^{-1}. Methanol of 500 ml was poured into it under pressure. The obtained resin was regarded to be fully converted into the CH_3O type.

Table 1 lists physical properties of the anion exchange resin used.

Table 1. Properties of anion exchange resin

Amberlite IRA904Cl	
Character	Anion
Group	-N≡$(CH_3)_3$X
Type	Macro reticular
Ion-exchange capacity [mol/kg-resin]	≥4.2
Specific surface area [m^2/g-resin]	46.9
Particle size [mm]	0.40–0.56

2.2 Transesterification

The process flow of the fixed bed reactor used is shown in Figure 1.

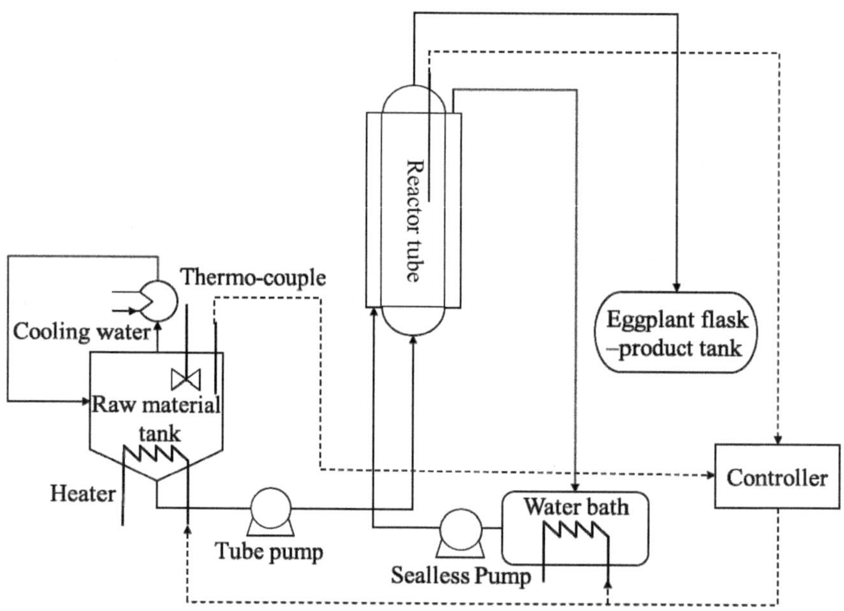

Figure 1. Process flow

The reaction tube (Length/diameter ratio (L/D) = 5, 15 or 45 keeping capacity constant at 400 ml) was filled with the CH_3O type resin. A thermocouple, a nitrogen supply nozzle, and a Dimroth condenser were installed in the feedstock tank (2 L capacity). The soybean oil, the methanol (90 wt% - soybean oil basis), and the methyl oleate (200 wt% - soybean oil basis) were put in the feedstock tank. The methyl oleate was used because it was one of the main component of FAME and in the actual operation, FAME was used as a mixuture. The soybean oil and the methanol were used as feedstocks, and the methyl oleate as a homogenizing solvent. There is a possibility that the transesterification doesn' occur when oils and fats and the methanol are not homogeneously mixed in the fixed-bed reactor (Ito, Kakuta, Sugano, & Hirano, 2011). The feedstock tank was kept warm at 50 °C with an immersed heater. Liquor in the feedstock tank was stirred by a magnetic stirrer. The liquid in the feedstock tank was sent to from the bottom of the reaction tube with a tube pump at the rate of 0.5 ml/min and flowed to an eggplant flask from the top and the flask was replaced with a fresh one every hour. The continuous operation was done for 60 h. The sample in the flask was analyzed by ^{1}H-NMR. The proton peak area from H in the methoxy group of FAME in the NMR is compared with that of H in the glycerin part of TG, and from the results, FAME yeld is calculated. (Pedro, Miguel, Lilian, & Maria, 2004).

The reaction tube was equipped with the outer tube in which the warm water in the water bath was circulated with a pump to keep the inner tube temperature at 50 °C.

2.3 Removal of Adsorbate From Resin

The column with the glass filter was filled with the spent resin after the above 60 h run. Two separate expriments for 15 and 30 h were also conducted to analyse the same as follows using the column of L/D = 5. First of all diethyl ether of 500 ml was poured into the column at the speed of SV = 10 h^{-1} to physically remove the oil adhering to the resin. Next, citric acid methanol solution of 2.3 L was poured into the column at the speed of SV = 4h^{-1}. The adsorbed material was recovered by evaporative removal of the methanol and diethyl ether from the effluent. The resin adsorbates were quantitatively analysed by column chromatography and GC-MS. The condition of column chromatography is shown in Table 2 and the condition of GC-MS is shown in Table 3.

Table 2. Condition of column chromatography

Column	5 wt% water content silica gel 65 g, dia. 24 × 160 mm	
Sample amount	1 g	
Dissolution medium		
Diethyl ether / n-Hexane (1:99, vol.)	200 ml	(FAME)
Diethyl ether / n-Hexane (13:87, vol.)	200 ml	(Triacyl glyceride)
Diethyl ether / n-Hexane (50:50, vol.)	150 ml	(Diacyl glyceride)
Diethyl ether	100 ml	(Monoacyl glyceride)

Table 3. Condition of GC-MS

Injector temperature [°C]	380	
Ion source temperature [°C]	200	
Interface temperature [°C]	345	
Pressure [kPa]	124.4	
Total flow rate [ml/min]	111.5	
Column flow rate [ml/min]	2.13	
Linear velocity [cm/s]	52.9	
Rate [°C/min]	Column temperature [°C]	Retention time [min]
-	50	1
15	180	0
7	230	0
30	380	10

3. Result and Discussion

3.1 Durability in Fixed Bed Reactor of CH_3O Type Resin

The FAME yield by the transesterification in the fixed bed reactor filled with the CH_3O type resin is shown in Figure 2. The recovered amounts of adsorbates from the spent resin are shown in Figure 3. The time variation of the amount of the glycerin absorbed to the resin is shown in Figure 4, comparing the amount of active sites of the resin (Hirano et al., 2010) in the column and the produced amount of glycerin calculated from integrated amount of produced FAME in Figure 2 assuming one mole of glycerin is produced for three moles of FAME production.

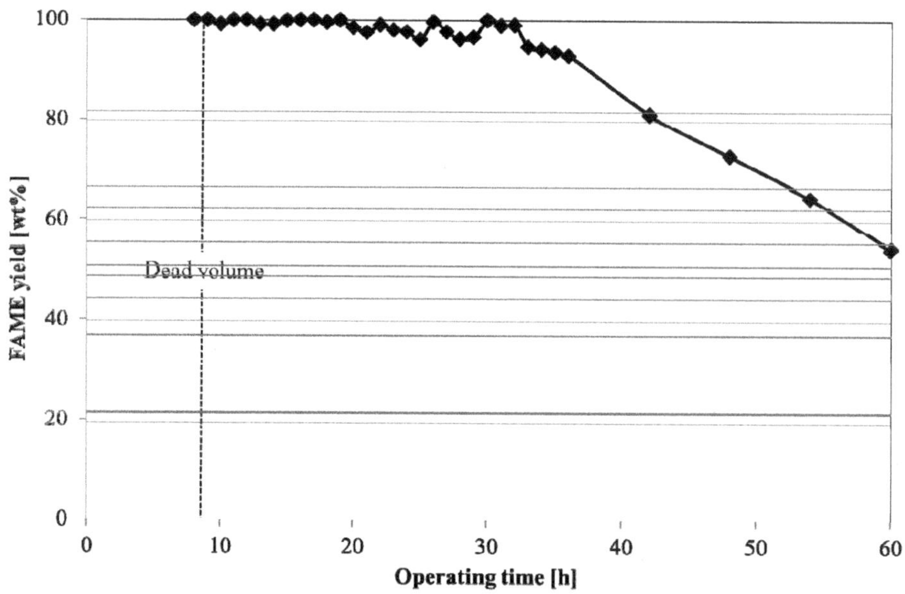

Figure 2. FAME yield of transesterification in fixed bed reactor (MeOH 90 wt%, Methyl oleate 200 wt%, L/D = 5)

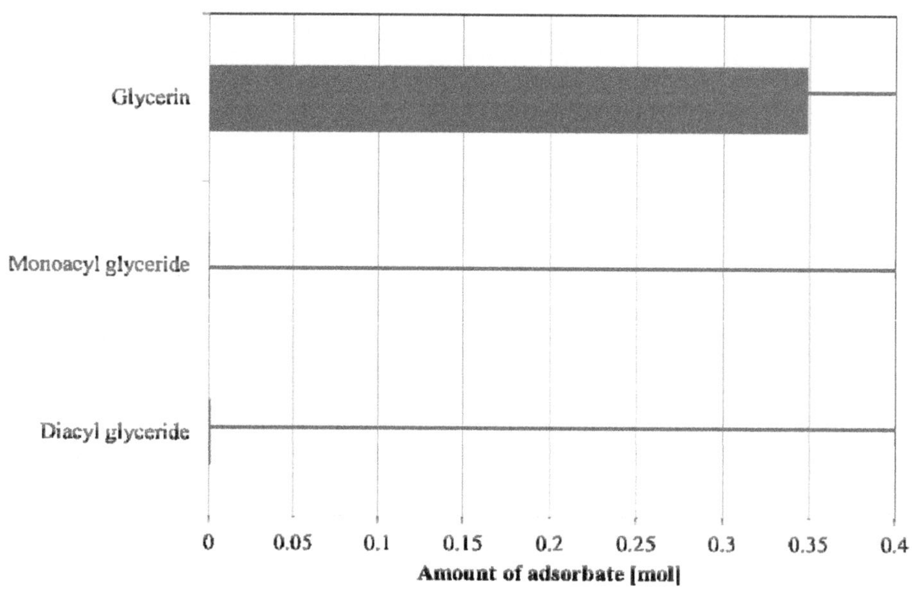

Figure 3. Amount of adsorbate of spent resin (MeOH 90 wt%, Methyl oleate 200 wt%, L/D = 5)

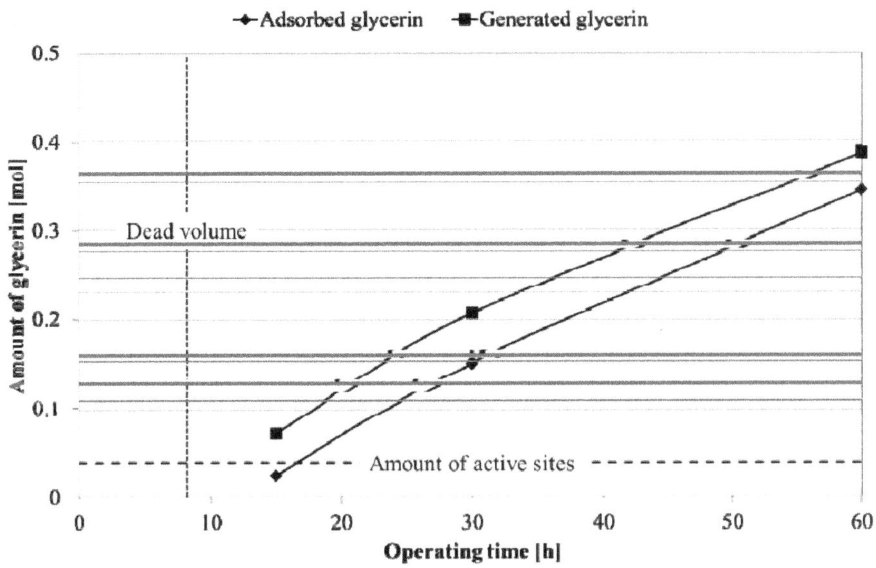

Figure 4. The variation with time of the amount of absorbed glycerin (MeOH 90 wt%, Methyl oleate 200 wt%, L/D = 5)

As shown in Figure 2, the FAME yield started to decrease around 30 h after. As shown in Figures 3 and 4, the main component of absorbed materials was found to be glycerin and most of the generated glycerin was adsorbed by the resin. Moreover, the amount of the adsorbed glycerin was much larger than the amount of active sites of the resin.

It was found that the deactivation occurred even when the CH_3O type resin that didn't generate fatty acid was used in the fixed bed reactor. It is thought that the cause of the resin deactivation is adsorption of the glycerin of the byproduct by the resin. Generally, chemisorption is known as the cause of the deactivation when the ion exchange resin is used (Oberhauser, Lavacchi, Vizza, Capozzoli, & Lee, 2013; Pyrlik et al., 2012). However, because the amount of adsorbed glycerin were much more than that of active sites of resins, it is thought that this phenomena is caused not by chemisorption on the active site of the resin but broad-sense physisorption in the pores of it.

3.2 Removal of Adsorbed Glycerin

Effect of L/D on amount of the glycerin adsorbed by resin is shown in Table 4. Effect of L/D on the FAME yield is shown in Figure 5.

Table 4. Effect of L/D on amount of the glycerin adsorbed by resin (MeOH 90 wt%, Methyl oleate 200 wt%, L/D = 5)

Length/diameter radio (L/D)	5	15	45
Amount of the glycerin [mol]	0.35	0.22	0.12

As shown in Table 4, the amount of the glycerin adsorbed by the resin decreases as L/D increases keeping total volume unchanged. The present results may be explained by the increased physical shear stress to the resin with increased linear velocity. As a result, it is thought that removing the glycerin adsorbed by the resin is promoted. As shown in Figure 5, the decreased tendency to the FAME yield was inhibited by increasing the L/D. It is thought that removing the glycerin adsorbed by the resin is promoted because of an increase in the linear velocity according to an increase in L/D and the deactivation of the resin was inhibited. It was found that in order to inhibit the deactivation of the resin, it was effective to remove the glycerin adsorbed by the resin by increasing the linear velocity.

Discussion for promotion of transesterification and its **verification by additional experiments**

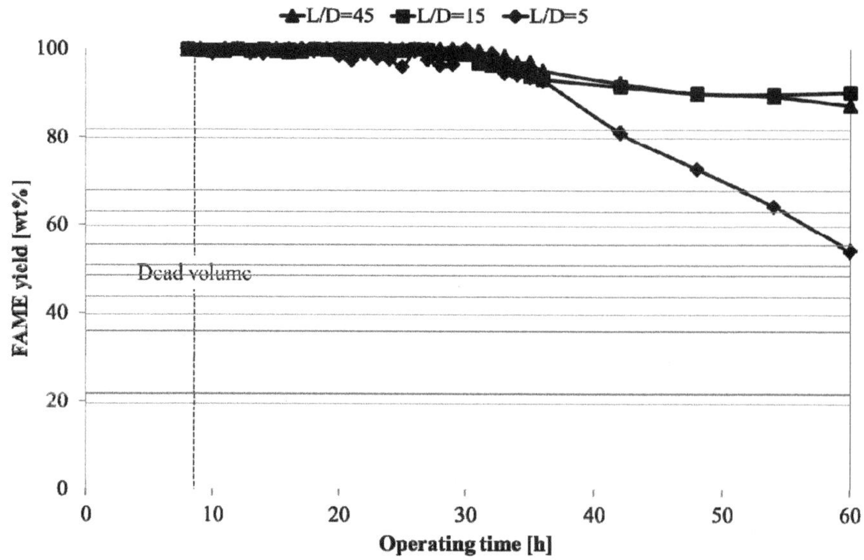

Figure 5. Effect of L/D on the FAME yield (MeOH 90 wt%, Methyl oleate 200 wt%)

3.3 Discussion on Factors for Increasing FAME Yield

Decrease of concentration of the methyl oleate used as homogenizing solvent and increase of that of the methanol are effective for equilibrium shift to the right. The problem of bypass upflow of phase-separated methanol without reaction occurs, when concentration of the methyl oleate is decreased. It was tried that the sufficient residence time of the methanol is secured by inserting the drift plates.

3.4 Modification of Reactor Configuration for Increasing FAME Yield

The same reaction tube (L/D = 5) used in Chapter 2.2 was also used, while 13 drift plates were installed in it at equal intervals as shown in Figure 7. The linear velocity in the modified reactor is equivalent to L/D = 45 and the removal effects of gryceline by the shear stress is also expected. Furthermore, the revision of the flow direction may be expected to inhibit the bypass upward flow of phase-separated methanol which may increase the retention time of methanol.

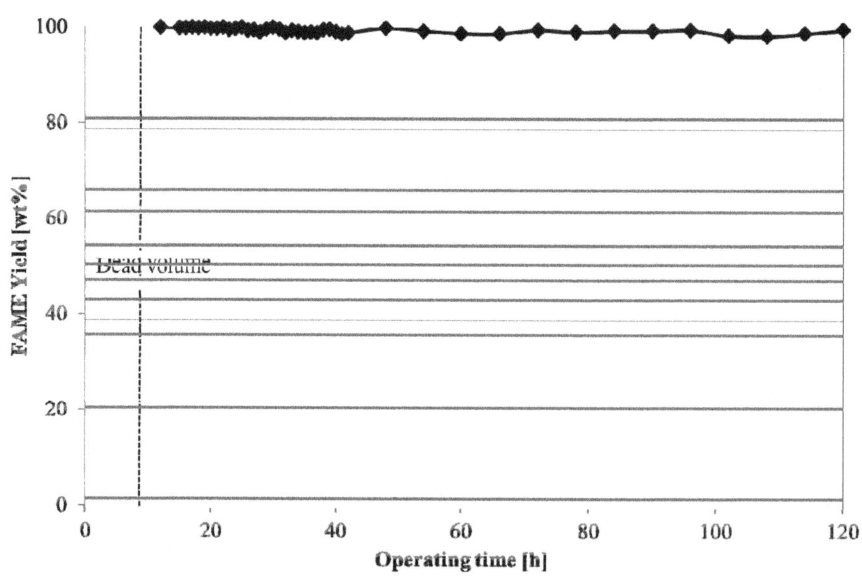

Figure 6. Effect of insertion of drift plates in reaction tube (MeOH 272.5 wt%, Methyl oleate 0 wt%, L/D = 5)

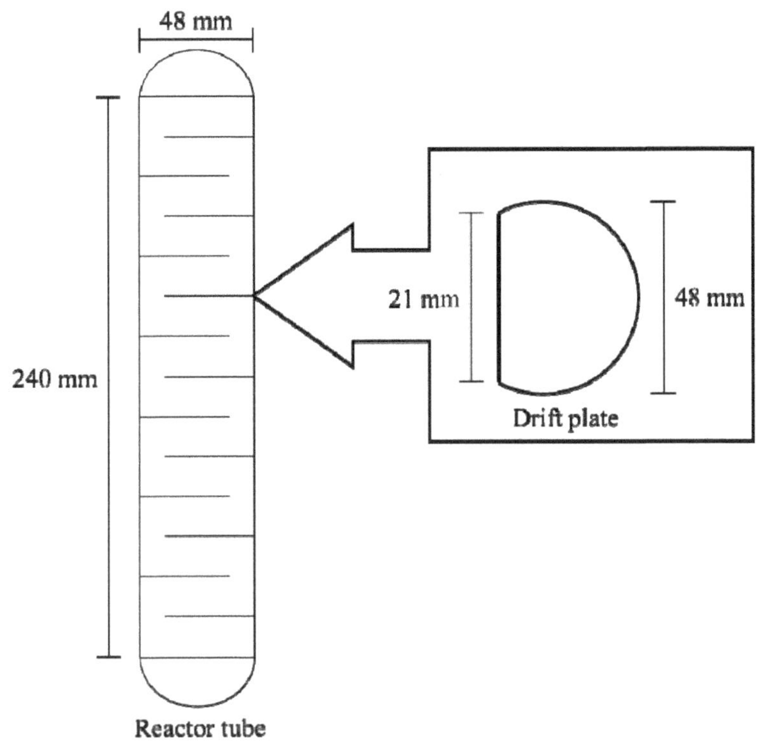

Figure 7. Configuration of drift plates

The soybean oil and the methanol (275.5 wt% - soybean oil basis) were put in the feedstock tank. The volume concentration of soybean oil in the feedstock tank of this time was equal to the condition in Chapter 2.2. The continuous operation was done for 120 h. The other experimental setups and conditions are same as those in Section 2.2.

3.5 Demonstration Results

The FAME yield when the drift plates were inserted in the reaction tube is shown in Figure 6. The FAME yield was maintained for 120 h. Glycerins adsorbing on the resin at this time were measured to be 0.023 mol. It was found to be possible that the FAME yield is maintained high enough by the methanol concentration increase, the methyl oleate concentration decreases, while inserting the drift plates in the reaction tube.

4. Conclusion

It was clarified that the contact of the resin with triglycerides of the reactant was obstructed by glycerins of the byproduct adsorbed by the resin even when the CH_3O type resin that doen't generate fatty acid was used in the fixed bed reactor. It was suggested that increasing the linear velocity in the reactor tube decreased the adsorption of glycerin, so that the deactivation of the resin could be inhibited. The durability of the CH_3O type resin has been improved by the methanol concentration increase, the FAME concentration decrease, and inserting the drift plates in the reaction tube.

Reference

Hirano, K., Ito, T., Kakuta, Y., & Sugano, M. (2010). Biodiesel Fuel Production from Waste Animal Fats Using Alternative Solid Catalysts. *Journal of the Japan Institute of Energy, 89*(1), 36-41. http://dx.doi.org/10.3775/jie.89.36

Horie, R., Sugano, M., Hirano, K., & Mashimo, K. (2006). Proceedings of the 15th Annual Conference of the Japan. *Institute of Energy, 15*, 211-212.

Ito, T., Kakuta, Y., Sugano, M., & Hirano, K. (2011). Proceedings of the 15th Annual Conference of the Japan. *Institute of Energy, 20*, 72-73.

Kouzu, M., Nakagaito, A., & Hidaka, J. S. (2011). Pre-esterification of FFA in plant oil transesterified into biodiesel with the help of solid acid catalysis of sulfonated cation-exchange resin. *Applied Catalysis A: General, 405*(1), 36-44. http://dx.doi.org/10.1016/j.apcata.2011.07.026

Matsumura, M. (2007). *Bio liquid fuel*, 27-239.

Neto, P. R. C., Caro, M. S. B., Mazzuco, L. M., & da Graça Nascimento, M. (2004). Quantification of soybean oil ethanolysis with 1H NMR. *Journal of the American Oil Chemists' Society, 81*(12), 1111-1114. http://dx.doi.org/10.1007/s11746-004-1026-0

Oberhauser, W., Lavacchi, A., Vizza, F., Capozzoli, L., & Lee, H. M. (2013). Aerobic diol lactonization by Au-nanoparticles supported onto an anion-exchange resin. *Applied Catalysis A: General, 451*, 58-64. http://dx.doi.org/10.1016/j.apcata.2012.11.012

Petroleum. (2013). *Association of Japan Petroleum Industry in Japan*, 3-8

Pyrlik, A., Hoelderich, W. F., Müller, K., Arlt, W., Strautmann, J., & Kruse, D. (2012). Dimethyl carbonate via transesterification of propylene carbonate with methanol over ion exchange resins. *Applied Catalysis B: Environmental, 125*, 486-491. http://dx.doi.org/10.1016/j.apcatb.2011.09.033

Saka, S. (2007). Recent Progress in biofuels for its production and Utilization. *Journal of Environmental Conservation Engineering, 36*(12), 842-846.

Shibasaki-Kitakawa, N., Honda, H., Kuribayashi, H., Toda, T., Fukumura, T., & Yonemoto, T. (2007). Biodiesel production using anionic ion-exchange resin as heterogeneous catalyst. *Bioresource Technology, 98*(2), 416-421. http://dx.doi.org/10.1016/j.biortech.2005.12.010

16

A Comparative Field-Scale Simulation Study on Feasibility of SAGD and ES-SAGD Processes in Naturally Fractured Bitumen Reservoirs

Yaser Souraki[1], Mohammad Ashrafi[1] & Ole Torsaeter[1]

[1] Department of Petroleum Engineering and Applied Geophysics, Norwegian University of Science and Technology – NTNU, Trondheim, Norway

Correspondence: Yaser Souraki, Department of Petroleum Engineering and Applied Geophysics, Norwegian University of Science and Technology – NTNU, S. P. Andersens vei 15A, Trondheim 7491, Norway. E-mail: yaser.souraki@ntnu.no

Abstract

Thermal processes are widely applied in heavy oil and bitumen reservoirs. Steam-Assisted Gravity Drainage (SAGD) and Expanding Solvent-SAGD (ES-SAGD) are the two most promising techniques for production of heavy oil and bitumen reservoirs. The feasibility and efficiency of the aforementioned processes have been studied extensively for sandstone reservoirs. As there are some naturally fractured heavy oil resources in the world similar thorough studies are necessary to find out the applicability of these processes in these types of reservoirs. In this work several numerical simulations have been performed to investigate the feasibility of these methods in naturally fractured reservoirs (NFR). Dual permeability option was applied in this study. This work tries to cover the effect of several parameters such as reservoir thickness, fracture permeability, matrix permeability, fracture spacing, steam quality and oil viscosity on SAGD and ES-SAGD processes. Results have elucidated that basically both methods are efficient and economically applicable in naturally fractured reservoirs. Recovery factor and cumulative steam-oil ratio (CSOR) were economical parameters to evaluate the performance of processes. However, in some cases net amount of injected energy and oil production rate were considered for evaluation too. Generally ES-SAGD had better performance based on the recovery factor, CSOR, oil production rate and net amount of injected energy. Effect of solvent type and concentration were studied for ES-SAGD process only. For this purpose, pentane, hexane and heptane were selected. Pentane displayed better performance in earlier periods of production; however the heavier solvents showed better recovery performance at later times.

Keywords: naturally fractured reservoirs, SAGD, ES-SAGD, heavy oil and bitumen

Nomenclature

Symbols		*scripts*	
CSOR	Cumulative Steam-Oil Ratio	g	Gas
ES-SAGD	Expanding Solvent-SAGD	o	Oil
GHG	Green-House-Gas	w	Water
k_r	*R*elative Permeability		
NFR	Naturally Fractured Reservoir		
S_w	*W*ater Saturation		
SAGD	Steam-Assisted Gravity Drainage		
SAS	Steam-Alternating-Solvent		
STB	Stock Tank Barrel		
VAPEX	Vapor Extraction		
THAI	Toe-to-Heal Air Injection		
CSS	Cyclic Steam Stimulation		

1. Introduction

Fractured reservoirs have complex structures, and are usually modeled based on dual porosity or dual permeability assumption. Production from fractured heavy oil reservoirs was not interesting for companies until a decade or so ago. Generally, researchers believed that high permeability of fractures provides an easy path for steam to bypass the oil without recovering significant amount of oil. However, results of several experimental works, simulation studies and some pilot tests in recent years indicated that thermal processes such as SAGD, Vapor Extraction (VAPEX) and hot-solvent injection are operationally feasible and economically efficient.

The fractures in reservoirs may have either negative or positive effects on flow of fluids inside the porous media (Aguilera, 1995). Naturally fractured reservoirs contain a remarkable portion (about one-third) of world-wide heavy oil and bitumen resources (Mollaie, 2010). Dusseault and Shafiei (2010) estimated around 2 trillion barrels of heavy oil to be deposited in carbonate fractured reservoirs. Canada, former Soviet Union and the Middle East have several fractured viscous oil reservoirs. The Alberta State of Canada has one of the largest carbonate bitumen resources which is contained in Grosmont formation and located in north-central Alberta. Among the several operated pilots on this formation, Cyclic Steam Stimulation (CSS) was the most prosperous method based on the recovery and CSOR (Edmunds et al., 2009). The estimated amount of Iranian heavy oil deposited in fractured reservoirs is about 86 billion STB (Kharrat et al., 2003).

To study the effect of fracture orientation towards well configurations on the SAGD process some experimental works have been done by Joshi (1986). The obtained data indicated that existence of vertical fractures perpendicular to the horizontal injector plane enable a fluent and clear path for steam to go upward resulting in high production of oil at earlier period of production. Pooladi-Darvish et al. (1994, 1995) analytically presented that in NFRs, heat conduction under gravity drainage is the primary heating mechanism. Azin et al. (2005) investigated feasibility of Vapor Extraction (VAPEX) method to Kuh-e-Mond field which is an Iranian heavy oil resource. They modeled the VAPEX process mathematically and to approve it they tried to history match the Butler and Jiang (2000) experimental data. Finally, they found good match and consistency between simulation results and experimental data. This model was applied to a fractured reservoir to study the effect of reservoir permeability, oil viscosity and solvent injection rate. They concluded that initial viscosity of oil affected the VAPEX process in fractured reservoirs. Also oil production rate increased with increase in matrix permeability.

Rostami et al. (2005) performed simulation study to evaluate the applicability of VAPEX process in high-pressure fractured heavy oil reservoirs. Results showed that optimum solvent injection rate should be found from economical and technical points of view. They investigated the effect of diffusion coefficient in fractured and non-fractured reservoirs and deduced that the amount of diffused solvent into the heavy oil was greater in a fractured system compared to a non-fractured system. Bagci (2006) has also performed several experiments to investigate the effect of fracture orientations towards well configurations on SAGD process. He found that performance of SAGD process was affected by location of fractures. Horizontal well pairs displayed the highest oil production in presence of fractures. He also deduced that SAGD performance was improved by vertical fractures.

3-D field scale SAGD simulation studies have been applied to one of Iran's viscous oil carbonate reservoirs by Sedaee Sola (2006). Results of simulation elucidated that SAGD is an efficient method for producing oil from this reservoir. Simulation analysis on the effect of vertical well spacing showed that the optimum vertical distance between wells should be 200 ft in that case. In addition, effect of some parameters like fracture permeability, matrix permeability and length of the horizontal wells were studied. Applicability of two most promising thermal techniques namely CSS and SAGD were studied by Das (2007). He performed several simulation analysis Using STARS module of Computer Modeling Group (CMG) software to investigate the effect of different parameters such as oil viscosity, fracture spacing, capillary pressure, wettability and matrix permeability on these two processes. Results demonstrated that better recovery of oil was obtained for SAGD in the case of higher oil viscosity whereas CSS showed better performance in reservoirs with lower viscous oil. Also he mentioned that oil expulsion from matrix and spontaneous water imbibitions are the two key mechanisms for the success of thermal processes.

Mollaei (2007) briefly presented a review of performed researches on the effect of important parameters on application of steam flooding in naturally fractured reservoirs (NFR). The effect of temperature on physical properties of rock and fluids such as viscosity, density, porosity and permeability was investigated. Moreover, effective parameters on two important oil recovery mechanisms in NFRs, namely oil expulsion from matrix to the fracture and oil displacement through the fracture networks, were introduced. Bagci (2007) performed comparative simulation study on SAGD and Steam-Alternating-Solvent (SAS) processes in fractured reservoirs under different injection schedules, bottom-hole pressures and fracture orientations. He concluded that SAS process showed better performance than SAGD, however, the proper rate of solvent injection is required. Maintaining the reservoir temperature is important in the SAS process while solvent is injected inside the reservoir. Fracture orientation had no effect on oil production for the various injection schedules.

In order to investigate the effect of fracture orientations on SAGD process, Chen et al. (2007) carried out several simulations. They found that average oil production rate in a model with vertical fractures was more than twice that in a model with horizontal fractures. Edmunds et al. (2009) comprehensively reviewed geological and engineering aspects of historical and current data of bitumen recovery from the Grosmont formation in Alberta. They found that presence of fractures and vugs provided a noteworthy combination of porosity, thickness and oil saturation for steam recovery. A new method called Steam-Over-Solvent has been applied to the fractured reservoirs by Al Bahlani and Babadagli (2009). They carried out series of laboratory experiments and then developed dimensionless parameters to upscale the lab data to the field data. They observed three phases namely fracture oil depletion phase, diffusion dominant and dispersion dominant phases during solvent injection period. The experimental data and simulation results indicated that this technique was successful and promising in the lab and field conditions.

Rezaei et al. (2010) performed some experiments to investigate bitumen recovery from vuggy porous media using VAPEX process. They found the same results as the previous works. Souraki et al. (2011) performed simulation study using Athabasca bitumen PVT experimental data on a core sample surrounded by fractures. They concluded that higher CSOR and faster oil recovery can be obtained for higher injection rates; however, the optimum steam injection rate should be obtained to have economic CSOR. They evaluated the effect of fracture permeability for different steam injection rates and results revealed that higher fracture permeability may be effective in lower rates of injection. The feasibility of Toe-to-Heel Air Injection (THAI) technique in Iranian fractured heavy oil reservoirs was investigated by Fatemi et al. (2011a). They performed experimental work and 3-D simulation study using Kuh-e-Mond oil characteristics. They investigated effect of several parameters such as fracture density, initial oil saturation, horizontal fracture, matrix thickness, matrix permeability and fracture permeability. Higher matrix permeability had positive effect on the process whereas higher fracture permeability reduced the efficiency of the process. Higher oil recovery was achieved for the reservoir with higher fracture density. Moreover, horizontal fractures reduced the ultimate oil recovery.

Fatemi et al. (2011b) investigated the geometrical properties of fractured systems such as discontinuity, length and orientation of fracture by simulating the SAGD process. Simulation results indicated positive and negative effects on oil production for vertical and horizontal fractures respectively. Longer vertical fractures increased oil recovery whereas longer horizontal ones decreased it. In terms of oil recovery, discontinuity displayed damping behavior on performance of both vertical and horizontal fractures. An experimental work on recovery of bitumen from fractured carbonates using hot-solvents was carried out by Pathak and Babadagli (2012). They prepared three original cores with original fluid saturations and tested them in three experiments using different solvents and different pressures and temperatures. They found that the dilution attained by butane was substantially greater than the dilution attained by propane. However, the experimental results revealed that using hot-solvent is a prosperous technique for producing bitumen from complex carbonate rocks.

Based on the aforementioned researches, mostly VAPEX and hot-solvent processes investigated by simulation study or experimental work in fractured reservoirs. Although some simulations and laboratory experiments have been performed on SAGD process but ES-SAGD process which is one of the most promising processes of heavy oil and bitumen recovery has not been conducted in NFRs so much. This process can improve the SAGD performance because it is benefiting from synergy of steam and solvent to reduce the high viscosity of bitumen crudes. In fact, vapor of solvent co-injected into the reservoir with steam. Steam and solvent in vapor phase contact cold bitumen and condense. Condensed solvent diffuses inside the oil and makes it lighter in the boundary of steam chamber. Afterwards, condensations of solvent and steam together with diluted oil are extracted from horizontal producer. Results of simulation studies on sandstone reservoirs revealed that ES-SAGD yield lower cumulative energy consumption and green-house-gas emissions (GHG) than SAGD process. Further, oil production rate, ultimate oil recovery and cumulative steam-oil ratio (CSOR) are significantly improved (Nasr & Ayodele, 2006). The main objective of this work was to investigate the feasibility of SAGD and ES-SAGD processes in naturally fractured heavy oil reservoirs. Sensitivity analysis has been done on the effect of geological and operational parameters such as fracture spacing, matrix permeability, fracture permeability, reservoir thickness, initial oil viscosity and steam quality. The results of this work have been described in details and demonstrated in related graphs.

2. Simulation Studies

In order to study the applicability of thermal processes in heavy oil naturally fractured reservoirs, several field-scale simulation models were built using STARS module of Computer Modeling Group software (CMG). Two thermal processes namely Steam-Assisted Gravity Drainage (SAGD) and Expanding-Solvent SAGD (ES-SAGD) were studied here. These two methods are the most promising methods in heavy oil and bitumen reservoirs. Athabasca bitumen was used as a fluid in these simulation studies. Some experimental properties of Athabasca bitumen measured in previous works by the authors were applied.

2.1 Model Description for SAGD Process

In this work a two dimensional model consisting of 51 x 1 x 17 grid blocks was built to represent a naturally fractured reservoir. The reservoir has dimensions of 306 ft (93 m) in length, 328 ft (100 m) in width and a thickness of 51 ft (15.5 m). Since SAGD process is symmetric, half of the reservoir was simulated. Dual permeability option was applied in all simulation runs, this indicates that in addition to the matrix to fracture fluid flow and inter-block fracture to fracture flows (convection and conduction) inter-block matrix to matrix flows were also accounted for and calculated. Identical fracture spacing of 1 foot in all directions was considered for the base model. The fracture network has permeability of 5 Darcies in each part of the reservoir whereas matrix system has permeability ratio of 0.04 to the fracture network (0.2 D). Matrix and fracture porosities are 0.3 and 0.02 respectively. Two horizontal wells were positioned with 15 ft (4.57 m) vertical distance at the bottom of the reservoir. The upper well is considered as injector whereas the lower well is acting as producer and located 1.5 ft above the base of model. Both wells were drilled horizontally through the whole J-direction of the model. Figure 1 shows the schematic view of simulation model. Additional information about rock-fluid, fluid properties and heat loss parameters are presented in Table 1. Figures 2 and 3 are presenting the matrix relative permeability curves and viscosity of bitumen which was obtained experimentally in previous works by Ashrafi et al. (2011). The internal liquid viscosity correlation for selected normal alkanes presented in STARS manual was used for the viscosities of the solvents used in ES-SAGD process simulations. This correlation is as follows (CMG-STARS user's guide, 2012):

$$VISC = AVISC \times exp(BVISC/T)$$

Where values of the coefficients *AVISC* and *BVISC* are given for selected components in STARS manual, *VISC* is viscosity and *T* is absolute temperature. Relative permeabilities of fracture network were considered as straight lines. This connotes that the fracture network is almost equivalent to a bundle of tubes where the residual oil and irreducible water saturations are equal to zero (Aguilera, 1995).

Table 1. Reservoir simulation input parameters

Item	Value
Initial oil saturation in matrix system	0.80
Initial oil saturation in fracture network	0.05
Rock thermal conductivity, Btu/(ft*day*°F)	106
Rock heat capacity, Btu/(ft^3*°F) (CMG, 2012)	35
Density of bitumen, lb/ft^3	63.26
Molecular weight of bitumen, lb/lbmole	534
Rock compressibility, 1/psi	1.4E-5
Three phase relative permeability model (CMG, 2012)	Stone's model 2
Initial reservoir temperature, °F	50
Initial reservoir pressure at the top of the reservoir, psi	400
Over/under burden heat capacity, Btu/(ft^3*°F) (CMG, 2012)	35
Over/under burden thermal conductivity, Btu/(ft*day*°F)	24
Reservoir has no gas cap and aquifer	

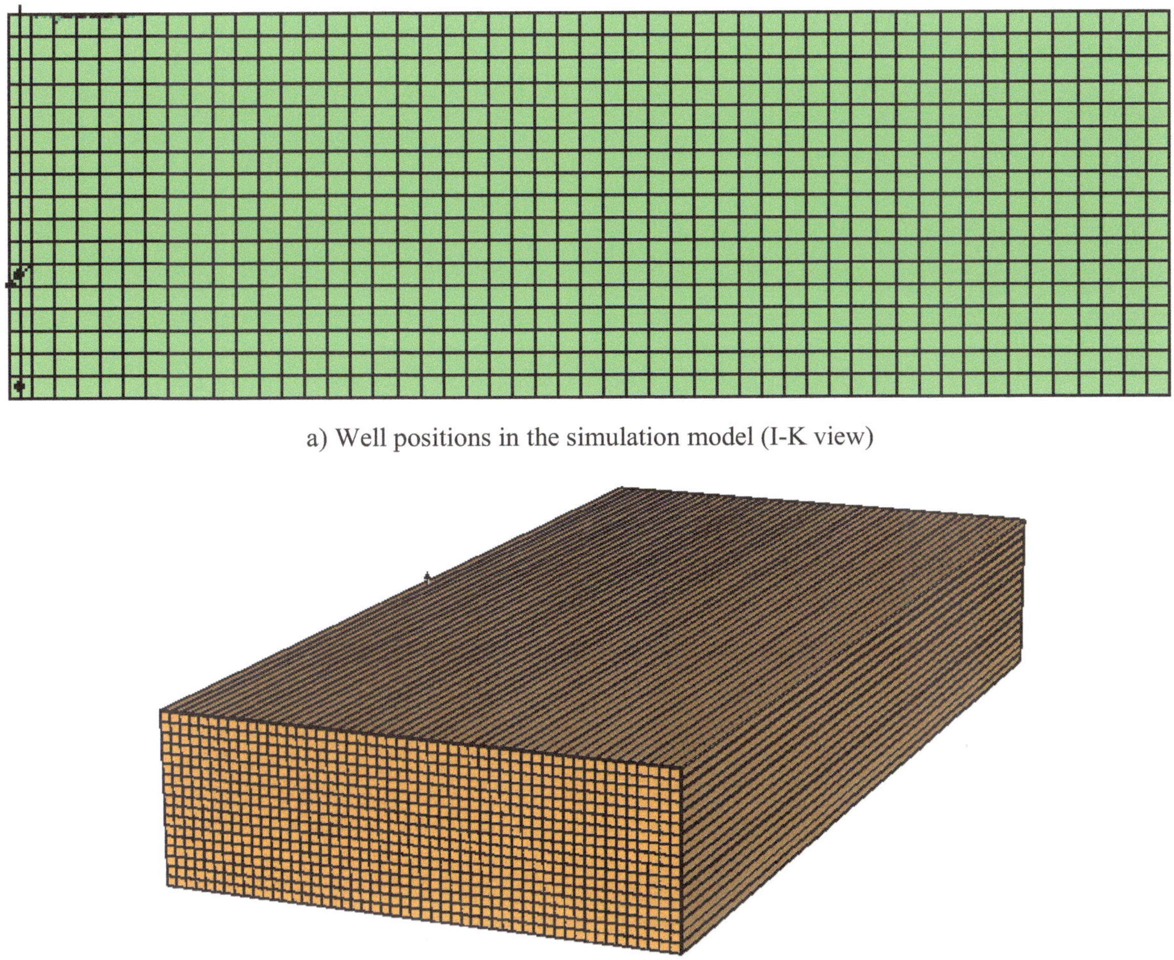

a) Well positions in the simulation model (I-K view)

b) 3-D schematic of the simulation model

Figure 1. Reservoir simulation model

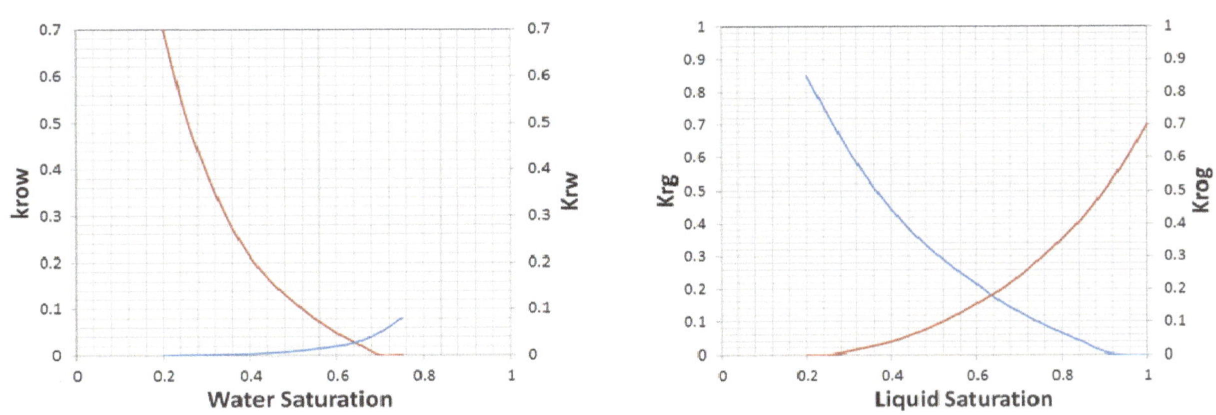

Figure 2. Sketch of the matrix relative permeability data (Chen, 2006)

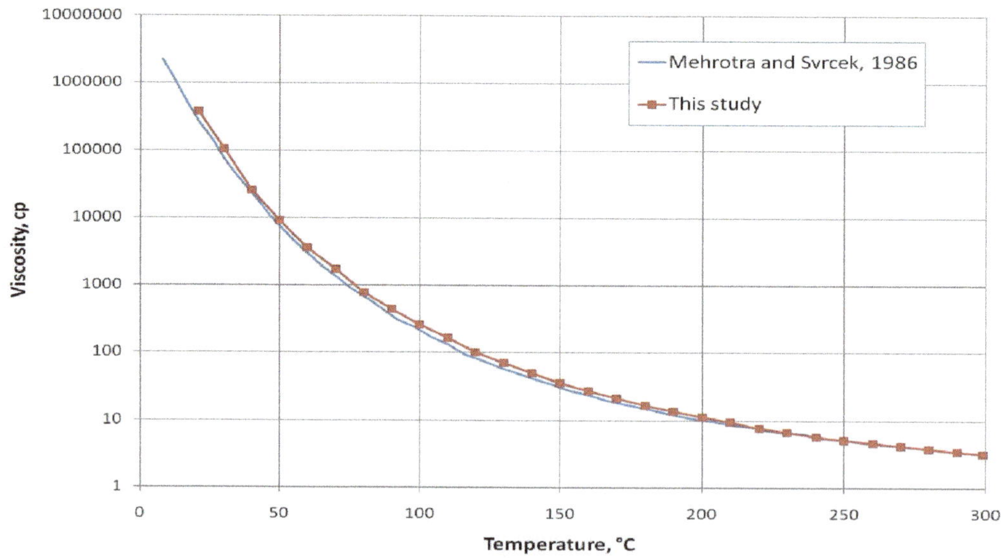

Figure 3. Viscosity of Athabasca bitumen versus temperature (Ashrafi et al., 2011)

Due to the high viscosity of bitumen at the reservoir conditions it is not possible to produce it by pressure depletion which is a primary production method in conventional oil reservoirs. Hence, the reservoir should be heated up to the extent that bitumen becomes mobile. In the SAGD process, in order to establish a good communication between injector and producer, steam is circulated through both horizontal wells or electrical heaters are used to transfer heat to the layers between wells before starting the operation. In this work, electrical heaters were applied to prepare the reservoir for operation. The set temperature of heaters was 500 °F and it was done for 80 days when the average temperature of layers between the two wells reached to 302 °F (150 °C). Afterwards, the electrical heaters were turned off. Minimum bottom-hole pressure of 100 psi was applied to the production well as a primary constraint. In addition, steam trap control option was used and set to 20 °F to hinder the steam production. Steam with quality of 70%, temperature of 490 °F and maximum injection rate of 400 STB/day was injected. The maximum bottom-hole pressure of 600 psi was imposed to injection well as a secondary constraint. Both wells have no skin factor.

2.2 Model Description for ES-SAGD Process

The same parameters were applied in this process except the steam injection constraint. Here 1% molar based of hexane was co-injected with steam. Co-injection was performed under the temperature of 490 °F and steam quality of 70%. The maximum fluid injection rate of 400 STB/day and bottom-hole pressure of 600 psi were imposed to the injection well as constraints. The ES-SAGD process benefits from advantages of steam and solvent together in bitumen's viscosity reduction. Steam conducts heat to the reservoir contains and therefore reduces the oil viscosity but solubility of solvent into the bitumen around the edges of steam chamber results in viscosity reduction of oil. Solubility of the solvents in bitumen is reflected in K-values of the solvents. K-values are the ratios of the mole fraction of each solvent in vapor phase to that in liquid phase at thermodynamic equilibrium condition. K-values are therefore temperature, pressure and concentration dependant. In this study the K-values of the solvents versus temperature and pressure were estimated using the following equation provided in STARS manual (CMG-STARS user's guide, 2012):

$$K = (KV1/P) \times exp(KV4/(T - KV5))$$

Where coefficients *KV1*, *KV4* and *KV5* are component dependant and presented in STARS manual for selected components. *P* and *T* are pressure and absolute temperature respectively.

Table 2. K-value coefficients of selected solvents

K-value coefficients	n-pentane	n-hexane	n-heptane
KV1 (psi)	1.4546e+5	1.4594e+5	1.5145e+5
KV4 (°F)	-4458.73	-4855.59	-5240.38
KV5 (°F)	-387.78	-371.87	-357.95

3. Results and Discussion

3.1 SAGD vs. ES-SAGD

Simulation results demonstrated that both processes in this work are efficient and practically applicable. However, ES-SAGD had a better performance from economical point of view. The most important economical parameters are recovery factor and cumulative steam-oil ratio (CSOR). Figure 4a illustrates that ultimate recovery factor in SAGD case is about 64 percent after 4400 days (12 years) which is economically acceptable whereas ES-SAGD shows a very promising recovery factor of 90 percent just after 2000 days (5.5 years). It should be noted that this is a net recovery factor, i.e., produced solvent has been deducted from total oil production. Considering CSOR, figure 4b remarks that ES-SAGD can be applied with a reasonable CSOR of 4 bbl/bbl. However, SAGD performance was inside the economical border of CSOR in the field conditions. Average oil production rate of 50 and 100 bbl/day were obtained up to 2000 days of production for SAGD and ES-SAGD processes respectively (Figure 4c). In order to generate steam, energy is required. Energy will be lost inside the reservoir via over/under burden. Figure 4d illustrates that energy lost to over/under burden is lower in ES-SAGD process than in the SAGD process.

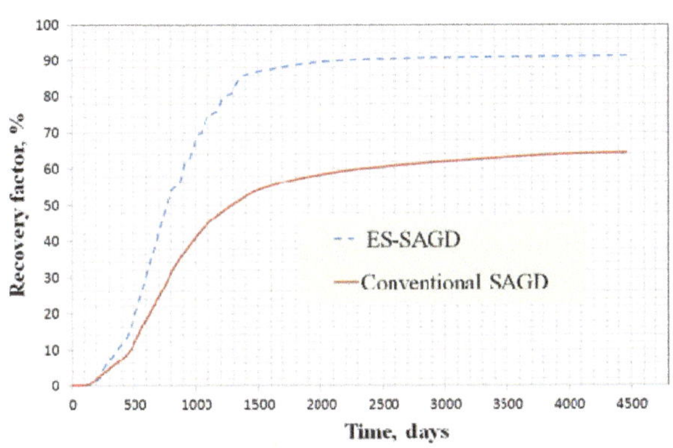

a) Recovery factor vs. time

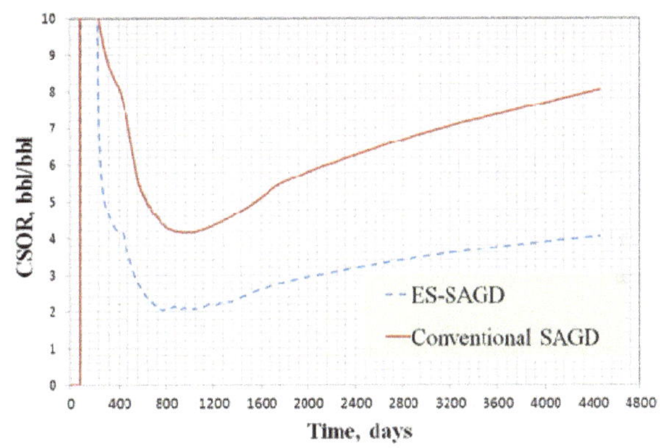

b) CSOR vs. time

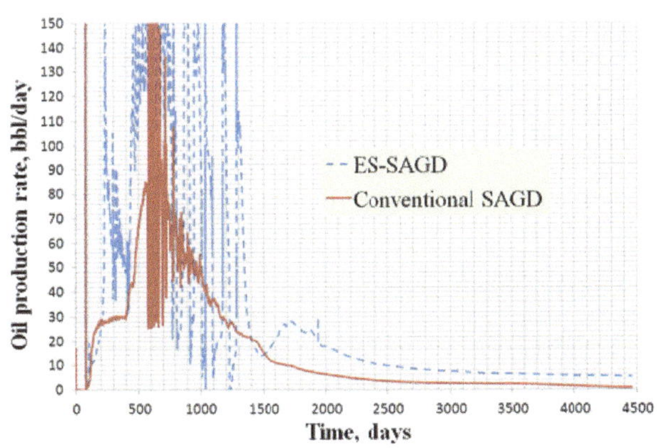

c) Oil production rate vs. time

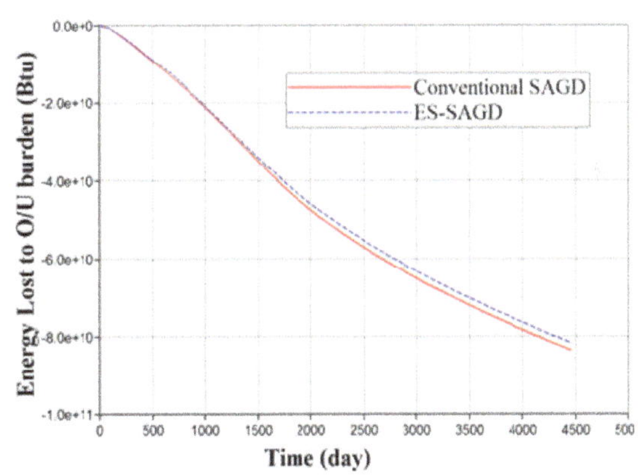

d) Energy lost to over/under burden vs. time

Figure 4. Illustration of SAGD and ES-SAGD performance comparison

Solvents are expensive; hence, injecting solvent into the reservoir should be evaluated economically. However, it is possible to recover most of the injected solvents as heavy oil and bitumen are upgraded at the surface. In SAGD process produced oil will be upgraded by adding solvent at surface and then transferred to the refinery. In fact, ES-SAGD is a method to upgrade heavy oil inside the reservoir. Figure 5 shows the injected and produced amount of hexane as solvent in this study. During production time 40914 bbls of hexane injected and 39029 bbls produced. In fact, only 1885 bbls of hexane remained inside the reservoir which is less than 5% of injected solvent volume.

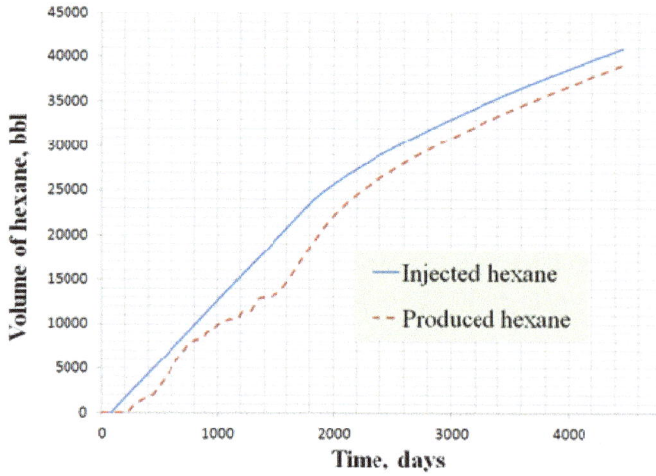

Figure 5. Illustration of injected and produced volume of hexane vs. time

3.2 Effect of Fracture Spacing

Impact of fracture spacing on SAGD and ES-SAGD processes was studied. The base case had fracture spacing of 1 foot and it was changed to 2 ft for deliberating the results. The results show that fracture spacing has no significant effect on recovery factor and CSOR in SAGD process. Larger fracture spacing resulted in higher oil recovery factor in ES-SAGD process and lower amount of CSOR in earlier period of production (Figure 6).

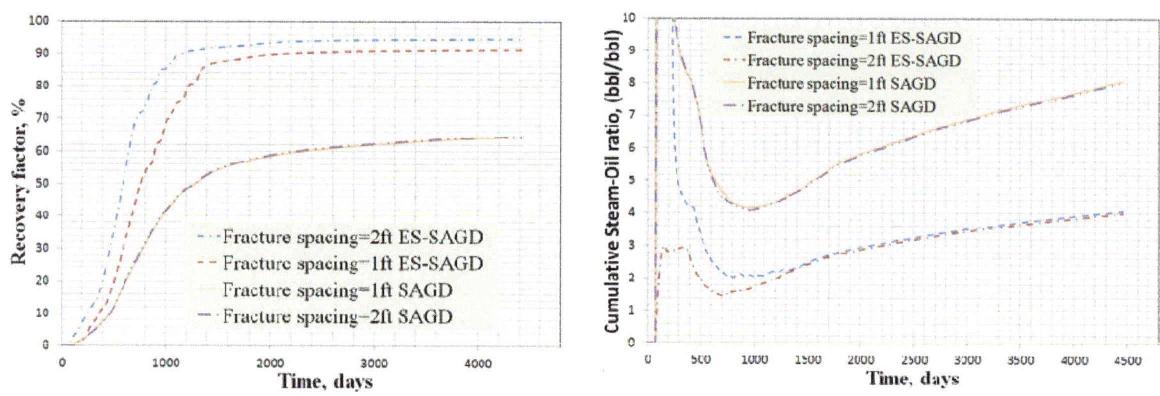

Figure 6. Illustration of the effect of fracture spacing

3.3 Reservoir Thickness

Conventional reservoirs with higher thickness are more suitable for conventional SAGD process as lower amount of steam will be lost due to the over/under burden. Hence, in this part of the study we tried to investigate the effect of reservoir thickness on SAGD performance in naturally fractured reservoirs. Reservoir thickness was increased to 81 ft (24.7 m), moreover production well was located 1.5 ft above the base of the reservoir and injector was placed at a vertical distance of 15 ft from producer. The other parameters and constraints were kept the same as before. Since the rate of injection was equal for all cases, the ultimate recovery factor was identical. However, reservoir with larger thickness has weaker performance at earlier period of production. It means that there is possibility to increase rate of injection as it can be seen based on the CSOR curves. Therefore, higher thickness can improve the performance of thermal processes such as conventional SAGD and ES-SAGD even in naturally fractured reservoirs. Figure 7 illustrates the reservoir thickness effects.

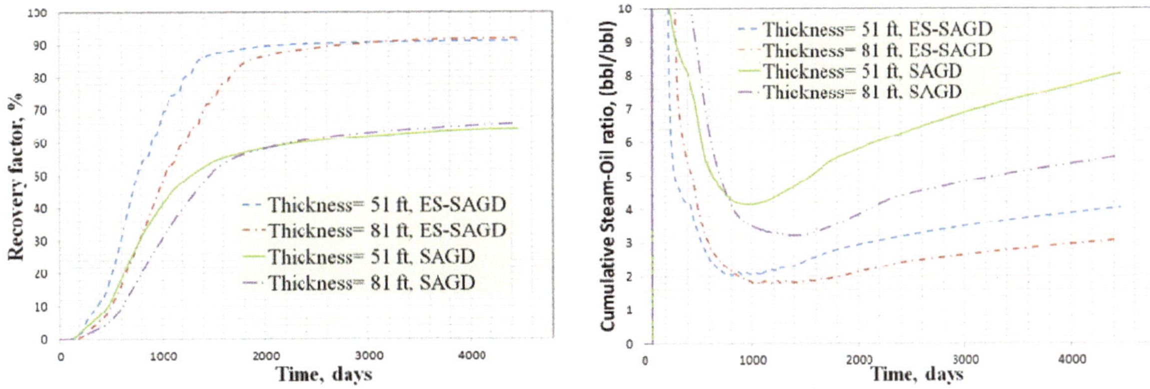

Figure 7. Illustration of the effect of reservoir thickness

3.4 Oil Viscosity

Athabasca bitumen properties were used for base cases simulations of SAGD and ES-SAGD processes. To investigate impacts of in-situ oil viscosity, Lloydminster oil properties were applied. Viscosities of these two types of heavy oils versus temperature are depicted in Figure 8.

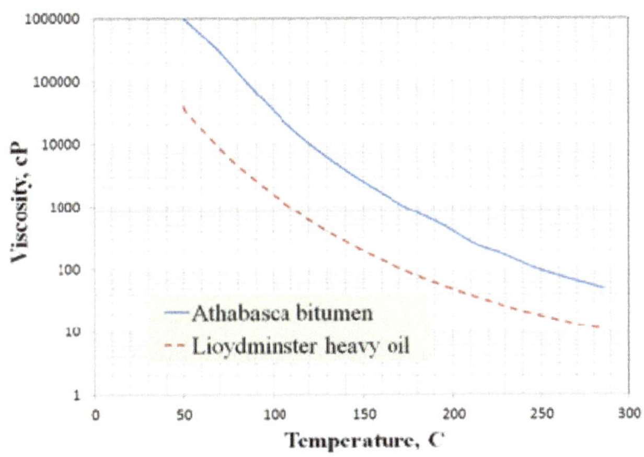

Figure 8. Illustration of different heavy oil viscosities vs. time

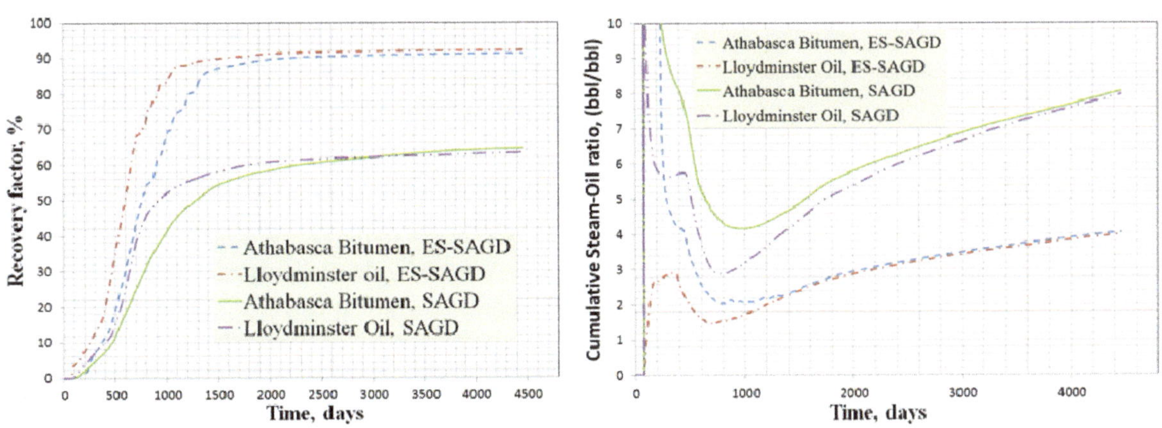

Figure 9. Illustration of the oil viscosity effect

The ultimate recovery factor of both processes for the aforementioned types of oils are almost the same, faster oil production was observed in the case of lighter oil. Hence, lighter oil will be mobile in lower temperatures and can be produced earlier than the heavier oil. This results in lower CSOR for earlier periods of production. When the reservoir matures both kinds of oil can be produced, thus the ultimate recovery is almost the same. Figure 9 displays the effect of oil viscosity.

3.5 Fracture Permeability

Evaluation of fracture permeability was performed by changing it from 5 Darcies to 10 Darcies. Figure 10 demonstrates that faster and higher recovery has been obtained for higher fracture permeability model even in SAGD process. Fracture transfers drained oil from matrices, so fracture properties govern flow of fluids. Hence, higher permeability of fracture means fluids can be transferred faster and easier. Due to the faster and higher recovery, CSOR will be lower which is more favorable for economy of the operation.

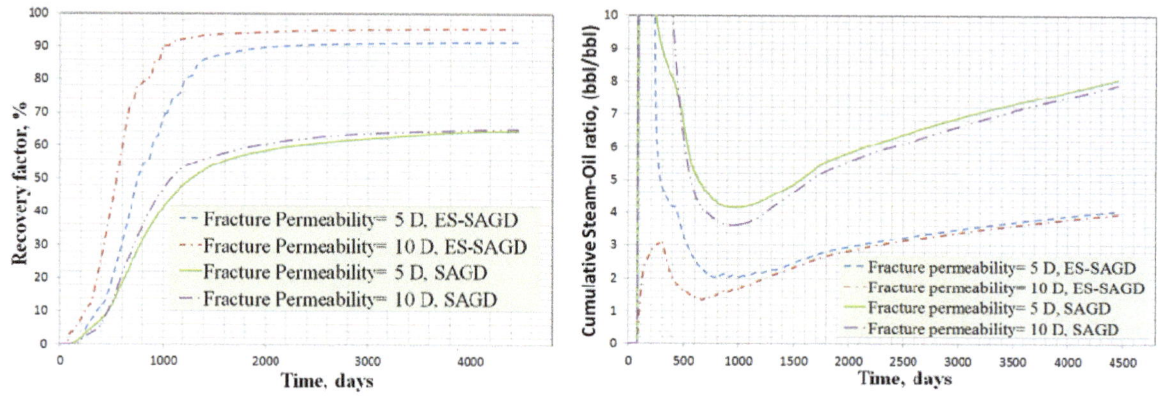

Figure 10. Illustration of the effect of fracture permeability

3.6 Matrix Permeability

Naturally fractured reservoirs are usually modeled using two different porosity and permeability properties due to matrix system and fracture network. Fracture permeability impacts were studied in the previous section. In this part, matrix permeability effects are evaluated. For this purpose permeability of matrix system was increased from 0.2 to 0.5 Darcies. Figure 11 shows that higher matrix permeability has significant effect on recovery factor and CSOR. In fact, higher matrix permeability resulted in faster and higher drainage of oil from matrices to the fractures and therefore improvement in oil recovery and CSOR and eventually the economy of the process.

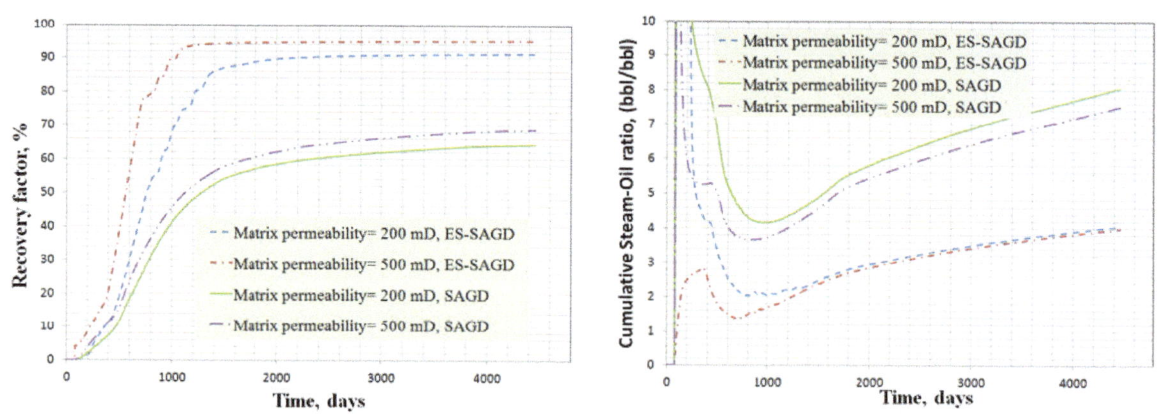

Figure 11. Illustration of matrix permeability impact

3.7 Steam Quality

In the field conditions it is impossible to have dry steam inside the reservoir due to the heat losses from surface to the reservoir inside the casing and liner. Two different steam qualities of 0.7 and 0.85 were studied in this work. Simulation results illustrated that faster recovery of oil can be obtained if steam has higher quality inside the reservoir (Figure 12). Obviously, steam has more heat capacity than water due to its latent heat. If steam has higher quality inside the reservoir, more heat content and energy can be transferred to the reservoir. It seems that steam quality effects are more pronounced in case of SAGD process where there is only steam injected to reduce the viscosity of heavy oil.

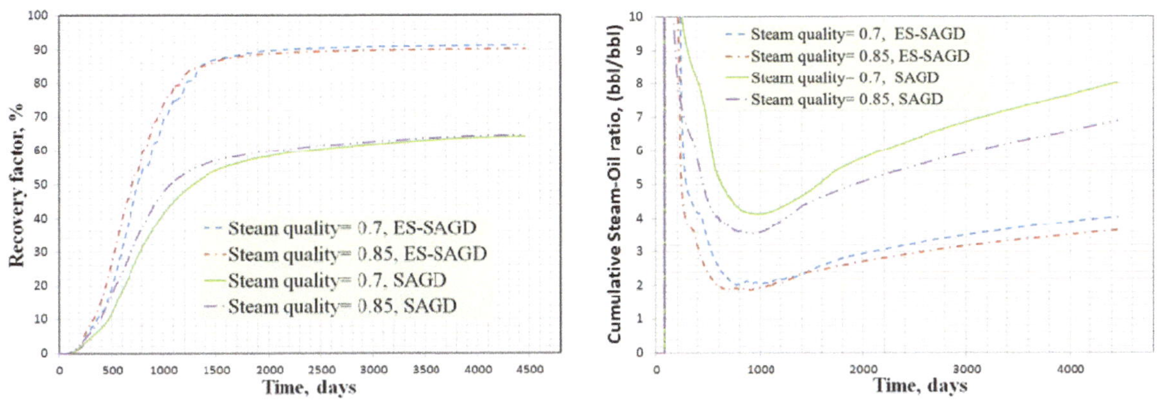

Figure 12. Illustration of steam quality effect

3.8 Solvent Type Effect

In order to find out the effective parameters on ES-SAGD process, solvent type effects were studied in this section. For this purpose, three different solvent types namely pentane, hexane and heptane were selected. They were injected with same mole percent of 1 in to the reservoir. Figure 13 illustrates the results of simulation studies for different types of solvents.

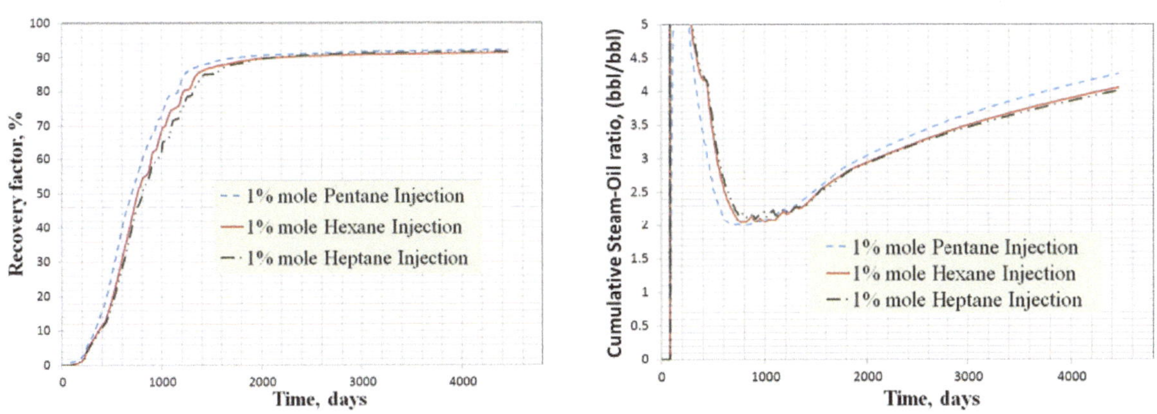

Figure 13. Illustration of the effect of solvent type

Figure 14 compares the net injected energy for each solvent until 1800 days of production. Pentane had a better performance than the others, because more energy was injected in this case. Obviously lighter solvent has lower volume based on the same mole fraction. Hence, more steam will be injected in lighter solvent case, and because steam has the main role in the mitigation of oil viscosity, higher recovery factor can be obtained. Considering CSOR it can be seen that more steam was injected in lighter solvent case which was pentane in this work. However, at later period of production heavier solvents had better performance based on CSOR.

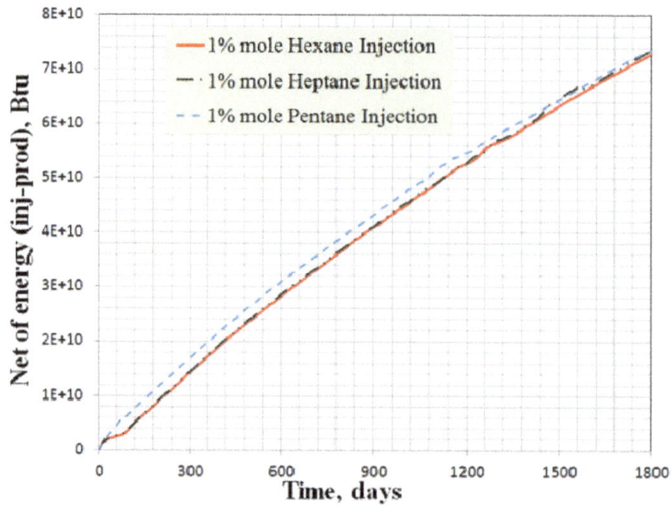

Figure 14. Comparison of net injected energy for different solvent types

3.9 Effect of Solvent Concentration

In the ES-SAGD process solvent concentration was increased to 2 mole percent in order to investigate solvent concentration effects. Results displayed significant impacts on recovery factor and CSOR. With only 1 mole percent increase in solvent concentration recovery factor increased from 91 % to 95% and CSOR improved from 4 to 3.1 bbl/bbl (Figure 15). Clearly higher solvent injection can help the process to reduce the viscosity of heavy oil and bitumen but it should be noted that steam has main role in ES-SAGD process as it can reduce viscosity faster than solvent. Solvent is expensive, so it is very important to find the optimum solvent concentration for each ES-SAGD operation based on the reservoir conditions.

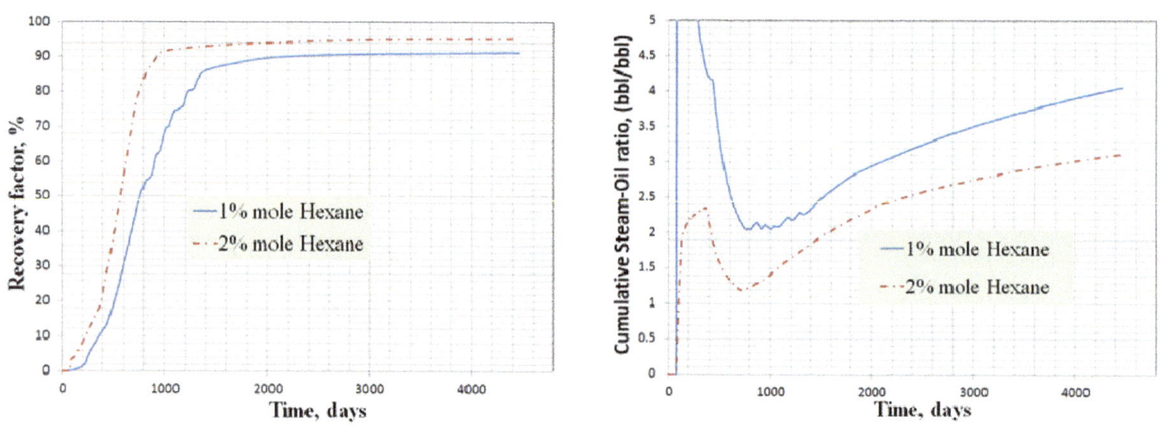

Figure 15. Illustration of solvent concentration effects

4. Conclusion

The results of this study showed that SAGD and ES-SAGD are feasible techniques to recover bitumen and heavy oil from naturally fractured reservoirs. ES-SAGD displayed extremely prosperous performance resulting in lower amount of CSOR and higher recovery than SAGD. Wider fracture spacing had no effect on SAGD process while faster recovery was obtained in the ES-SAGD process. Thicker reservoirs provide better conditions for thermal processes such as SAGD and ES-SAGD. Initial oil viscosity and steam quality can affect the mentioned processes. In fact, lower initial oil viscosity and higher steam quality can help the processes to be faster in the early periods of production. SAGD performance was more sensitive to the changes in the permeability of matrix than fracture. However, ES-SAGD was affected by changes in both the matrix and fracture permeabilities. Since solvents are expensive oil products, the type and optimum concentration of injected solvent should be chosen based on thorough study of the reservoir properties and operational conditions. High consumption of energy and water plus emission of green-house-gases like CO_2 are the main deficiencies of SAGD process, while ES-SAGD can yield higher recovery by using lower amount of water and energy. Another advantage of this process is lower amount of green-house-gas emission which is important from environmental point of view.

Acknowledgment

The authors would like to thank Statoil ASA for financial supports. We are also thankful to CMG Ltd. for providing us the license to STARS software.

References

Aguilera, R. (1995). *Naturally fractured reservoirs* (2nd ed.). Tulsa, Oklahoma: Penn Well Books.

Al Bahlani, & Babadagli, T. (2009). *Laboratory and Field Scale Analysis of Steam Over Solvent Injection in Fractured Reservoirs (SOS-FR) for Heavy-Oil Recovery*. Paper SPE 124047 presented at the 2009 SPE Annual Technical Conference and Exhibition held in New Orleans, Louisiana, USA, 4-7 October.

Ashrafi, M., Souraki, Y., Karimaie, H., Torsaeter, O., & Bjorkvik, B. J. A. (2011). *Experimental PVT Property Analyses for Athabasca Bitumen*. Paper CSUG/SPE 147064 presented at the Canadian Unconventional Resources Conference held in Calgary, Alberta, Canada, 15-17 November. http://dx.doi.org/10.2118/147064-MS

Azin, R., Kharrat, R., Ghotbi, C., & Vossoughi, S. (2005). *Applicability of the VAPEX Process to Iranian Heavy Oil Reservoirs*. Paper SPE 92720 presented at the 14th SPE Middle East Oil & Gas Show and Conference held in Bahrain International Exhibition Center, Bahrain, 12-15 March. http://dx.doi.org/10.2118/92720-MS

Bagci, A. S. (2006). *Experimental and Simulation Studies of SAGD Process in Fractured Reservoirs*. Paper SPE 99920 presented at the 2006 SPE/DOE Symposium on Improved Oil Recovery held in Tulsa, Oklahoma, U.S.A., 22-26 April. http://dx.doi.org/10.2118/99920-MS

Bagci, A. S., Samuel, O. M., & Mackay, M. (2007). Paper 2007-040 presented at the Petroleum Society's 8th Canadian International Petroleum Conference (58th Annual Technical Meeting) held in Calgary, Alberta, Canada, 12-14 June. http://dx.doi.org/10.2118/2007-040

Butler, R. M., & Jiang, Q. (2000). Improved Recovery of Heavy Oil by VAPEX with Widely Spaced Horizontal Injectors and Producers. *JCPT, 39*(1), 48-56. http://dx.doi.org/10.2118/00-01-04

Chen, Q. (2006). *Assessing and Improving Steam-Assisted Gravity Drainage: Reservoir Heterogeneities, Hydraulic Fractures and Mobility Control Foam, Dissertation for Doctor of Philosophy Degree, Stanford University, USA*. Retrieved from http://pangea.stanford.edu/ERE/pdf/pereports/PhD/ChenQing09.pdf

Chen, Q., Gerritsen, M. G., & Kovscek, A. R. (2007). Paper SPE 109873 presented at the 2007 SPE Annual Technical Conference and Exhibition held in Anaheim, California, USA, 11-14 November. http://dx.doi.org/10.2118/109873-MS

CMG-STARS User's Guide. (2012). *Computer Modeling Group Ltd.*

Das, S. (2007). *Application of Thermal Processes in Heavy Oil Carbonate Reservoirs*. Paper SPE 105392 presented at the 15th SPE Middle East Oil & Gas Show and Conference held in Bahrain International Exhibition Center, Bahrain, 11-14 March. http://dx.doi.org/10.2118/105392-MS

Dusseault, M. B., & Shafiei, A. (2010). Ullmann's Encyclopedia of Industrial Chemistry, Wiley-VCH. http://dx.doi.org/10.1002/14356007

Edmunds, N., Barrett, K., Solanki, S., Cimolai, M., & Wong, A. (2009). Prospect for Commercial Bitumen Recovery from the Grosmont Carbonate, Alberta. *JCPT, 48*(9), 26-32. http://dx.doi.org/10.2118/09-09-26

Fatemi, S. M., Ghotbi, C., Kharrat, R., & Badakhshan, A. (2011a). *Application of Toe-to-Heel Air Injection (THAI) Process in Fractured Carbonate Systems: 3D Simulation of the Effect of Fractures Geometrical properties, Reservoir and Operational Parameters*. Paper SPE 143434 presented at the SPE EUROPEC/EAGE Annual Conference and Exhibition held in Vienna, Austria, 23-26 May. http://dx.doi.org/10.2118/143434-MS

Fatemi, S. M., Kharrat, R., & Vossoughi, S. (2011b). *Investigation of Steam Assisted Gravity drainage (SAGD) and Expanding Solvent-SAGD (ES-SAGD) Processes in Complex Fractured Models: Effects of Fractures' Geometrical Properties*. Paper SPE 150082 presented at the SPE Heavy Oil Conference and Exhibition held in Kuwait City, Kuwait, 12-14 December. http://dx.doi.org/10.2118/150082-MS

Joshi, S. D. (1986). *A Laboratory Study of Thermal Oil recovery Using Horizontal Wells*. Paper SPE 14916 presented at the SPE/DOE Symposium on Enhanced Oil Recovery held in Tulsa, Oklahoma, USA, 20-23 April. http://dx.doi.org/10.2118/14916-MS

Kharrat, R., Vossoughi, S., Azin, R., & Razzaghi, S. (2003). *The Investigation of Using Thermal Methods for Heavy Oil Recovery*. Phase I, Internal Document, Petroleum University of Technology, Iran.

Mollaei, A., Maini, B., & Jalilavi, M. (2007). *Investigation of Steam Flooding in Naturally Fractured Reservoirs*. Paper presented at the International Petroleum technology Conference held in Dubai, UAE, 4-6 December. http://dx.doi.org/10.2523/11087-MS

Mollaei, A., & Maini, B. (2010). Steam Flooding of Naturally Fractured Reservoirs: Basic Concepts and Recovery Mechanisms. *JCPT, 49*(1), 65-70. http://dx.doi.org/10.2118/132485-MS

Nasr, T. N., & Ayodele, O. R. (2006). *New Hybrid Steam-Solvent Processes for the Recovery of Heavy Oil and Bitumen*. Paper SPE 101717 presented at the 2006 Abu Dhabi International Petroleum Exhibition and Conference held in Abu Dhabi, UAE, 5-8 November. http://dx.doi.org/10.2118/101717-MS

Pathak, V., & Babadagli, T. (2012). Experimental Investigation of Bitumen Recovery from Fractured Carbonates Using Hot-Solvents. Paper SPE 159439 presented at the SPE Annual Technical Conference and Exhibition held in San Antonio, Texas, U.S.A., 8-10 October. http://dx.doi.org/10.2118/159439-MS

Pooladi-Darvish, M., & Farouq, A. (1994). *Steam Heating of Fractured Formations Containing Heavy Oil: Basic Premises and a Single-Block Analytical Model*. Paper SPE 28642 presented at the SPE 69th Annual Conference and Exhibition held in New Orleans, LA, U.S.A., 25-28 September. http://dx.doi.org/10.2118/28642-MS

Pooladi-Darvish, M., Tortike, W. S., & Farouq, A. (1995). *A New Semi-Analytical Model for Thermal Recovery Processes*. Paper SPE 29660 presented at the SPE Western Regional Meeting held in Bakersfield, CA, U.S.A., 8-10 March. http://dx.doi.org/10.2118/29660-MS

Rezaei, N., Mohammadzadeh, O., & Chatzis, I. (2010). Experimental Investigation of Vapor Extraction Process in the Recovery of Bitumen from Vuggy Porous Media. Paper CSUG/SPE 137829 presented at the Canadian Unconventional Resources & International Petroleum Conference held in Calgary, Alberta, Canada, 19-21 October. http://dx.doi.org/10.2118/137829-MS

Rostami, B., Azin, R., & Kharrat, R. (2005). *Investigation of the Vapex Process in High-Pressure Fractured Heavy Oil Reservoirs*. Paper SPE/PS-CIM/CHOA 97766 presented at the 2005 SPE International Thermal Operations and Heavy Oil Symposium held in Calgary, Alberta, Canada, 1-3 November. http://dx.doi.org/10.2118/97766-MS

Sedaee, S. B., & Rashidi, F. (2006). *Application of the SAGD to an Iranian Carbonate Heavy Oil Reservoir*. Paper SPE 100533 presented at the 2006 SPE Western Regional/AAPG Pacific Section/GSA Cordilleran Section Joint Meeting held in Anchorage, Alaska, U.S.A., 8-10 May. http://dx.doi.org/10.2118/100533-MS

Souraki, Y., Ashrafi, M., Karimaie, H., & Torsaeter, O. (2011). *Experimental Investigation and Numerical Simulation of Steam Flooding in Heavy Oil Fractured Reservoir*. Paper SPE 144552 presented at the SPE Western North American Regional Meeting held in Anchorage, Alaska, U.S.A., 7-11 May. http://dx.doi.org/10.2118/144552-MS

Mechanistic Study of the Wettability Modification in Carbonate and Sandstone Reservoirs during Water/Low Salinity Water Flooding

Zohreh Jalili[1] & Vahid Alipour Tabrizy[2]

[1] Department of Petroleum Engineering and Applied Geophysics, Norwegian University of Science and Technology, N7491, Trondheim, Norway

[2] Department of Reserve Replacements, Åsgard Petroleum Technology, Statoil ASA, Norway

Correspondence: Zohreh Jalili, Department of Petroleum Engineering and Applied Geophysics, Norwegian University of Science and Technology, N7491, Trondheim, Norway. E-mail: zohrehj@stud.ntnu.no

Abstract

The paper pertains to the analysis of the chemical interaction between sea water ions, asphaltene colloids and silicate /calcite mineral as a substrate during water/low salinity water flooding. The work tackles modeling of salinity dependent relative permeability and capillary pressure functions from contact angle to estimate oil recovery during water/low salinity water. The paper has two main parts. In the first part, static contact angle is calculated based on disjoining pressure and compared to the experimental values, reported in the literature. In the second part, the model is used to demonstrate that water film is more stable in presence of low salinity water compared to distilled water and sea water, for carbonate and silicate minerals. Increasing temperature enhances the stability of the water film around the substrate for both types of minerals. This could be future interpreted as an indication for extra oil recovery applying low salinity water injection at elevated temperature. It is interesting to observe from the model that, increasing the Mg^{2+} ion concentration enhances the hydrophilicity characteristics for calcite mineral modified by asphaltene while for silicate surface modified by asphaltene, SO_4^{2-} ion enhances the hydrophobicity behaviour.

Keywords: wettability alteration, calcite, quartz, magnesium, sulphate, low salinity water

Nomenclature

A	Hamaker constant (J)
A_s	Structural force coefficient (Pa)
e	electronic charge ($=1.602\times10^{-19}$ C)
h	Distance of separation (m)
h_s	Decay length (m)
J	Mean surface curvature
k	Boltzmann constant (1.38×10^{-23} m^2 kg s^{-2} K^{-1})
n	Refractive index of media
P_c	Capillary pressure (pa)
pH_0	Isoelectric point of the surface
r	Radius (m)
T	Absolute Temperature (K)
T_o	Reference Temperature (K)
V	Interaction potential (J)
v_ρ	Plasma frequency of the free electron gas (3×10^{15} sec^{-1})
z	Ion valence

Greeks

δ	Collision diameter (m)
ε	Permittivity or dielectric constant of medium
$\hbar$	Plank constant (6.63×10^{-34} J/sec)
Φ	Surface potential (J)

κ Debye length (m)
λ Wave length (m)
Π Disjoining pressure (atm)
σ Surface/interfacial tension (N m^{-1})
θ Contact angle
ζ Zeta potential (V)

Subscripts

01 Mineral surface
02 Mineral surface
1 Solid
2 Oil phase
3 Aqueous phase (water film)
BR Born short range repulsive force
DLR Double layer electrostatic repulsive force
LVA London -Van der Waals attractive force
o Oil/original condition
w Water
p Particle

Abbreviations

DLVO Derjaguin, Landau, Verwey, Overbeek
DW Distilled water
LSW Low salinity water
Mg^{2+} Magnesium
SO_4^{2-} Sulfate
SSW Synthetic sea water
TDS Total dissolved solid

1. Introduction

Low salinity water-based enhanced oil recovery (EOR) has been recognized to contribute for extra oil recovery from sandstone reservoirs containing clay minerals in the literature (Lager et. al 2007 and Loahardjo et. al, 2007). In 2011, a laboratory study was documented for reservoir limestone cores from aqueous section of reservoir with two to five percentage of extra oil recovery by tertiary low salinity water injection (Austad et. al 2011). The experiments were performed by injection of formation water with salinity of 200000 ppm as a secondary recovery while the tertiary recovery was observed by injection of 100×diluted formation water or 10×diluted seawater (from Persian Gulf) at 110 °C (Austad et. al 2011). Migration of fines, double layer expansion and ionic exchange were seems to be the main hypothetical explanations for the efficiency of low salinity water flooding in sandstone reservoirs (Tang and Morrow1999; McGuire et al. 2005; Lager et al. 2007). Rezaei Doust et al. (2010) proposed a new chemical mechanism for Low salinity EOR based on wettability alteration by desorption of organic basic material form silicate minerals at specific range of pH. An increase of pH associated with flooding with low-salinity water (LSW) has been observed in the experimental work reported by Hamouda and Valderhaug (2014). It was observed by Hamouda and Valderhaug (2014) also that the pressure drop across the flooded cores with LSW is almost doubled of the pressure drop compared to cores flooded with SSW. They argued also low salinity brine will expand the electrical double layer as well as dissolution which will lead to fine detachment and mobilization; hence the sweep efficiency will be increased (Hamouda and Valderhaug 2014). Austad et al. (2011) also claimed that presence of anhydrite in limestone is necessary for enhanced oil recovery by low salinity water flooding in carbonates as the main mechanism is dissolution of anhydrite from mineral to ionic phase.

Hamouda and collogues in series of articles dealing with carbonate/silicate minerals, water and oil documented that the parameters that contributes the initial and final wetting condition, before and after imbibition, could be addressed as oil composition, ionic composition of irreducible and imbibing water, pH, rock surface characteristics and temperature (Alipour Tabrizy et al. 2011a; Alipour Tabrizy et al. 2011b; Petrovich and Hamouda 1998; Rezaei Gomari et al. 2006a; Rezaei Gomari et al. 2006c; Karoussi and Hamouda 2007; Hamouda and Karoussi 2008; Chukwudeme and Hamouda 2009). Several articles have been reported also in the literature on improving oil recovery by spontaneous and forced imbibition of seawater at high temperature

(>70 °C) (Cuie and Morrow 1991; Akin et al. 2000; Petrovich and Hamouda 1998; Babadagli 2003; Hirasaki and Zhang 2004; Rezaei Gomari et al. 2006a; Rezaei Gomari et al. 2006b; Karoussi and Hamouda 2007; Hamouda and Karoussi 2008; Chukwudeme and Hamouda 2009). It was reported also as the temperature increases, the degree of exchange of Mg^{2+} and SO_4^{2} ions enhances (Karoussi and Hamouda 2007 and Petrovich and Hamouda 1998). Mahani et al. (2014) used a model of sandstone rock that consists of Na- Montmorillonite clay over glass substrate to observe the kinetic of oil droplet detachment in presence of low and high salinity water. They reported a wettability alteration process toward less oil-wet condition when the system is exposed to low salinity brine with slow kinetic reaction (Mahani et al. 2014). The field response to low salinity water flooding as secondary and tertiary modes were also addressed in another research work by Mahani et al. (2011). For one field, it was documented that dual step water cut is developed and the viscous force is the dominant mechanism which can be interpreted as a characteristic of change in wetting state; in contrast for another case study, the low salinity water flooding is immature and the breakthrough of low salinity brine was not yet observed due to strong bouncy effect caused by high vertical permeability of reservoir (Mahani et al. 2011).

In spite of the fact that several observations were reported for the application of low salinity water flooding in sandstone and carbonate rocks as a promising EOR technique, there are few reports in the literature to study oil recovery mechanism(s) associated to low salinity water injection applying universal mathematical model. This paper addresses the calculation of disjoining pressure and total interaction potentials of mineral/brine/asphaltene colloids linked to the effect of temperature, to study the influence of seawater ions and low salinity seawater on wettability modification of calcite and quartz surfaces. Static contact angle is calculated based on disjoining pressure and compared to the experimental values, reported previously in the literature by Alipour Tabrizy et al. (2011b). It is attempted also to generate oil/water relative permeability curves from calculated contact angle to investigate, the influences of sea water ions/low salinity water flooding on recovery process.

2. Theory

2.1 Fines Detachment by DLVO (Derjaguin Landau Verwy Overbeek) Theory

DLVO theory and experimental observations could indicate that fine detachment is the mechanism that may alter the wettability of the rock at high temperatures (Hamouda and Karoussi 2008). However, Hamouda and Karoussi (2008) did not provide any calculation of contact angle based on DLVO theory and simply used variation of disjoining pressure versus water film thickness to compare with imbibition and flooding experimental results. In DLVO theory, the interaction between particles (sphere-shape) and a pore surface is described by London van der Waals (LVA) attraction, double-layer repulsion (DLR) and Born repulsion (BR). The DLVO model may not be able to fully describe colloidal behavior in aqueous media. Grasoo et al. (2002) reviewed many of the interactions that play a role in environmental systems and are not commonly included by the traditional DLVO model: e.g., hydrogen bonding and the hydrophobic effect, hydration pressure, non-charge transfer Lewis acid base interactions, and steric interactions. They documented also that there are conflicting opinions in the literature as to whether steric interactions and other extended DLVO forces should be considered or not. While several studies have concluded that these contributions to be additive (Van Oss 1994; Jucker et al. 1998; Giasson et al. 1998 and Freitas et al. 2001); others disagree with this approach (e.g. Rijnaarts et al. 1999). It seems that the complexity of aqueous, hydrocarbon and oil system will most likely preclude the development of a unique approach that can be used for all conditions. In this work, hydration and structural forces are not taken into account for simplicity. Therefore the total interaction potentials (between a fine particle and mineral surface) may be expressed as follows (Israelachvili 1992):

$$V_t(h) = V_{LVA}(h) + V_{DLR}(h) + V_{BR}(h) \tag{1}$$

For a spherical particles and flat mineral surfaces, the interactions potential are expressed for the different interactions by the following equations (Israelachvili 1992):

$$V_{LVA}(h) = -\frac{A}{6}\left[\frac{2(H+1)}{H(H+2)} + \ln(\frac{H}{H+2})\right] \tag{2}$$

where $H=h/r_p$; A is Hamaker constant; h is distance of separation; r_p is particle radius; ε is permittivity of medium (water); ε_0 is permittivity of free space; κ is Debye length and Φ_{01} and Φ_{02} are surface and particle surface potentials. Hamaker constant is calculated using Lifshitz theory as a function of temperature (Israelachvili 1992):

$$A = \frac{3}{4}kT\left(\frac{\varepsilon_1-\varepsilon_3}{\varepsilon_1+\varepsilon_3}\right)\left(\frac{\varepsilon_2-\varepsilon_3}{\varepsilon_2+\varepsilon_3}\right)+\frac{3\hbar v_e}{8\sqrt{2}}\frac{\left(n_1^2-n_3^2\right)^2\left(n_2^2-n_3^2\right)^2}{\left(n_1^2+n_3^2\right)^{1/2}\left(n_2^2+n_3^2\right)^{1/2}\{\left(n_1^2+n_3^2\right)^{1/2}+\left(n_2^2+n_3^2\right)^{1/2}\}} \quad (3)$$

Where, ε_1, ε_2 and ε_3 are the static dielectric constants of the three media, while n_1, n_2 and n_3 are values for the refractive index of media. K is Boltzmann constant (1.38×10^{-23} m^2 kg s^{-2} K^{-1}), T is temperature in Kelvin, $\hbar$ is Plank constant (6.63×10^{-34} J/sec) and v_e is plasma frequency of the free electron gas (3×10^{15} sec^{-1}). The Born repulsion interaction potential can be calculated as follows:

$$V_{BR}(h) = -\frac{A\delta^6}{7560}\left[\frac{8r_p+h}{(2r_p+h)^7}+\frac{6r_p-h}{h^7}\right] \quad (4)$$

Where δ is collision diameter which in this research the collision diameter is assumed 5 Å to compare with the work done by Karoussi and Hamouda (2007) and Schembre et al. (2006). The double layer repulsion force can be calculated as:

$$V_{DLR}(h) = -\frac{\varepsilon\varepsilon_0 r_p}{4}(\Phi_{01}^{\ 2}+\Phi_{02}^{\ 2})\{\left[\frac{2\Phi_{01}\Phi_{02}}{\Phi_{01}^{\ 2}+\Phi_{02}^{\ 2}}\right]\ln\left[\frac{1+\exp(-kh)}{1-\exp(-kh)}\right]-\ln\left[1-\exp(-2kh)\right]\} \quad (5)$$

In this equation, Φ_{01} and Φ_{02} are surface and particle surface potentials. The Debye length (κ) is calculated on the basis of (Israelachvili 1992):

$$\kappa = \sqrt{\frac{e^2 z^2 n_b}{\varepsilon\varepsilon_0 kT}} \quad (6)$$

Where, e is electron charge; n b is total ion density in bulk solution; z is ion valence; k is Boltzmann constant and T is temperature in K. The reported permittivity values for the water at different temperatures by Schembre et al. (2006) are interpolated to the corresponding temperatures. The surface potential (i.e. $\zeta1$) for mineral surface ($\Phi01$) is determined by Nernst equation and extrapolated to the corresponding temperature (Israelachvili 1992).

$$\zeta_1 = \Phi_{01} = -2.3\left(\frac{kT}{e}\right)(pH-pH_{oi}) \quad (7)$$

$$\zeta(T) = \Phi_{01} = -0.01712\left[(T-T_o)+1\right]\zeta(T_o) \quad (8)$$

In this equation, pH_{0i} is the iso-electric point of calcite and quartz and T is the temperature in Kelvin.

2.2 Disjoining Pressure

Basu and Sharma (1996) described the stability of water films on the basis of disjoining pressure. Disjoining pressure was expressed by Young-Laplace equation:

$$P_c = \Pi + 2\sigma_{ow}J \quad (9)$$

Where *Pc* is the capillary pressure between wetting (water) and non-wetting (oil) phases (for water wet mineral), Π is the disjoining pressure in thin film, σ_{ow} is interfacial tension between oil and water, and *J* is the mean surface curvature. *J* is positive for concave surfaces and negative for convex surfaces. If the applied capillary pressure exceeds the net forces due to disjoining pressure and surface curvature, the wetting films will be unstable (Basu and Sharma 1996). The disjoining pressure is directly related to the contact angle by the following equation:

$$\cos(\theta) = 1+\frac{1}{\sigma^{\alpha\gamma}}\int_{\infty\ 0}^{P_c} h\,d\Pi \quad (10)$$

The detail of mathematical calculation is shown in appendix. Controversial equation was reported by Karoussi and Hamouda (2007) where they claimed that $\cos(\theta) = 1 + \frac{1}{\sigma^{\alpha\gamma}} \int_{\infty\ 0}^{P_c} \Pi dh$, without any mathematical derivation or justification. The reported equation in their work is dimensionally inconsistent as the dimension of derivate part is length (h or thickness of film) while the dimension of upper and lower limits of integration is pressure (p_c or capillary pressure). The stability of wetting phase can be examined by calculation of $I = \int_0^{P_c} h d\Pi$ and is discussed in appendix. Disjoining pressures could be derived from intermolecular or inter ionic forces such as van der Waals, double layer electrostatic and structural forces, which are estimated by the following equations (Israelachvili 1992):

$$\Pi_t(h) = \Pi_{LVA}(h) + \Pi_{DLR}(h) + \Pi_{BR}(h) \tag{11}$$

$$\Pi_{LVA}(h) = -\frac{A(15.96\frac{h}{\lambda} + 2)}{12\pi h^3 (1 + 5.32\frac{h}{\lambda})^2} \tag{12}$$

$$\Pi_s(h) = -A_s e^{-\frac{h}{h_s}} \tag{13}$$

$$\Pi_{DLR}(h) = nkT \left[\frac{\Phi_{r1}\Phi_{r2}\cosh(\kappa h) - \Phi_{r1}^2 - \Phi_{r2}^2}{\sinh^2(\kappa h)} \right] \tag{14}$$

$$\Phi_{ri} = \frac{e\zeta_i}{KT} \tag{15}$$

Where, h is film thickness; λ is wave length (=100 nm); A_s is structural force coefficient (1.5×10^{10} Pa); h_s is decay length (0.05 nm) and Φ_{r1} and Φ_{r2} are reduced potential for surface and particle, respectively (Israelachvili, 1992).

2.3 Relative Permeability and Capillary Pressure Model

Several experiment on core flooding showed that the maximum oil recovery occurs in neutral or slightly water-wet cores (Anderson 1985). In strongly oil-wet media, oil occupies the smallest pores, which cross ponds to high capillary pressure and high residual oil saturation. In contrast, for strongly water-wet porous media, oil become disconnected from continues oil film (snap off effect); herby trapped in pores. This results to lower oil recovery and early water breakthrough (Zhao et al. 2010). The maximum oil recovery is archived in intermediate-wet porous media, where water invades the smallest pores and oil production will continue through film flow (Hui and Blunt 2010). It is possible to speculate theoretically that, the oil saturation decreases during water/low salinity water flooding exponentially with respect to rock wettability according to Equation (16):

$$S_{or} = ae^{b\cos(\theta)} \tag{16}$$

In this equation, $S_{or}u$ is residual oil saturation after water flooding, a and b are constants and can be calculated from boundary conditions. θ is the contact angle between oil, water and mineral. If it is assumed that the rock is initially strongly oil-wet and after following with low salinity water, wettability will be altered to intermediate wetting, one can calculate the constants in Equation (16). Substituting the boundary conditions to Equation (17) yields to the following equation:

$$S_{or} = (S_{or})_f \left(\frac{(S_{or})_f}{(S_o)_i}\right)^{\cos(\theta)} \tag{17}$$

In this equation, S_{or} is the residual oil saturation, $(S_{or})_f$ is the final residual oil saturation after low salinity water flooding and $(S_o)_i$ is the initial oil saturation in the rock. Assuming the definition of normalized oil saturation and Corey correlation for calculation of normalized relative permeability and capillary pressure, it is possible to write:

$$S_{on} = \frac{S_o - S_{or}}{1 - S_{or} - S_{wi}} \tag{18}$$

$$k_{ro} = \left[\frac{S_o - (S_{or})_f \left(\frac{(S_{or})_f}{(S_o)_i}\right)^{\cos(\theta)}}{1 - (S_{or})_f \left(\frac{(S_{or})_f}{(S_o)_i}\right)^{\cos(\theta)} - S_{wi}} \right]^{\frac{2+\lambda}{\lambda}} \tag{19}$$

$$k_{rw} = \left[1 - \left(\frac{S_o - (S_{or})_f \left(\frac{(S_{or})_f}{(S_o)_i}\right)^{\cos(\theta)}}{1 - (S_{or})_f \left(\frac{(S_{or})_f}{(S_o)_i}\right)^{\cos(\theta)} - S_{wi}} \right) \right]^{\frac{2+3\lambda}{\lambda}} \tag{20}$$

$$P_C = P_C^{th} \left[1 - \left(\frac{S_o - (S_{or})_f \left(\frac{(S_{or})_f}{(S_o)_i}\right)^{\cos(\theta)}}{1 - (S_{or})_f \left(\frac{(S_{or})_f}{(S_o)_i}\right)^{\cos(\theta)} - S_{wi}} \right) \right]^{\frac{-1}{\lambda}} \tag{21}$$

Where P_C^{th} and λ are capillary threshold and pore size distribution, respectively. Similar approach were reported in the literature to model continues wettability alteration during immiscible CO_2 flooding (Al-Mutairi et al. 2012).

3. Results and Discussion

This section is dividing in three subsections. The first part addresses the calculation procedures of contact angle and discusses the influences of ions and temperature on wettability alteration. Total potential interactions between modified minerals and different ions are discussed in the second part, while the last section is dealing with mathematical modeling of relative permeability and capillary pressure from contact angle.

3.1 Modeling of Contact Angle and Subsequent Effects of Ions and Temperature

Before addressing the stability of water film for modified minerals by asphaltene in presence of different ions, the Hamaker constant is computed for different temperatures in accordance with Lifshitz theory for ternary system of mineral/water/oleic. In Figure 1, the variation of Hamaker constant versus temperature is shown for quartz and calcite minerals. The used dielectric properties of three media and refractive index corresponding to the media are shown in Table 1.

Table 1. Dielectric properties and refractive index values for mineral /water/ toluene systems (Israelachvili 1992)

Calcite/Water/Toluene					
Media 1	Calcite	n_1	1.638	ε_1	8
Media 2	Toluene	n_2	1.496	ε_2	2.38
Media 3	Water	n_3	1.323	ε_3	80.1
Quartz/Water/Toluene					
Media 1	Quartz	n_1	1.536	ε_1	4.29

Media 2	Toluene	n_2	1.496	ε_2	2.38
Media 3	Water	n_3	1.323	ε_3	80.1

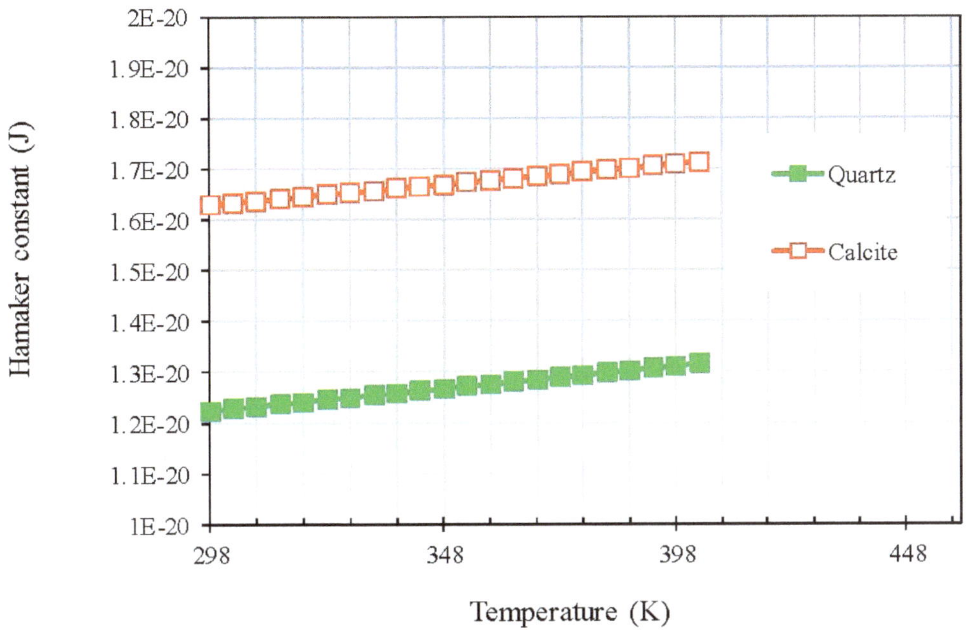

Figure 1. The variation of Hamaker constant versus temperature for quartz/water/toluene and calcite/water/toluene systems

From Figure 1, it is clear that the Hamaker constants for calcite and quartz are increasing linearly with the temperature. However these trends for calcite (≈7.98E-24 J/K) and quartz (≈8.77E-24 J/K) are slightly different. This can be also observed from Equation (22) in which the derivate of Hamaker coefficient to the temperature is a constant and is equal to:

$$\frac{dA}{dT} = \frac{3}{4}k\left(\frac{\varepsilon_1 - \varepsilon_3}{\varepsilon_1 + \varepsilon_3}\right)\left(\frac{\varepsilon_2 - \varepsilon_3}{\varepsilon_2 + \varepsilon_3}\right) \tag{22}$$

This is in contrast of the trend reported by Karoussi and Hamouda (2007) where they showed that Hamaker coefficient for calcite/decane/water system is decreasing by increasing temperature. Lifshitz theory has been used in their work to calculate the Hamaker constant; however the method of calculation (or simplification) for the second term ($\frac{3\hbar}{4\pi}\int_{v_1}^{\infty}\left(\frac{\varepsilon_1(v)-\varepsilon_3(v)}{\varepsilon_1(v)+\varepsilon_3(v)}\right)\left(\frac{\varepsilon_2(v)-\varepsilon_3(v)}{\varepsilon_2(v)+\varepsilon_3(v)}\right)dv$) was not mentioned. Equation (22) shows that sign of $\frac{dA}{dT}$ is dependent to the values of ε_1, ε_2 and ε_3. As the dielectric constant of water (ε_3) is larger than the dielectric constant of calcite or normal decane (ε_1 and ε_2) so $\frac{dA}{dT}$>0; then the Hamaker constant could not decrease by increasing temperature.

The stability of water film in calcite/silicate system (modified with 0.35 wt. % asphaltene) in presence of distilled water (DW), 0.1 M NaCL, 0.1 M Na_2SO_4 and 0.1 M $MgCl_2$ is addressed in this section. Jada et al. (2002) and Rezaei Gomari et al. (2006a) reported the zeta potential charge of quartz and calcite as a function of pH. Parra- Barraza et al. (2003) also reported the zeta potential of asphaltene versus pH when the asphaltene is precipitated from crude oil in presence of n-heptane with the ratio of 1 to 40. In Figure 2, calculated disjoining pressure as a function of water film thickness at 25 °C for quartz and calcite is demonstrated. From Figure 2a) it can be seen that water film over silicate mineral in presence of asphaltene dissolved in toluene and DW or 0.1 M Mg^{2+} is unstable since the area under disjoining capillary pressure is less than $-\sigma^{\alpha\gamma}_{\infty}$ (Type 3 according to appendix); hence the system is strongly oil-wet. The water film for quartz/0.1 M SO_4^{2-}/asphaltene or quartz/0.1 M NaCl/ asphaltene is meta-stable (Type 2 according to appendix) and the system is more water-wet respect to quartz/0.1 M Mg^{2+}/asphaltene or quartz/DW/ asphaltene. As it can be seen in Figure 2 b, for calcite, the water film is unstable regardless to the type of ions; however the instability of water-film over calcite in presence of 0.1 M Mg^{2+} and 0.1 M SO_4^{2-} is in lesser extent compared to 0.1 M NaCl and DW.

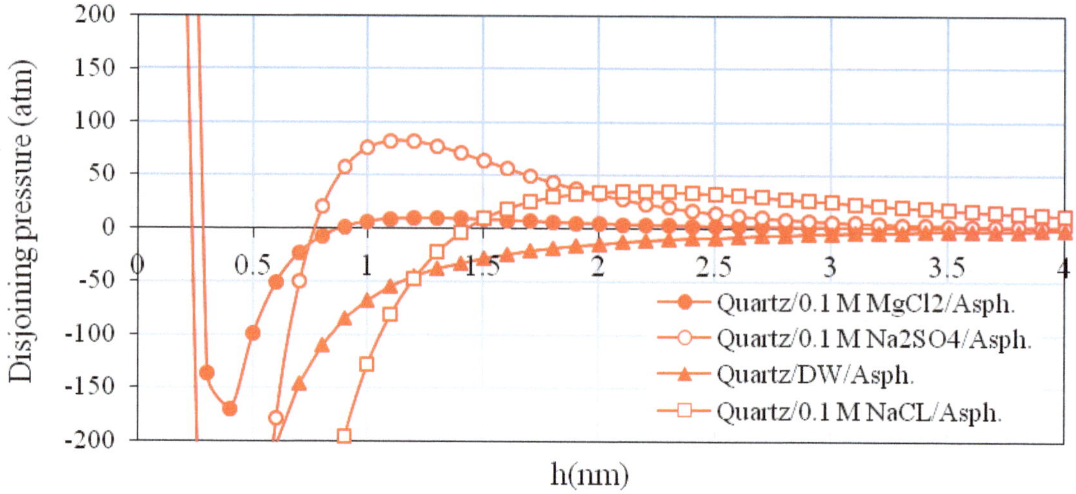

a) Quartz

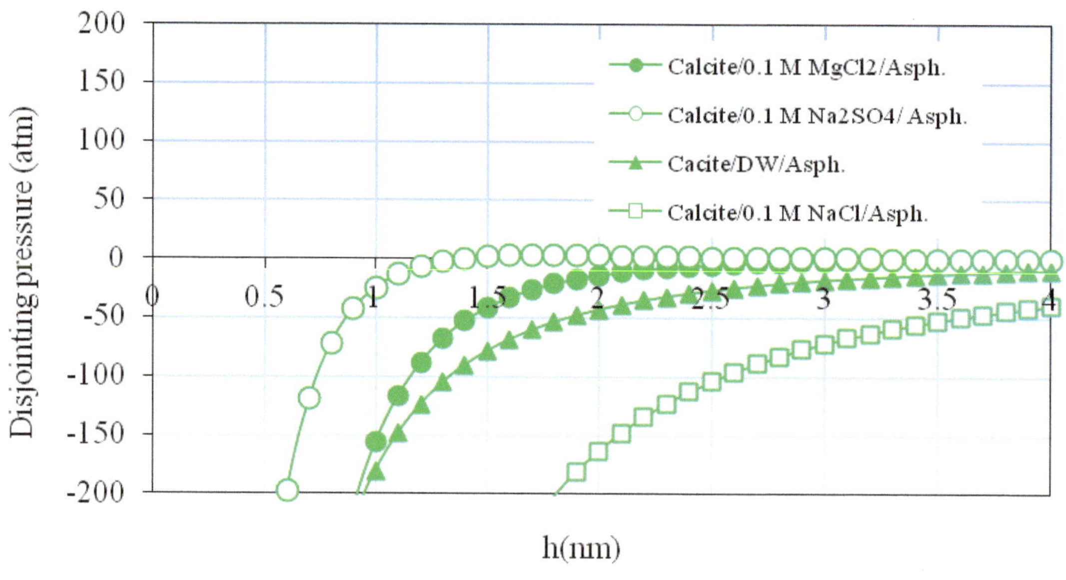

b) Calcite

Figure 2. Calculated disjoining pressure as a function of water film thickness at 25 °C for a) quartz and b) calcite in presence of distilled water (DW), 0.1 M NaCl, 0.1 M Na_2SO_4 and 0.1 M $MgCl_2$

In Figure 3, disjoining pressure as a function of water film thickness at 25 °C for quartz and calcite in presence of low salinity and synthetic sea water is shown. The ionic composition of low salinity seawater is obtained by 1000 times diluting the concentration of ions in seawater. In Table 2, the ionic composition of synthetic seawater and low salinity seawater is shown.

Table 2. Ionic composition of synthetic sea water and low salinity sea water

Ions	Concentration (mol/l)	Concentration (mol/l)
	Synthetic Sea Water	Low Salinity Water
NaCl	0.4	0.0004
KCl	0.01	0.00001
$CaCl_2\,(2H_20)$	0.013	0.000013
$MgCl_2(6H_20)$	0.045	0.000045
$NaHCO_3$	0.002	0.000002
$Na_2(SO_4)$	0.024	0.000024
TDS	0.494	0.000494

From Figure 3, it can be seen that water film over silicate and calcite minerals in presence of asphaltene dissolved in toluene and synthetic sea water (SSW) is unstable as the area under disjoining capillary pressure is less than $-\sigma^{\alpha\gamma}_{\infty}$ (Type 3); hence the system is strongly oil-wet. The water film for both types of minerals (quartz and calcite) in presence of low salinity water (LSW) and asphaltene is meta-stable (Type 2) and the contact angle forms between zero and 90 °. This shows that LSW may alter the wettability of calcite and silicate minerals toward water wet condition. Mahani et al. (2014) also observed a wettability alteration process toward less oil wet condition when clay coated silica glass is exposed to low salinity brine with slow kinetic reaction (Mahani et al. 2014). They addressed that the time scale of wettability alteration is ten to twenty times longer than expected time by diffusion (Mahani et al. 2014). It seems to be that the kinetic of the process is consistent to the electro kinetic ion diffusion mechanism involving DLVO forces between oil layer, water film and clay minerals.

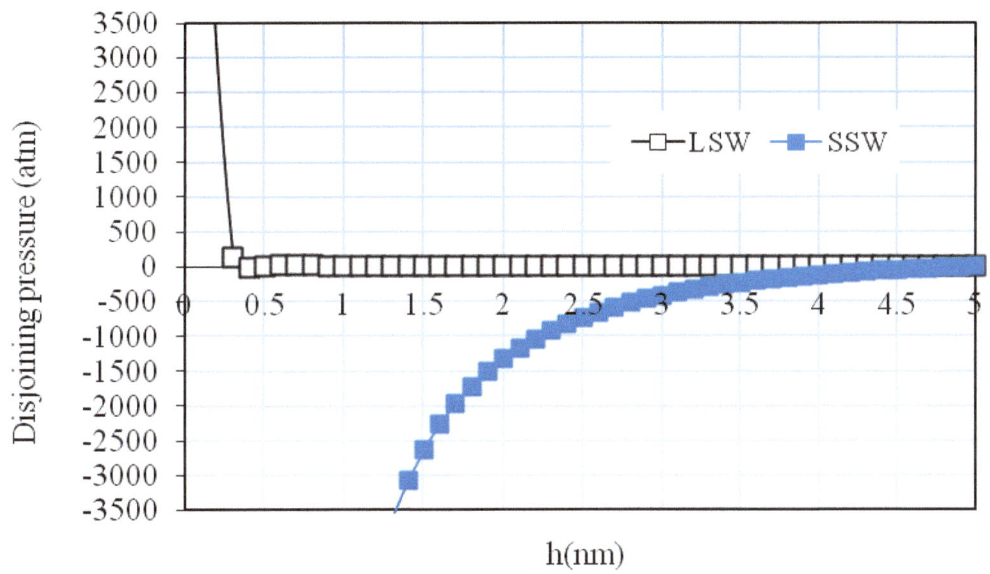

a) Quartz

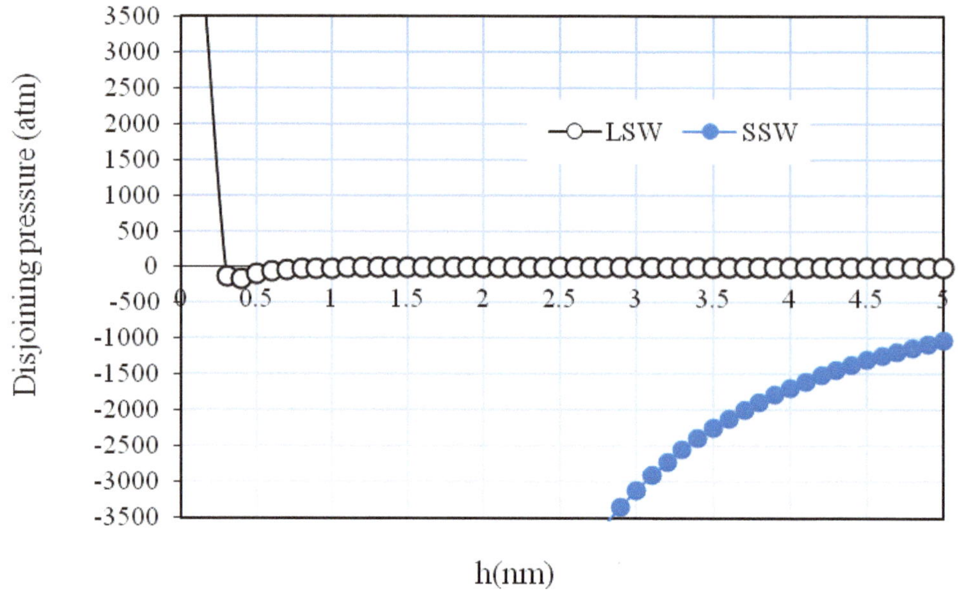

b) Calcite

Figure 3. Calculated disjoining pressures as a function of water film thickness at 25 °C for a) quartz and b) calcite in presence of low salinity water (LSW) and synthetic sea water (SSW)

Static contact angle between n-decane, distilled water and aged mineral (prewetted with distilled water (DW), 0.1 M NaCL, 0.1 M Na_2SO_4 and 0.1 M $MgCl_2$ and 0.35 wt. % dissolved asphaltene in toluene, are calculated based on Equation (16) and shown and compared in Table 3 with experimental values reported by Alipour Tabrizy et al. (2011b) and Mahani et al. (2014). The experimental error measurement was reported within 3° and 5° by Alipour Tabrizy et al. (2011b) and Mahani et al. (2014), respectively.

Table 3. Contact angle between oil (n-decane), distilled water and aged mineral (pre wetted with distilled water (DW), 0.1 M NaCL, 0.1 M Na_2SO_4 and 0.1 M $MgCl_2$ and 0.35 wt. % dissolved asphaltene in toluene at 25 °C

Modification brine	0. 1 M Mg^{2+}	DW	0. 1 M SO_4^{2-}	0. 1 M NaCL	LSW	SSW
Quartz						
Theoretical calculation	120	109	93	97	60	93
Experimental **	133	95	90	93		
Experimental *					67	90
Calcite						
Theoretical calculation	78	105	120	95	80	140
Experimental *	65	100	127	95		

*From Alipour Tabrizy et al. (2011b).

**From Mahani et al. (2014).

The effect of temperature on disjoining pressure has been shown in Figure 4. for quartz/0.1 M Mg^{2+}/asphaltene and calcite/0.1 M Mg^{2+}/asphaltene systems for temperatures 25 °C and 130 C°. Temperature has two main effects on calculated disjoining pressure, the first effect is related to Hamaker constant and the second effect is related to zeta potential of surface mineral and asphaltene colloids which they control the double layer interaction force. For both systems, increasing the temperature is enhancing the stability of water film over mineral surface; hence it will improve the hydrophilicity of system. The contact angle is decreasing from 77.78° to 31.78° for calcite mineral and from 124.75° to 16.25° for quartz mineral by increasing of temperature from 25 C° to 130 C°. This indicates the influences of ions on water film stability are accelerated by increasing temperature both for carbonate and silicates mineral.

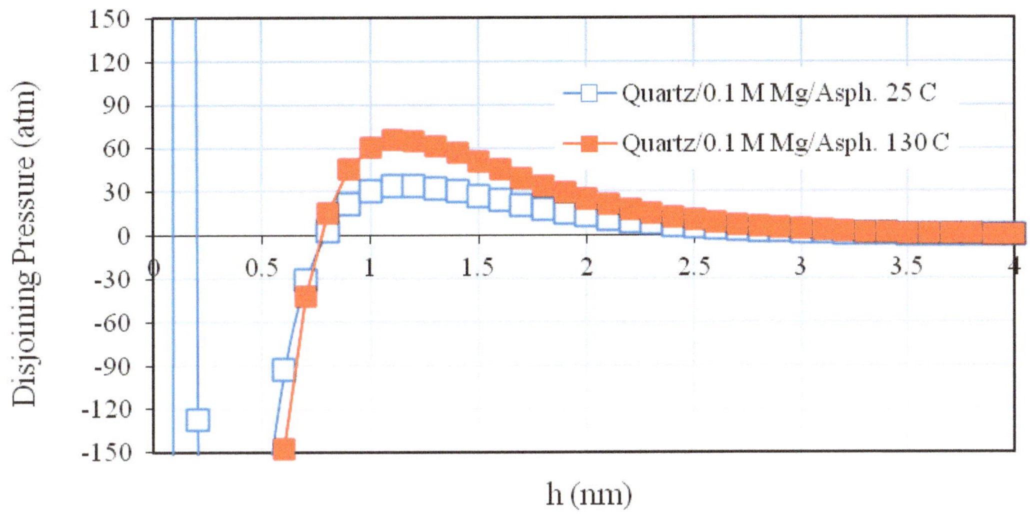

a) Quartz

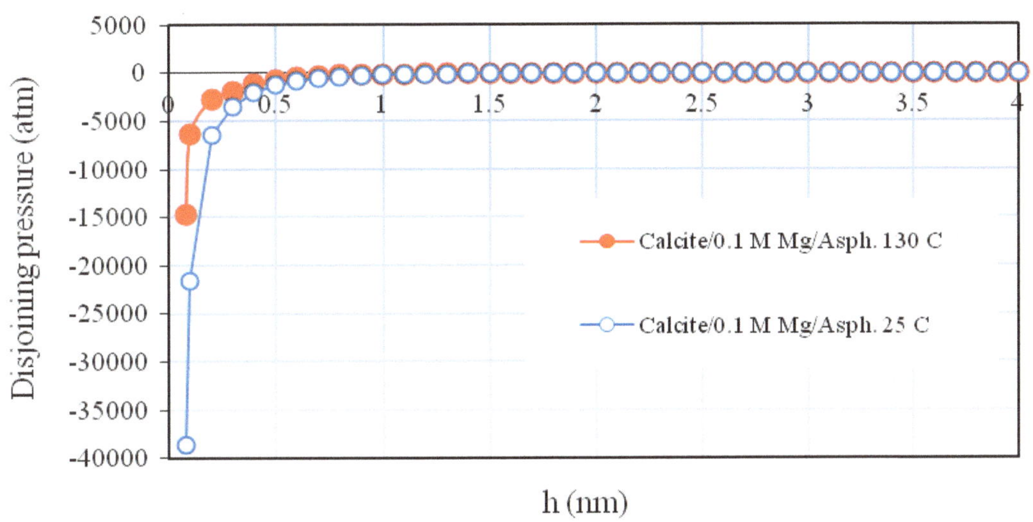

b) Calcite

Figure 4. Calculated disjoining pressure as a function of water film thickness and temperature for a) quartz and b) calcite in presence 0.1 M $MgCl_2$

3.2 Total Potential Interactions between Modified Minerals and Different Ions

Total interaction potential can be used to order the affinity of different ions/polar components to adsorb to the mineral surface. The calculation of total interaction potential for quartz and calcite are shown in Figure 5. From this figure it is possible to see that for modified quartz with asphaltene in presence of distilled water (DW), 0.1 M NaCL, 0.1 M Na_2SO_4 and 0.1 M $MgCl_2$, the total interaction potential between asphaltene colloids and mineral surface are always positive; hence the interaction force between asphaltene colloids and mineral surfaces is repulsive. However the magnitude of this force is dependent to the type of brine, where for 0.1 M NaCL and 0.1 M Na_2SO_4 this repulsion is more pronounced; hence the modified quartz with asphaltene in presence of 0.1 M NaCl and 0.1 M Na_2SO_4 are more water-wet comparing to the modified quartz with asphaltene in presence of

DW and 0.1 M $MgCl_2$. For modification of quartz by asphaltene in presence of 0.1 M $MgCl_2$, Mg^+ ions may act as bridge between asphaltene and quartz surfaces. This calculation is supported by water vapor adsorption and contact angle measurements by Alipour Tabrizy et al. (2011b) and is in agreement with pervious research works reported by Chukwudeme and Hamouda (2009) and Morrow et al. (2000). They explained in different experiments, possible bridging as an explanation. In other flooding experiments in micro model tests (Liu et al. 2010) for enhanced heavy oil recovery by alkaline flooding; it was found that oil recovery was greatly affected by wettability alteration of water-wet grains to preferentially oil-wet due to magnesium ion bonding. For calcite, the potential interaction between calcite and asphaltene in presence of DW and 0.1 M NaCL are negative which supports the high affinity of calcite and carbonate mineral to adsorb asphaltene. Calcite and carbonate minerals modified by asphaltene are strongly oil-wet compared to silicate minerals since the interaction potential is negative and the interaction force between asphaltene and calcite mineral surfaces is attractive unlike asphaltene and quartz minerals. Figure 5b) shows 0.1 M Mg^{2+} ions can enhance strongly the hydrophilicity of modified calcite with asphaltene as the affinity of asphaltene to interlock with Mg^{2+} is much less than Ca^{2+}. The total interaction potential between asphaltene colloids and Ca^{2+} is smaller that in two orders of magnitude compared to asphaltene colloids and Mg^{2+}. It has been reported previously that the affinity of cations to adsorb negatively charged colloidal particles (e.g. asphaltene) has the following order: (Tang and Morrow 1999)

$Li^+ < Na^+ < K^+ < Mg^{2+} < Ca^{2+} < H^+$

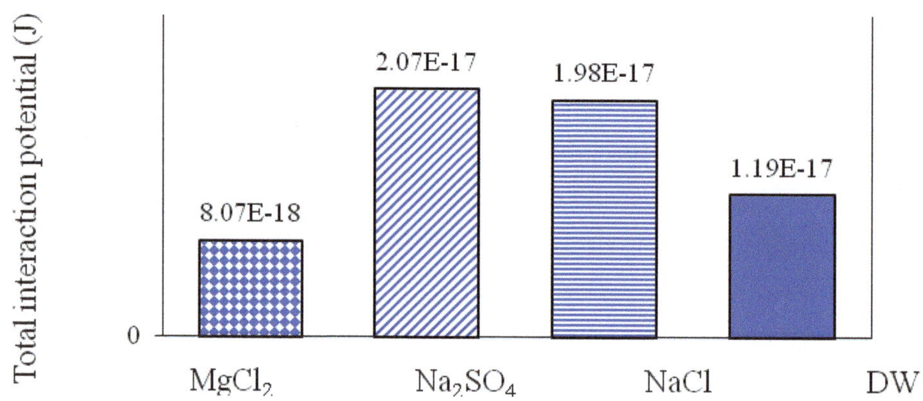

a) Quartz

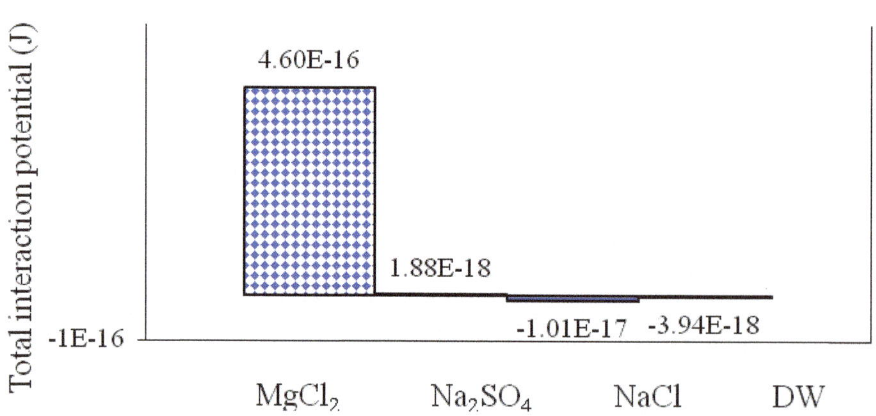

b) Calcite

Figure 5. Calculated total interaction potential between asphaltene colloids, quartz and calcite in presence of distilled water (DW), 0.1 M NaCl, 0.1 M Na_2SO_4 and 0.1 M $MgCl_2$

The calculations of total interaction potential are shown also for synthetic seawater and low salinity seawater for quartz, calcite and asphaltene colloidal systems in this subsection. The total interaction potential of quartz and /or calcite, synthetic sea water (SSW) and low salinity sea water (LSW), asphaltene colloids are demonstrated in Figure 6. In quartz/brine/asphaltene system, the total interaction potential is slightly positive for DW and SSW. For LSW, the total interaction potential is considerably larger than DW and SSW (10^3 order of magnitude larger that SSW and 10^2 order of magnitude larger than DW). In calcite/brine/asphaltene system, the total interaction potential is negative (attractive) for DW and SSW. For LSW, the total interaction potential is positive (repulsive). This clearly shows the applicability of low salinity seawater for tertiary improved oil recovery after sea water injection both for sandstone and carbonate reservoir rocks. The low salinity seawater flooding makes the crude oil/brine/rock (COBR) system more water-wet; hence it can produce additional oil between 5% up to 15% from water flooded sandstone and carbonate rocks (Tang and Morrow 1999).

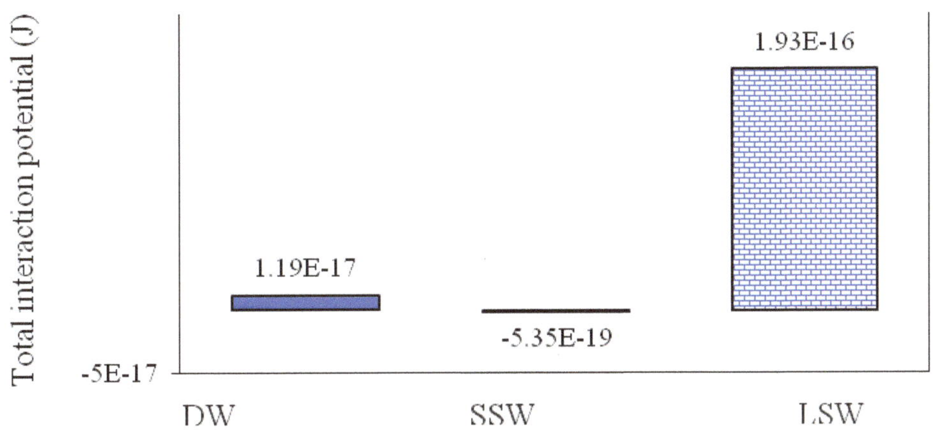

a) Quartz

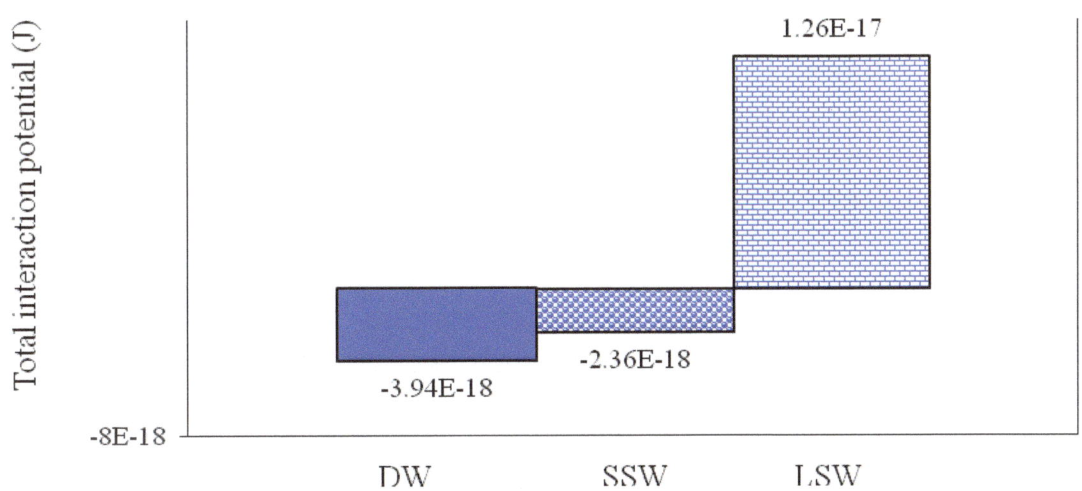

b) Calcite

Figure 6. Calculated total interaction potential between asphaltene colloids, quartz and calcite in presence of distilled water (DW), Synthetic Sea Water (SSW) and Low Salinity Sea Water (LSW)

3.3 Modeling of Relative Permeability and Capillary Pressure

Figure 7 shows the relative permeability and capillary pressure curves from estimated contact angles by applying Equation (19) till Equation (21) for a sandstone rock saturated initially with 0.35 wt. % dissolved asphaltene in toluene and flooded with low salinity water (LSW), synthetic sea water (SSW) and 0.1 M $MgCl_2$. Capillary threshold and pore size distribution for this rock sample is assumed as 2.5 bar and 0.5 respectively.

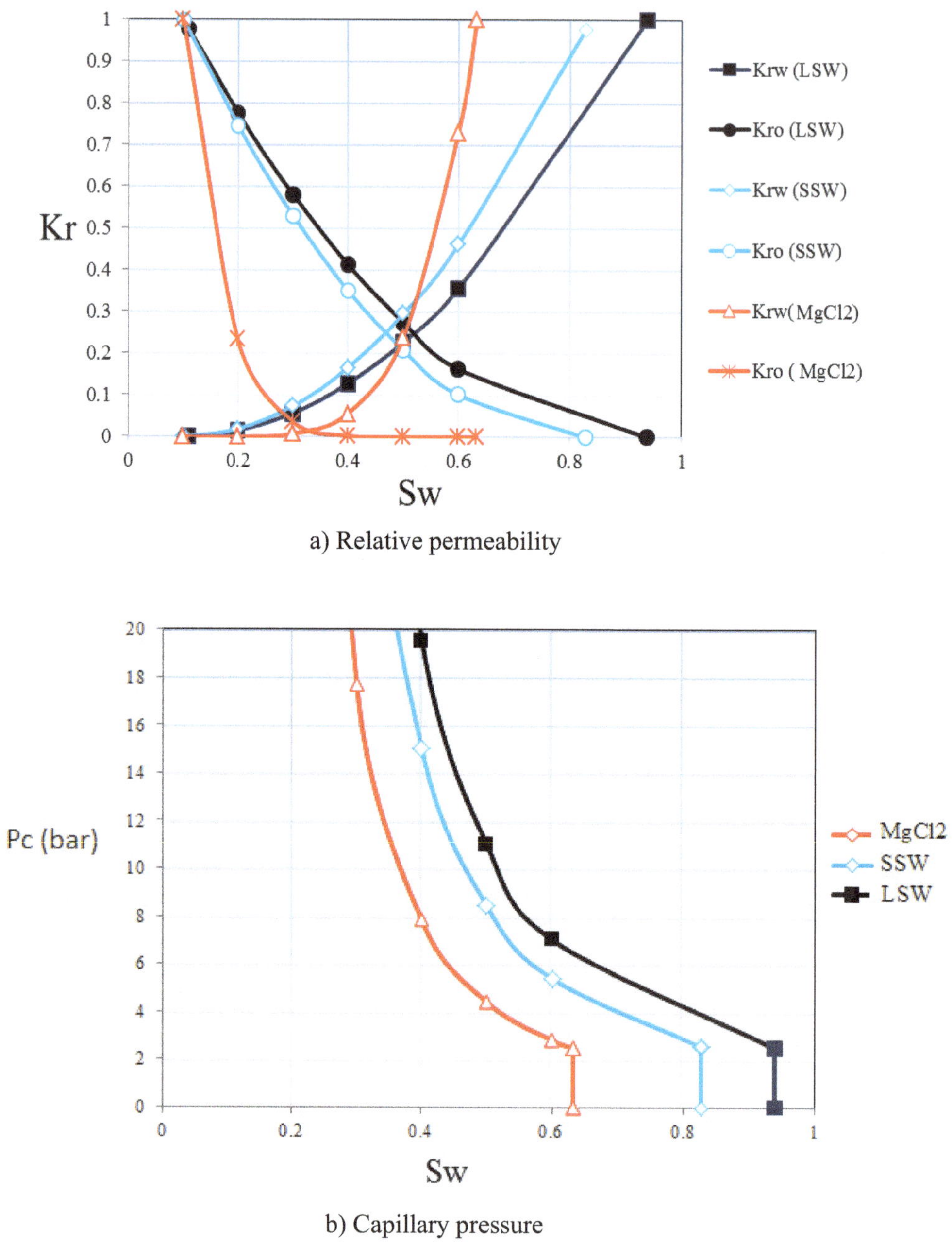

Figure 7. Calculated relative permeability and capillary pressure curves from contact angle for sandstone rock saturated with 0.35 wt. % dissolved asphaltene in toluene and flooded with low salinity water (LSW), synthetic sea water (SSW) and 0.1 M $MgCl_2$

Two observations can be seen from Figure 7. The first observation is the shift of the oil relative permeability toward the right hand side and a decrease in residual oil saturation with reduction of contact angle for the core flooded with low salinity water (LSW). This indicates oil becomes confined to smallest pores as the rock is altered from strongly oil wet to intermediate wet condition. On the other hand, water relative permeability decreases as the rock wetting state is altered towards intermediate wet condition during low salinity water (LSW) injection. The crossover point for oil/water relative permeability curves is located at 0.5 of S_w for the core flooded with low salinity water, whereas for the core flooded with synthetic sea water and 0.1 M $MgCl_2$, the crossover points are located at 0.4 and 0.3 respectively. The second observation is related to the area of the capillary pressure curves for the sandstone rock saturated with 0.35 wt. % dissolved asphaltene in toluene and flooded with different brines. Figure 7b demonstrates the area below capillary pressure curve (work per unit volume of rock to displace water by oil) is higher for flooded core with low salinity water (LSW) compared to that for flooded core with sea water or 0.1 M $MgCl_2$. This also shows the wettability alteration of sandstone rock towards intermediate wetting condition during low salinity water flooding.

4. Conclusions

This paper provides insights to a more fundamental understating of oil recovery by water/low salinity water flooding in silicates and carbonate rocks. An approach is discussed to assess the reservoir rock wettability (contact angle) from oil and mineral surface chemistry properties, ions concentration, pH and temperature. A model for evaluating relative permabilities and capillary pressure is used also to describe the wettability alteration mechanism and estimates the flooding/imbibition recovery by different ions.

The wettability change to a less oil-wet condition was shown for both types of sandstone and carbonate reservoirs during low salinity water; however the wettability alteration is more pronounced for modified silicate mineral compared to modified calcite mineral. Increasing the temperature also enhances the stability of the water film around the mineral surface and increases the efficiency of low salinity water flooding. The model shows also Mg^{2+} ion enhances the hydrophilicity characteristics for oil-wet calcite mineral while for oil-wet silicate surface, SO_4^{2-} ion enhances the hydrophobicity behaviour. This is consistent to the experimental observations reported in the literature; however it is worth mention that the physics underlying water/low salinity flooding is more complex than the simple model presented in this work, among them geomechanical considerations of rock during water/low salinity flooding (such as dissolution, precipitation, compaction and clay swelling) should receive considerable attention.

Acknowledgments

Authors declare no other competing interests. The paper reflects only the viewpoints of authors and not their corresponding organizations by any means.

References

Akin, S., Schembre, J. M., Bhat, S. K., & Kovscek, A. R. (2000). Spontaneous imbibition characteristics of diatomite. *Journal of Petroleum Science and Engineering, 25*, 149-165. http://dx.doi.org/10.1016/S0920-4105(00)00010-3

Alipour, T. V., Denoyel, R., & Hamouda, A. A. (2011a). Characterization of wettability modification of calcite, quartz and kaolinite: Surface energy analysis. *Colloid and Surface A., 384*, 98-108. http://dx.doi.org/10.1016/j.colsurfa.2011.03.021

Alipour, T. V., Hamouda, A. A., & Denoyel, R. (2011b). Influence of sulfate and magnesium ions on wettability alteration of calcite, quartz and kaolinite: Surface energy analysis. *Energy & Fuels, 25*, 1667-1680. http://dx.doi.org/10.1021/ef200039m

Al-Mutairi, S. M., Abu-khamsin, S. A., & Hossain, M. E. (2012). A Novel approach to handle continuous wettability alteration during immiscible CO2 flooding process. *Abu Dhabi International Petroleum Conference and Exhibition*, 11-14 November, Abu Dhabi, UAE. http://dx.doi.org/10.2118/160638-MS

Anderson, W. G. (1986). Wettability literature survey: Part 1. Rock /oil/ brine interactions and the effects of core handling on wettability. *Journal of Petroleum Technology, 38*, 1125-1144. http://dx.doi.org/10.2118/13932-PA

Austad, T., Shariatpanahi, F., Starnd, S., Black, C. J. J., & Webb, K. J. (2012). Conditions for a low-salinity enhanced oil recovery (EOR) effect in carbonate oil reservoirs. *Energy & Fuels, 26*, 569-575. http://dx.doi.org/10.1021/ef201435g

Babadagli, T. J. (2000). Evaluation of EOR methods for heavy oil recovery in naturally fractured reservoirs. *Journal of Petroleum Science and Engineering, 37*, 25-37. http://dx.doi.org/10.1016/S0920-4105(02)00309-1

Basu, S., & Sharma, M. M. (1996). Measurement of critical disjoining pressure for dewetting of solid surfaces.

Colloid and Surface A., 181, 443-455. http://dx.doi.org/10.1006/jcis.1996.0401

Chukwudeme, E. A., & Hamouda, A. A. (2009). Oil recovery from polar components (asphaltene and SA) treated chalk rocks by low salinity water and water containing SO_4^{2-} and Mg^{2+} at different temperatures. *Colloids Surface A., 336*, 174-182. http://dx.doi.org/10.1016/j.colsurfa.2008.11.051

Freitas, A., & Sharma, M. (2001). Detachment of particles from surfaces: An AFM study. *Journal of Colloid and Interface Science, 233*, 73-82. http://dx.doi.org/10.1006/jcis.2000.7218

Giasson S, Kuhl, T., & Israelachvili, J. N. (1989). Adsorption and interaction forces of micellar and microemulsion solutions in ultrathin films. *Langmuir, 14*, 891-898. http://dx.doi.org/10.1021/la960747w

Grasso, D., Subramaniam, K., Butkus, M., Strevett, K., & Bergendahl, J. (2002). A review of non-DLVO interactions in environmental colloidal systems. *Reviews in Environmental Science and Bio Technology, 1*, 17-38. http://dx.doi.org/10.1023/A:1015146710500

Hamouda, A. A., & Karoussi, O. (2008). Effect of temperature, wettability and relative Permeability on oil recovery from oil-wet chalk. *Energies, 1*, 19-23. http://dx.doi.org/10.3390/en1010019

Hamouda, A. A., & Rezaei G. K. A. (2006). Influence of temperature on wettability alteration of carbonate reservoirs. *SPE/DOE Symposium on Improved Oil Recovery*, Tulsa, OK, April 22–26. http://dx.doi.org/10.2118/99848-MS

Hamouda, A. A., & Valderhaug, O. M. (2014). Investigating enhanced oil recovery from sandstone by low salinity water and fluid/rock Interaction. *Energy & Fuels, 28*, 898-908. http://dx.doi.org/10.1021/ef4020857

Hui, M. H., & Blunt, M. J. (2000). Effects of wettability on three-phase flow in porous media. *Journal of Physical Chemistry B., 104*, 3833-3845. http://dx.doi.org/10.1021/jp9933222

Israelachvili, J. N. (1998). *International and Surfaces Forces* (2nd ed.). Academic Press.

Jada, A., Choau, A., Bertrand, Y., & Moreau, O. (2002). Adsorption and surface properties of isolating oils. *Fuel, 81*, 1227-1232. http://dx.doi.org/10.1016/S0016-2361(02)00019-4

Jucker, B., Zehnder A., & Harms, H. (1998). Quantification of polymer interactions in bacterial adhesion. *Environmental Science and Technology, 32*, 2909-2915. http://dx.doi.org/10.1021/es980211s

Karoussi, O., & Hamouda, A. A. (2007). Imbibition of sulfate and magnesium ions into carbonate rocks at elevated temperatures and their influence on wettability alteration and oil recovery. *Energy & Fuels, 21*(4), 2138-2146. http://dx.doi.org/10.1021/ef0605246

Lager, A., Webb, K. J., & Black, C. J. J. (2007). Impact of brine chemistry on oil recovery. *14th European Symposium on Improved Oil Recovery*, Cairo, Egypt.

Legens, C., Toulhoat, H., Cuiec, L., Villieras, F., & Palermo, T. (1998). Wettability change related to the Adsorption of organic acids on calcite: Experimental and Ab initio computational studies (pp. 27-30). *SPE Annual Technical Conference and Exhibition*, New Orleans, September. http://dx.doi.org/10.2118/49319-MS

Liu, Q., Dong, M., Asghari, K., & Tu, Y. (2010). Wettability alteration by magnesium ion binding in heavy oil/brine/chemical/sand systems - analysis of hydration forces. *Natural Science, 2*(5), 450-456. http://dx.doi.org/10.4236/ns.2010.25055

Loahardjo, N., Xie, X., Yin, P., & Morrow, N. R. (2007). Low salinity water flooding of a reservoir rock, *SCA*, 29.

Mahani, H., Berg, S., Ilic, D., Bartels , W. B., & Joekar-Niasar V. (2014). Kinetics of low salinity flooding effect. *SPE Journal,* 1-13.

Mahani, H., Sorop, D., Ligthelm, D. J., Brooks, D., Vledder, P., Mozahem, F., & Ali, Y. (2011). Analysis of field responses to low-salinity water flooding in secondary and tertiary mode in Syria, SPE EUROPEC/EAGE Annual Conference and Exhibition, 23-26 May, Vienna, Austria. http://dx.doi.org/10.2118/142960-MS

McGuire, P. L., Chatham, J. R., Paskvan, F. K., Sommer, D. M., & Carini, F. H. (2005). Low salinity oil recovery: An exciting new EOR opportunity for Alaska's North Slope. *SPE Western Regional Meeting*, Irvine, CA. http://dx.doi.org/10.2118/142960-MS

Morrow, N. R., Lim, H. T., & Ward, J. S. (1986). Effect of crude-oil-induced wettability changes on oil recovery. *SPE Formation and Evaluation*, 89-103. http://dx.doi.org/10.2118/13215-PA

Oss, C. V. (1994). Interfacial forces in aqueous media. Marcel Dekker, New York.

Parra, B. H., Hernandez, M. D., Lizardi, J., Herna´ndez, J., Herrera, U. R., & Valdez, M. (2004). The zeta potential and surface properties of asphaltene obtained with different crude oil/n-heptane proportions. *Fuel, 82*, 869-874. http://dx.doi.org/10.1016/S0016-2361(03)00002-4

Petrovich, R., & Hamouda, A. A. (1989). Dolomitization of Ekofisk oil field reservoir chalk by injected seawater. In Arehart, G. B., & Hulston, J. R. (Eds.), *Water-Rock Interaction* (pp. 345-348), Balkema, Rotterdam.

Rezaei, D. A., Puntervold, T., & Austad. T. (2011). Chemical verification of the EOR mechanism by using low saline/smart water in sandstone. *Energy & Fuels, 25*, 2151–2162. http://dx.doi.org/10.1021/ef200215y

Rezaei, G. K. A., Denoyel, R., & Hamouda, A. A. (2006b). Wettability of calcite and mica modified by different long chain fatty acids (C18 acids). *Journal of colloid and interface science, 297*, 470-479. http://dx.doi.org/10.1016/j.jcis.2005.11.036

Rezaei, G. K. A., & Hamouda, A. A. (2006c). Effect of fatty acids water composition and pH on the wettability alteration of calcite surface. *Journal of Petroleum Science and Engineering, 50*, 140-150. http://dx.doi.org/10.1016/j.petrol.2005.10.007

Rezaei, G. K. A., Hamouda, A. A., & Denoyel, R. (2006a). Influence of sulfate ions on the interaction between fatty acids and calcite surfaces. *Colloids Surface A., 287*, 29-35. http://dx.doi.org/10.1016/j.colsurfa.2006.03.018

Rijnaarts, H., Norde, W., Lyklema, J., & Zehnder, A. (1999). DLVO and steric contributions to bacterial deposition in media of different ionic strengths. *Colloids and Surfaces B: Bio interfaces, 14*, 179-195. http://dx.doi.org/10.1016/S0927-7765(99)00035-1

Schembre, J. M., Tang, G. Q., & Kovscek, A. R. (2006). Wettability alteration and oil recovery by water imbibition at elevated temperatures. *Journal of Petroleum Science and Engineering, 52*, 131-148. http://dx.doi.org/10.1016/j.petrol.2006.03.017

Tang, G. Q., & Morrow, N. R. (1999). Influence of brine composition and fines migration on crude oil/brine/rock interactions and oil recovery. *Journal of Petroleum Science and Engineering, 24*, 99-111. http://dx.doi.org/10.1016/S0920-4105(99)00034-0

Zhao, X., Blunt, M. J., & Yao, J. (2010). Pore-scale modeling: Effects of wettability on water flood oil recovery. *Journal of Petroleum Science and Engineering, 71*, 169-178. http://dx.doi.org/10.1016/j.petrol.2010.01.011

Zhou, X., Morrow, N. R., & Ma, S. (2000). Interrelationship of wettability, initial water saturation, aging time, and oil recovery by spontaneous imbibition and water flooding. *SPE Journal, 5*(2), 199-207. http://dx.doi.org/10.2118/62507-PA

Appendix

The disjoining pressure is expressed in terms of the change in the sum of interfacial tensions of the surrounding surfaces with respect to its thickness (Basu and Sharma, 1996). This can be expressed as:

$$\Pi = -\frac{\partial(\sigma^{\alpha\gamma} + \sigma^{\beta\gamma})}{\partial h}\Big|_{@T} \tag{A.1}$$

Where $\sigma^{\alpha\gamma}{}_{\infty}, \sigma^{\beta\gamma}{}_{\infty}$ the interfacial tensions for thin fluid films and h is the film thickness. By integration of this equation, the critical pressure is calculated as:

$$(\sigma^{\alpha\gamma} + \sigma^{\beta\gamma}) - (\sigma^{\alpha\gamma}{}_{\infty} + \sigma^{\beta\gamma}{}_{\infty}) = -\int_{\infty}^{h(P_c)} \Pi dh \tag{A.2}$$

Where $\sigma^{\alpha\gamma}{}_{\infty}, \sigma^{\beta\gamma}{}_{\infty}$ are bulk interfacial tensions between liquids and solid surface. $h(P_c)$ is the film thickness at capillary pressure. The final film tension is:

$$\sigma^{f} = \sigma^{\alpha\gamma} + \sigma^{\beta\gamma} + \Pi h \tag{A.3}$$

Where, Π is disjoining pressure at film thickness h. By applying Equation (A.2) the following relation is obtained:

$$\sigma^{f} = (\sigma^{\alpha\gamma}_{\infty} + \sigma^{\beta\gamma}_{\infty}) - \int_{\infty}^{h(P_c)} \Pi dh + \Pi h \tag{A.4}$$

Using the derivative chain rule, the following equation can be written:

$$\sigma^{f} = (\sigma^{\alpha\gamma}_{\infty} + \sigma^{\beta\gamma}_{\infty}) - \int_{0}^{P_c} h d\Pi \tag{A.5}$$

The contact angle can be expressed as applying force balance calculation:

$$\cos(\theta) = \frac{\sigma^{f} - \sigma^{\beta\gamma}_{\infty}}{\sigma^{\alpha\gamma}_{\infty}} \tag{A.6}$$

Hence; the disjoining pressure is directly related to the contact angle by the following equation:

$$\cos(\theta) = 1 + \frac{1}{\sigma^{\alpha\gamma}_{\infty}} \int_{0}^{P_c} h d\Pi u \tag{A.7}$$

One can divide the variation of disjoining pressure curve versus film thinness in three cases. In the first case (Type 1), the area under disjoining pressure versus film thinness curve is positive or zero ($I \geq 0$); accordingly the wetting phase is stable and cannot create an angle (or zero contact angle) between solid surface and two liquid phases. In the second case (Type 2), the area under disjoining pressure versus film thinness curve is between zero and $-\sigma^{\alpha\gamma}_{\infty}$ ($-\sigma^{\alpha\gamma}_{\infty} \leq I \leq 0$) where the wetting phase is meta-stable and the contact angle between zero and 90° is created. In the third case (Type 3), the area under disjoining pressure versus film thinness curve is less than $-\sigma^{\alpha\gamma}_{\infty}$ ($-2\sigma^{\alpha\gamma}_{\infty} \leq I \leq -\sigma^{\alpha\gamma}_{\infty}$) and the wetting phase is unstable with the contact angle between 90° and 180°.

Energy Input in the Production of Cassava

A. Isaac Bamgboye[1] & Babajide S. Kosemani[1]

[1] Department of Agricultural and Environmental Engineering, University of Ibadan, Ibadan, Oyo State, Nigeria

Correspondence: A. Isaac Bamgboye Department of Agricultural and Environmental Engineering, University of Ibadan, Ibadan, Oyo State, Nigeria. E-mail: isaacbam22@yahoo.com

Abstract

Energy is one of the largest components of the production cost in Agricultural activities. The efficiency of its use will often be compromised in favour of other equally important factors. Data were collected in 10 cassava farms by using face to face questionnaire method to determine the energy input in cassava production. Mathematical expressions were used to evaluate the energy requirement for each of the defined unit operations. Energy requirement in land preparation, planting, crop maintenance (fertilization and weed control) and harvesting were determined. It was observed that 78.67% of the total energy input used in cassava production was indirect, while 21.33% was direct. The average energy input in the production of cassava was 8571.26 MJ/ha, while 9960.00 kg was the average yield obtainable per hectare. Energy input in fertilizer was the highest with 64.0% of the total energy input; followed by diesel fuel with 19.50%. The net energy and energy productivity value were estimated to be 46,655.77 MJ/ha and 1.18 MJ/kg respectively. The ratio of energy output to energy input was 7.1.

Keywords: Energy, cassava, renewable fuel, energy ratio, energy productivity

1. Introduction

Agriculture is both a producer and consumer of energy. It uses large quantities of locally available noncommercial energy, such as seed, manure and animate energy, as well as commercial energies, directly and indirectly, in the form of diesel, electricity, fertilizer, plant protection, chemical, irrigation water, machinery etc. Efficient use of these energies helps to achieve increased production, productivity and contributes to the profitability and competitiveness of agricultural sustainability in rural living (Singh *et al*., 2002).

Energy inputs of different forms in agricultural production in almost every country are on the high levels. The reduction of these inputs by using agricultural products and residues as renewable energy sources is and will remain an important issue in future. It is closely related to the environmental problems arising from agricultural production on large surfaces. Therefore, the rational and efficient use of energy is essential for sustainable development. Chamsing, et al., 2006 in their work on some crops stated that for irrigated rice, rainfed rice, maize, wet-season soybean and cassava, energy input varied between 1.79-18.49, 10.09-13.11, 9.79-12.79, 5.21-10.03 and 4.95-9.13 GJ/ha respectively. They also observed that about 62% and 38% of energy inputs in farm operations was from indirect and direct energy inputs respectively. Energy from fertilizer contributed the highest followed by energy from seed, pesticide and herbicide.

Average energy ratio of selected crop production of Thailand was 5.20. This ratio is higher than the ratio of Turkish (1.18) in crop year 2000 (Ozkan et al., 2004) but lower that the ratio of Bangladesh (8.11) in crop year 2000/01 (Alam et al., 2005). Energy ratio for rice, maize and sugarcane production were depended on area and were nearly of the same range as India which were 2.15-12.75 for rice, 4.83-17.02 for maize and 5.82-6.67 for sugarcane production (Singh et al., 1997).

Both direct energy use for crop management and indirect energy use for fertilizers, pesticides and machinery production have contributed to the major increases in food production seen since the 1960s (Woods, 2010). However, the relationship between energy inputs and yields is not linear. Low-energy inputs can lead to lower yields and perversely to higher energy demands per tonne of harvested product. At the other extreme, increasing energy inputs can lead to ever-smaller yield gains.

Cassava (*Maninot esculenta*) is a perennial woody shrub with up to 32% (fresh) starch content which is cultivated extensively as a food crop in Africa, is the third largest source of carbohydrate in food for human consumption in the world. Cassava roots play an important role in the African diet and they are processed, using simple traditional methods, into products such as *Gari*, *Fufu* and *Lafun* flour. It is estimated that about ten million tonnes of cassava is processed for *Gari* annually in Nigeria alone. Because of recent characterization of cassava starch properties, market demand for cassava for the production of starch-based products has increased rapidly. This has led to an increase in the cassava growing area, which has reached about 500,000 ha in the past two years. Cassava production system is also changing from being a small-scale subsistence crop to a large-scale commercial crop. This paper determines the energy requirement in the production of cassava in each of the stages of production. This involves land preparation, planting, crop maintenance (fertilization and weed control) and harvesting.

2. Materials and Methods

Ten farms were surveyed to collect data on fuel, natural gas, fertilizer, pesticides and chemicals used on the farm for cassava production. The areas of study covered Oyo, Ogun, Osun and Kwara States of Nigeria. Four farms were surveyed in Oyo States, while two farms were surveyed in Ogun, Osun, and Kwara states respectively. The data for energy input resources in all the selected farms during cassava production from land preparation to transportation to the market or house was collected using structural questionnaire and oral interviews. The information obtained was extracted personally from the available documents in the surveyed areas. The input energy was divided into direct and indirect energy. The direct energy consists of diesel, human power and electricity, while the indirect energy contains seeds, fertilizers, farmyard manure, chemicals and machinery.

3. Processing Technology and Method of Energy Evaluation of Cassava Production

The type and magnitude of the energy consumed is a function of the process and the technology employed. In order to quantify the energy demands of each unit operation, quantitative data on operating conditions was required for each unit operation. Table 1 summarizes the production technologies under study of the cassava production.

Table 1. Measured Parameters for Evaluating Energy Input in Cassava Production

S/N	Operation	Required Parameter
1	d preparing	me taken for preparing the land, h Number of person involved fuel consumed, l Calorific value of fuel used, J/l
2	Stem harvesting	Time taken for preparing the stem, h
3	Planting	Number of person involved. Time taken to plant the stem, h
4	Herbicide application	Number of person involved. Time taken to apply the herbicide, h
5	Fertilizer application	Number of person involved. Time taken to apply the fertilizer, h
6	Inter – row Weeding	Number of person involved. Time taken for the weeding, h
7	Harvesting	Number of person involved. Time taken for the harvesting, h
8	Transportation	Number of person involved. Time taken for the harvesting, h

The types and magnitude of the parameters required for the energy evaluation of each unit operation are presented in Table 2. Jekayinfa and Bamgboye (2005, 2006) used similar procedure in the energy audits of palm-kernel oil mills and cashew-nut mills in Nigeria.

Table 2. Production Techniques at Each Stages of Production of the Cassava

S/N	Operation	Equipment and Principle Adopted
1	Land preparing	Mechanized. Tractor 4W, Disc plow Disc harrow Ridger
2	m harvesting and Gathering	Manual with the use of Cutlass
3	nting	nual with the use of cutlass.

4	Herbicide and Pesticide Application	Manual knapsack sprayer. Boom sprayer-hydraulic
5	Fertilizer application	Manual
6	Inter – row Weeding	Manual with the use of cutlass.
7	Complete Harvesting	De-stumping Manual
	De-stumping axe or machete	Use of truck
	Transportation	Transportation to factory, Tractor 4W – Trailer

The energy evaluation methods for each unit operation are as follows.

Land Preparation

Initial land clearing of field was done using tractor coupled with a disc plow for primary tillage, while secondary tillage was done using a disk harrow or rotary tiller. The time and fuel consumed by the tractor per hectare during each stage was obtained and used to compute the energy. The energy consumed for land clearing was obtained from the expression:

$$E_{lp} = 47.8\,D + 3.6(0.075\,N\,Ta) \quad \text{MJ} \tag{1}$$

Where

D = Amount of diesel consumed per unit operation, L.

0.075 = Energy input of an average adult male, MJ/h.

Ta = Useful time spent by a male worker per unit operation,

N is the number of persons involved in an operation

Stem harvesting and gathering

The amount of bundle required for planting one hectare, the time and the number of people required was obtained from the farmer and was used in the computation of the energy. The energy consumed was obtained from the expression:

$$E_{shg} = 3.6(0.075\,N\,Ta) \quad \text{MJ} \tag{2}$$

Planting

The amount of time required for planting and the number of people that performed the operation was obtained from the survey and used in the computation of the energy. The energy consumed was obtained from the expressing:

$$E_{p} = 3.6(0.075\,N\,Ta) \quad \text{MJ} \tag{3}$$

Pesticide Application

Pre emergence herbicide spraying was done using a tractor-mounted boom sprayer immediately after the planting operation in some farms. While both pre emergence herbicide spraying and post emergence herbicide spraying was done manually in most of the farms with knapsack sprayer in the farm surveyed. The amount of fuel used, time consumed, number of people involved in the operation, and amount of the pesticide and herbicide used was obtained through the questionnaire during the survey. The energy consumed was obtained from the expression:

When the operation is carried out manually

$$E_{[illegible]} = 3.6(0.075\,N\,Ta) \quad \text{MJ} \tag{4}$$

Pesticide energy input (MJ/ha)

Total pesticide input (MJ/ha) = $E_{[illegible]} \cdot P_{eav}$

$E_{[illegible]}$ = Applied rate (kg or lit/ha) of pesticide 1th for applied time

P_{eav} = Energy equivalent MJ/kg or litres of pesticide

When tractor-mounted boom sprayer is used

$$E_{[illegible]} = 47.8\,D + 3.6(0.075\,N\,Ta) \quad \text{MJ} \tag{5}$$

Fertilizer Application

This is done manually in all farms considered. The time required by the farmers to apply the fertilizer, the number of people that applied it, and quantity of fertilizer used per hectare was obtained and the energy input obtained with the following expression:

$$E_{fa} = 3.6(0.075\ N\ Ta) \quad \text{MJ} \qquad (6)$$

To calculate the labour input

While

Total fertilizer input (MJ/ha)

$$= \sum_{n=1}^{n} \frac{N.N_{eqv}}{AP} + \frac{P_2O_5.P_{eqv}}{AP} + \frac{K_2O.K_{eqv}}{AP} \qquad (7)$$

N_{eqv} = Energy equivalent value of N = 78.1MJ/kg

P_{eqv} = Energy equivalent value of P_2O_5 = 17.4MJ/kg

K_{eqv} = Energy equivalent value of K_2O = 13.7MJ/kg

N = Compound fertilizer rate applied × percentage of N ingredient (kg)

P_2O_5 = Compound fertilizer rate applied × percentage of P_2O_5 ingredient (kg)

K_2O = Compound fertilizer rate applied × percentage of K_2O ingredient (kg)

n = Compound fertilizer for applied time 1th

Weed control

Pre-emergence herbicide spraying was done immediately after planting to control the weeds and a contact herbicide was also sprayed in between planting rows to control the weeds after planting in some farms. Some farmers weed their farms by cutting the weed with cutlass and hoe. For herbicide application, the energy input was obtained by measuring the amount of herbicide used and the time it took to apply the herbicide. Energy input was obtained by determining the time required and the numbers of people required for cutting using cutlass and hoe. The follow expressions were used to obtain the energy input.

When herbicide spraying was used:

$$E_{cw} = H_{hav}.\ H_{eqv} + 3.6(0.075\ N\ Ta) \quad \text{MJ} \qquad (8)$$

Where

Total herbicide input (MJ/ha) = $H_{hav}.\ H_{eqv}$

H_{hav} = Applied rate (kg or lit/ha) of herbicide 1th for applied time

H_{eqv} = Energy equivalent MJ/kg or litres of herbicide

When weed is done by cutting:

$E_{cw} = 3.6(0.075\ N\ Ta)$ MJ

Biological energy input

The stem introduced into the soil was considered in the energy computation. The amount of stem (bundle), the amount of a stem stalk in a bundle and the weight of the stem was obtained. Biological energy input was obtained from the expression:

$$E_{bi} = \text{Amount of stem applied (kg/ha)} \times \text{Energy equivalent of stem (MJ/kg)} \qquad (9)$$

Where

Energy equivalent of cassava stem = 5.6 MJ

Weight of each stalk input = 0.075 kg.

No of stalk in a bundle = 30

Complete Harvesting

Harvesting was done manually in all the farms. The time taken by the farmer to harvest the cassava root, the number of people involved, and the yields were obtained from the farms. Energy input in this operation was obtained from the expression.

$$E_{hw} = 3.6(0.075\ N\ Ta) \quad \text{MJ} \tag{10}$$

$$\text{Energy use efficiency} = \frac{\text{energy output (MJ/ha)}}{\text{[illegible]}} \tag{11}$$

$$\text{Energy productivity} = \frac{\text{cassava output (t/ha)}}{\text{[illegible]}} \tag{12}$$

$$\text{Net energy} = \text{Energy output (MJ/ha)} - \text{Energy Input (MJ/ha)} \tag{13}$$

$$\text{The energy output} = \text{yield} \times \text{energy equivalent} \tag{14}$$

Table 3. Energy Equivalent of Inputs and Outputs in Agricultural Production

Input (unit)	Energy equivalent (MJ)	References
Chemicals (kg)	120	Singh 2002
Human labour (h)	1.96	Singh, 2002
Machinery (kg)	62.71	Singh, 2002
Nitrogen (kg)	78.1	Demerirean et al., 2006
Phosphorus (kg)	17.4	Demerirean et al., 2006
Potassium (kg)	13.7	Demerirean et al., 2006
Cassava (kg)	5.6	Demerirean et al., 2006
Diesel fuel (L)	56.3	Singh, 2002

5. Results and Discussion

5.1 Result and Discussion

Energy input in the production of cassava varied from 7388.6 – 10888.66 MJ/ha as shown in Table 4. This is relatively higher than 4950 – 9130 MJ/ha obtained from cassava in Thailand (Chamsing et al., 2006). From the table, energy varied from one farm to another. The variation is caused majorly by the different amount of biological energy input, chemical energy input and difference in method of equipment acquisitions. Tractor and equipment used in the cassava production in this region were obtained through rental, private and partnership. The distance the tractor has to travel to get to the farm and the fuel used in transporting the tractor down to the farm was accounted for. The fertilizer application to the farm also varies, thus contributed to the variation.

Human labour varied from 90.56 – 421.5 MJ/ha, which was higher than 56.5MJ/ha from machinery, but lower than the input fuel of 239 – 2485.6 MJ/ha. Farm 2 was observed to have the highest energy input of 10,888.66 MJ/ha with the yield of 12,000 kg/ha, which happens to be the highest cassava yield in this study. It has been observed that low-energy inputs can lead to lower yields and conversely to higher energy demands per kg of harvested product (Woods, et al., 2010). High energy input was attributed to high amount of fuel used in transporting the tractor to the farm from the research institute and the high yield was due to high amount of fertilizer usage. Farm 10 has the lowest energy input of 6634.95 MJ/ha with a corresponding yield of 7400 kg/ha. This is as a result of low energy input in fertilizer which resulted in low yield.

The input and output energy values used in cassava production are illustrated in Table 5. The average total energy input of 8560.03 MJ/ha was required in the production of cassava. This is similar to what was obtained in Thailand from cassava (Chamsing, et al., 2006).

The pattern of energy use as shown in Fig. 1 are chemical fertilizers 64.00%, 19.50% from diesel oil and machinery, human labour 2.20% and 6.67 % of biological energy (stem). About 77.5% of the total energy inputs used in cassava production was indirect (stem, fertilizers, chemicals, machinery) and 22.5% was direct (human labour, diesel). Mean cassava yield was about 9,960.00 kg/ha. This is close to the national average yield for cassava in Nigeria which varies from 10,000 kg/ha to 15,000 kg/ha (Phillips et al., 2004).

The net energy and energy productivity value were 46,655.77 MJ/ha and 1.18 kg/MJ, respectively (Table 5). Energy use efficiency of 7.01 was obtained in this work indicates an intensive use of energy inputs in the cassava production. This was within 6.3 to 9.1 obtained by Chamsing, et al (2006) in Thailand. Higher energy ratio in Thailand is as a result of higher yield of cassava 14300 (kg/ha) as compared to 9960 (kg/ha) obtained from this study.

Table 4. Energy Input and Output in the Production of Cassava (MJ/ ha)

Item	farm 1	farm 2	farm 3	farm 4	farm 5	farm 6	farm 7	farm 8	farm 9	farm 10
Human Labour	142.7	90.56	158.25	149.25	104.62	141	330	285	421.5	176.25
Machinery	56.5	56.5	56.5	56.87	56.87	56.5	-	56.5	-	56.5
Fuel	1957.4	2485.6	2151	2103.2	2390	1764.6	239	1912	239	2045.2
Pesticide and Herbicide	1200	1200	1200	480	600	1200	480	960	600	600
Nitrogen Fertilizer	2343	4686	2343	3514.5	4686	3514.5	5857.5	3514.5	4100	857.5
Phosphorus Fertilizer	522	1044	522	783	1044	783	1305	783	913.5	1305
Potassium Fertilizer	411	822	411	615	822	615	1027.5	615.5	512.5	1027.5
Biological Energy(Cassava Stem)	756	504	567	562.50	567	693	630	441	754	567
Total Energy(MJ/ha)	7388.6	10888.66	7408.75	8264.32	10270.49	8767	9869	8567.5	7540.5	6634.95
Yield (kg/ha)	10000	12000	11300	9500	11200	8500	11500	9200	9000	7400
Energy Output(MJ/ha)	56000	67200	63280	53200	62720	47600	64400	51520	50400	41440

Table 5. Energy output – input ratio and energy forms in cassava production

Items	Unit	Quantity
Energy ratio	-	7.01
-		
Energy productivity	kg/MJ	1.18
-		
Net energy	MJ/kg	46,655.77
Direct energy	MJ/kg	1,928.61
22.5		
Indirect energy	MJ/kg	6,642.75
77.5		
Total energy input	MJ/kg	8,571.26
-		
Total energy output	MJ/kg	55,776
-		

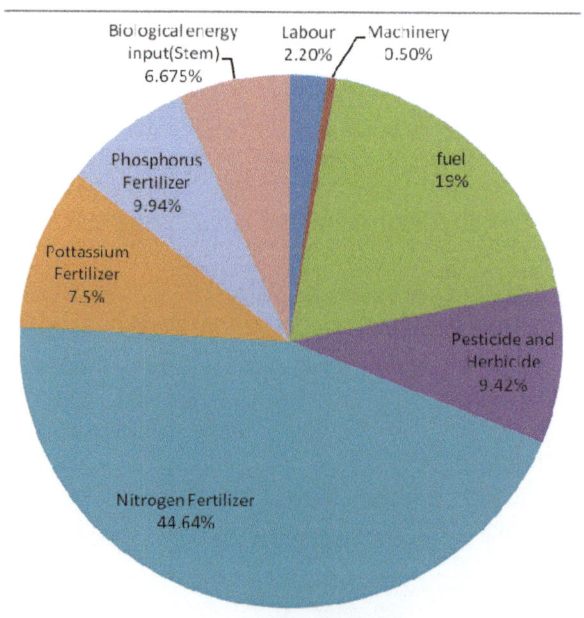

Figure 1. Pie Chat of the Energy Input in the Production of Cassava

Conclusion

The energy input in each of the cassava production stages was classified into direct and indirect energy. The indirect total energy input used in cassava production was 77.5%, while 22.5% was direct. The average energy input in the production of cassava was 8,571.26 MJ/ha, while the average yields obtainable per hectare was 9,960.00 kg.

References

Alam, M. S., Alam, M. R., & Islam, K. K. (2005). Energy flow in agriculture: Bangladesh.*American Journal of Environmental Sciences, 1*(3), 213-220.

Bamgboye, A. I., & Jekayinfa, S. O. (2006). Energy Consumption Pattern in Palm Kernel Processing Operations. Agricultural Engineering International: the CIGR Ejournal. *Manuscript EE05013, VIII.*

Chamsing, A., Salokhe, V., & Singh, G. (2006). "Energy Consumption Analysis for Selected Crops in Different Regions of Thailand". Agricultural Engineering International: the CIGR Ejournal. Manuscript EE 06013. VIII.

Jekayinfa, S. O., & Bamgboye, A. I. (2005). Estimating Energy Requirement in Cashew (*Anacardium Occidentale* L.) *Nut Processing operations Energy, 31*(8-9), 1305-1320

Ozkan, B., Akcaoz, H., & fert, C. (2004). Energy input-output analysis in Turkish agriculture. *Renewable energy, 29*(2004), 39-51

Phillips, T. P., Taylor, D. S., Sanni, L., & Akoroda, M. O. (2004). A cassava industrial revolution in Nigeria: The potential for a new industrial crop. Retrieved from http://www.fao.org/docrep/007/y5548e/y5548e00.htm (March, 2011).

Singh, H., Mishra, D., & Nahar, N. M. (2002). Energy use pattern in production agriculture of typical village in arid zone", *India—part–I. Energ Convers Manage, 43*, 227-586.

Woods, J., Williams, A., Hughes, J. K., Black, M., & Murphy, R. (2010). Energy and the food system. Phil. Trans. R. Soc. B365, 2991–3006.

Using Simulation to Test the Reliability of Regression Models

Fred J. Rispoli[1] & Vishal Shah[2]

[1] Department of Mathematics and Computer Sciences, Dowling College, Oakdale, The United States

[2] Department of Biology, Dowling College, Oakdale, The United States

Correspondence: Professor Fred J. Rispoli, Department of Mathematics, Dowling College, Oakdale, NY 11769, The United States. E-mail: rispolif@dowling.edu

Abstract

In many sciences, it is standard laboratory practice to use a statistical design of experiment and a regression model to study the influence of multiple parameters under a wide range of conditions. The current study aims at investigating the reliability of regression models by examining recently published models. Of particular interest are the assumptions that are not robust to violation such as the reliability of measurements, constant variation of residuals, and sample size. To test regression models simulation is used to model potential measurement error and the importance of sample sizes on parameter estimation. The randomly perturbed designs are then used together with associated mathematical models obtained from the original designs to simulate experiments and obtain new regression models. A comparison of the original model to the new model, and various statistical tests are performed to determine how accurate the original parameters have been predicted when exposed to simulated measurement error.

Keywords: design of experiments, environmental science, multiple regression analysis, simulation

1. Introduction

Scientists perform experiments in virtually all areas of study, often to determine a relationship between numerous input factors and either one or multiple output factors. The area known as the Design of Experiments (DOE) is concerned with planning and conducting of experiments, as well as analyzing the resulting data so that valid and objective conclusions are obtained (Montgomery, 2009). Factorial experiments are often used as an experimental strategy in which inputs are varied together. These experiments are an important class of experiments because they may be used to accomplish a variety of different goals, such as perform factor screening, or to determine optimum factor levels. In this study we focus on experiments in environmental sciences, where factorial experiments are primarily used to study the influence of multiple parameters (physical, chemical or biological.) Upon performing experiments dictated by a factorial design, a multiple regression model is created to make predictions and inferences. This is a widely used method, in fact more than 4,000 hits were obtained with the keywords "multiple regression analysis" in the Science Citation Index just within the areas of Environmental Sciences.

Typically, when evaluating a regression model one uses the coefficient of determination (R^2) to confirm the goodness of fit of the model to that of experimental data. Coefficients of each variable and its associated *p* value are also used to help assess the influence of the variable on the process under study. But R^2 values can be made artificially large by including an excessive number of terms, and p-values only indicate if a term is statistically significant and do not assess the accuracy of parameter estimation. Statisticians have studied the reliability of regression models and have identified a "reliability matrix" (Gleser, 1992) to help assess the model. It is widely known that measurement errors influence regression models (Pagano & Anoke, 2013). However, a reliability matrix is rarely used in an environmental science study.

For a multiple regression model to be reliable, a necessary condition is that it does not violate the regression model assumptions (Kahane, 2008). However, a literature search reveals that in environmental sciences, multiple regression models are often not tested for their robustness to regression assumptions prior to interpretation and drawing conclusions. Two of the key model assumptions that are made are:

1. The independent and dependent variables do not contain measurement errors.

2. Residuals of the model are independent over time, normally distributed, have mean zero, and exhibit constant variation.

To further exacerbate the problem, scientists have known for a long time that when carrying out experimental studies two types of errors are introduced during input and output analysis: precision errors and accuracy errors (Taylor, 1982). Precision errors are related to the random errors associated with an experiment (e.g. measurement error), whereas accuracy errors are related to the systematic differences observed between laboratories (e.g. different calibration of the instruments). A regression model developed based on experimental data should also be robust to these experimental errors. Compounded with a limited data set available due to practical considerations, a regression model developed could provide a false sense of parameter efficiency and output predictability.

The current study is aimed at developing a method to evaluate the reliability of regression models when the input and output parameters are subjected to small random perturbations created using simulation. It is our assumption that the coefficients of the variables and their p values should not significantly change in reliable models, even when the variables are subjected to simulated random perturbations. Also addressed is how residual analysis of the models can help to evaluate the reliability of the model prediction.

2. Methodology

Twenty studies within the field of environmental sciences were selected to determine if the published models were robust enough to withstand simulated random perturbations of the input and output values uniformly distributed between ±5%. The studies examined along with errors measures (as described in this section later) are given in Table 1. The different types of experimental designs considered are: full factorial, fractional factorial, mixture design, Box-Behnken and central composite. The perturbations are used to assess how sensitive the model is to small changes in the input and output data. The changes could be a result of measurement error or possibly due to other types of process variation. To introduce a random perturbation of some value between ±5%, every coordinate of the design points and the output values as well, are multiplied by a simulated random number between 0.95 and 1.05. Table 2a provides an example of design points used in a 2^3 full factorial design that have been modified by multiplying by a simulated random number between 0.95 and 1.05 to obtain the design matrix given in Table 2b.

Table 1. The list of 20 studies examined along with error measures of the regression models published

Reference	**Type**	**MAPE**	**APE_{90}**
(Bhunia & Ghangrekar, 2007)	Full Factorial	11%	24%
(Prasad & Srivasta, 2009)	Full Factorial	54%	125%
(Lima, et. Al., 2007)	Full Factorial	5%	9%
Saadat & Karimi-Jashni, 2010)	Full Factorial	135%	444%
(Srinivasan & Viraraghavan, 2010)	Fractional Factorial	18%	39%
(Mobilia, Scipioni, Veglio & Sciano, 2010)	Fractional Factorial	27%	67%
(Dobrev, Pishtiyski, Stanchev & Mircheva, 2007)	Fractional Factorial	5%	9%
(Chen, Lin, Jones, Fu & Zhan, 2009)	Fractional Factorial	37%	109%
(Rispoli, et. al, 2010)	Mixture Design	3%	6%
(Abdullah & Chin, 2010)	Mixture Design	3%	6%
(Chen, Huang, Hsiao & Tsai, 2010)	Mixture Design	7%	14%
(Santafe-Moros, et. Al., 2005)	Mixture Design	38%	113%
(Gurkok, Cekmecelioglu & Ogel, 2011)	Box-Behnken	5%	8%
(Anunziata & Cussa, 2008)	Box-Behnken	674%	1679%
(Baskan & Pala, 2010)	Box-Behnken	33%	53%
(Dopar, Kusic & Koprivanac, 2011)	Box-Behnken	7%	19%
(Djoudi, Aissani-Benissad & Bourouina-Bacha, 2007)	Central Composite	21%	42%
(Landaburu-Aguirre, Pongracz, Peramaki & Keiski,2010)	Central Composite	57%	151%
(Mohajeri, Aziz, Isa & Zahed, 2010)	Central Composite	10%	17%
(Imandi, Bandaru, Somalanka, Bandaru & Garapati, 2008)	Central Composite	14%	27%

Table 2. (a) Full factorial 23 experimental design. (b) The factorial design with 5% random perturbations

x_1	x_2	x_3	x_1	x_2	x_3
+1	+1	−1	+1.04	+1.04	−1.02
−1	−1	+1	−1.03	−0.99	+1.00
−1	+1	+1	−1.01	+1.00	+0.99
+1	+1	+1	+1.00	+1.03	+1.05
+1	−1	+1	+0.98	−0.97	+0.96
+1	−1	−1	+1.03	−0.96	−0.97
−1	−1	−1	−0.97	−1.02	−1.03
−1	+1	−1	−0.96	+1.05	−1.04
		(a)		**(b)**	

Let **X** denote the entire set of original experimental design points stored as a matrix, and let **Y** denote the vector of experimental outputs. To introduce a random perturbation of some value between ±5%, every coordinate of the design points in **X**, and the output values as well, are multiplied by a simulated random number uniformly distributed between 0.95 and 1.05. This procedure yields the modified design matrix denoted by $\hat{\mathbf{X}}$. Once $\hat{\mathbf{X}}$ has been obtained we use the initial polynomial model together with $\hat{\mathbf{X}}$ to compute predicted output values. The predicted output values are then multiplied by a random number between 0.95 and 1.05 to yield $\hat{\mathbf{Y}}$. Next, perform a regression analysis using $\hat{\mathbf{X}}$ and $\hat{\mathbf{Y}}$, which generates a new polynomial model. Let β_i denote the coefficients of the original model. Observe that these are the parameters we are trying to estimate. We let $\hat{\beta}_i$ denote coefficients of the model obtained from $\hat{\mathbf{X}}$ and $\hat{\mathbf{Y}}$. For every parameter β_i, we compute the absolute percentage error (APE) given by $\left|\frac{\beta_i-\hat{\beta}_i}{\beta_i}\right|$. Using these values we then compute the mean absolute percentage error (MAPE). However, since the MAPE can be greatly influenced by one or two extreme terms, we also calculate the 90th percentile absolute percentage error (APE_{90}). Observe that 90% of the absolute percentage errors will be below APE_{90} and 10% will be larger. Additional nonparametric statistical tests such as the Sign Test can also be performed to test the reliability of the coefficients. A summary of the complete simulation process is given in Figure 1.

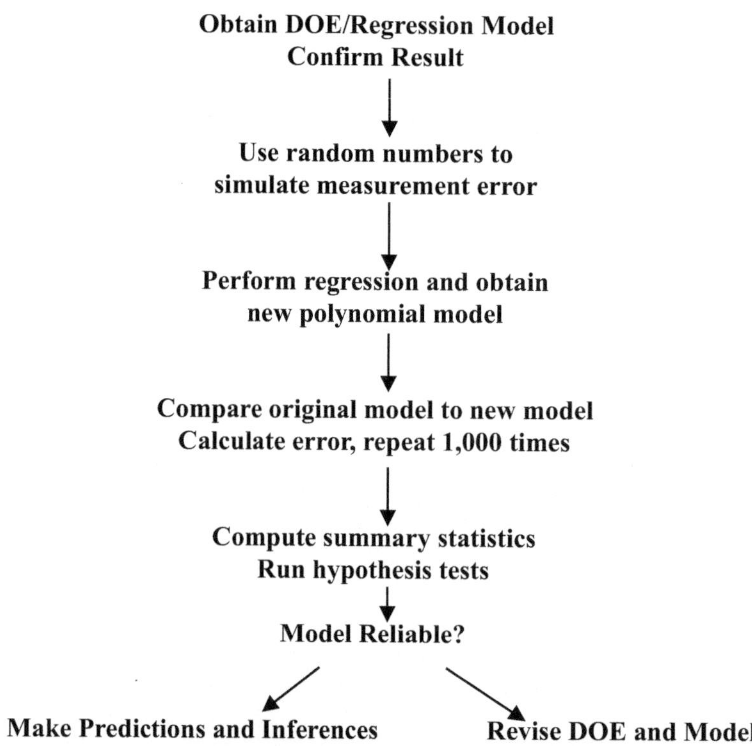

Figure 1. The simulation process to test regression model reliability

3. Results and Discussion

We propose through this study that once a regression model has been developed using a DOE, a random perturbation of ±5% should be introduced into the design points and the output values to help assess the model performance. The perturbation is intended to represent the measurement and systematic errors introduced when performing experiments with limited measurement resolution.

We begin with an analysis of how the type of experimental design influences the robustness with respect to measurement errors. A summary of the 20 studies is given in Table 3. The data illustrates a significant potential of errors influencing the results with Box-Behnken designs which produced results with an average MAPE of 180% and an average APE_{90} of over 400%. In contrast, mixture design seems to be more resilient to error with minimum % errors in both of the measures used. The percentages in Table 3 were obtained by finding the average MAPE and APE_{90} of the four studies in each design group shown in Table 1.

Table 3. A summary of the simulation results for the twenty studies examined

Design Type	MAPE	APE_{90}
Basic Factorial	51%	151%
Fractional Factorial	22%	56%
Mixture Design	13%	35%
Box-Benken	180%	440%
Central Composite	26%	59%

A second factor considered is the sample size measured by the ratio of number of design points divided by the number of predictor variables used in the final model. These ratios varied from a low of 1.25 to a maximum of 4.33. Surprisingly, the increase in this ratio does not necessarily decrease the MAPE. Sample size alone does not seem to be critical with respect to the robustness of the model. What is more important than sample size, is the location of the design points within the design space. In particular, the design types with points located throughout the design space, especially at or near the center, are the designs with the lowest MAPE. To test this rigorously we identified designs that have included design points in the interior of the design space vs. those that contain points only on the boundary. The results indicate a significant improvement in reducing the MAPE when interior points are included in the design.

4. Design Specific Analysis

The full factorial designs studies cited in Table 1 all had either 3 or 4 input variables, and between 8 and 16 predictor variables. The R^2 values for the models constructed were typically good and all above 0.95. When looking at the full factorial designs as a group we see that the MAPE's vary considerably with a minimum of 5% (Lima et al., 2007), to a maximum of 135% (Saadat & Karimi-Jashni., 2011). If one averages the MAPE's over the 4 studies we obtain a grand mean of 51% for the full factorial designs (Table 2). The design used in (Bhunia & Ghangrekar, 2007) is a full factorial 2^3 design with 11 design points which includes 3 center points, and yielded the model given in equation (1). The model constructed in (Saadat & Karimi-Jashni., 2011) is based on a full factorial 2^4 design, no additional interior or exterior points, but with repeated trials and is given below in equation (2).

$$Y = -26.1 + 190.78x_1 + 0.09x_2 + 159.65x_3 - 0.19x_1x_2 - 676.56x_1x_3 \tag{1}$$

$$Y = 57.4 + 18.0x_1 + 22.7x_2 - 1.2x_3 + 0.7x_4 - 12.8x_1x_2 + 4.9x_1x_3 - 0.1x_1x_4 - 6.0x_2x_3 + 0.1x_2x_4 - 0.1x_3x_4 - 2.3x_1x_2x_3 - 0.03x_1x_2x_4 + 0.08x_1x_3x_4 - 0.05x_2x_3x_4 - 0.2x_1x_2x_3x_4 \tag{2}$$

The residual plot for the model described in equation 1 suggests that despite low number of sample size, the assumptions are not violated. When introducing measurement error into the model constructed in (Bhunia & Ghangrekar, 2007), we obtained an MAPE of 11% and an APE_{90} of 24%.

The model (2) was obtained with a sample of 32 design points for 16 predictor variables, so the sample size is only twice the number of predictor variables. A potential source of error that arises in the model given in equation (2) is the inclusion of 3-way and 4-way cross terms with relatively small coefficients. Terms such as these are very sensitive to small measurement errors. Unless there is extreme confidence in the measurement system, we believe that a high degree term should not be included. Indeed, this is consistent with the "sparsity-of-effects-principle" which states that a system is usually dominated by main effects and low order interactions. The principle has been explained in depth by (Wu & Hamada, 2000). The residuals are essentially normally distributed with mean zero. However, the constant variation assumption is violated with a large variation in the vertical spread, and there is a nonrandom pattern with decreasing residuals illustrated in the residuals versus observation order plot. Hence time is related to decreasing residuals indicating a violation of time independent residuals. After introducing random perturbations we obtained an MAPE of 135% and an APE_{90} of 444%. Hence, the model is neither reliable nor robust to the input data variations. The study (Prasad & Srivasta, 2009) exhibits similar problems. There are too many variables artificially inflating R^2 and the constant variation of residuals requirement is violated.

Mixture designs have recently been developed and are included as an option in most computer aided experimental design software. For a comprehensive reference on mixture designs, see (Cornel, 1981). Unlike the previous two types of designs, mixture design have more than half of their points in the interior of the design space, including one point in the center. In full factorial designs the design space is an n-dimensional hypercube where n is the number of input variables. However, in a mixture design each point gives the proportion of each input into the mixture. Hence, there is a constraint that requires the sum of the proportions to be one. This implies that the design space is now an n-dimensional simplex.

The study (Abdullah & Chin, 2010) utilizes a simplex-centroid design for optimizing the composting of kitchen waste. The response is the carbon to nitrogen ratio Y, and the model developed is shown in equation (3). Observe that the model contains 3 input variables and 3 interaction terms for a total of 6 predictor terms.

$$Y = 14.9x_1 + 8.2x_2 + 281.6x_3 + 17.7x_1x_2 - 273x_1x_3 - 509.3x_2x_3 \quad (3)$$

The experimental design consists of 13 design points which yields a ratio of roughly 2 times as many sample points as predictor variables. The model shown in (3) had a high goodness of fit with a R^2 of 0.98. The residuals do not violate any of the residual assumptions. Perhaps a better assessment of accurate parameter estimation is indicated by the MAPE being only 2.7% and the APE_{90} at 5.8% (Table 1). The other studies in Table 1 based on a mixture design all produced remarkably similar results with respect to the error measures MAPE and APE_{90}, even when the residuals appeared to be violated. The mixture design is an example of a design that is very robust to small perturbations in the input data.

The next group considered are the Box-Behnken designs listed in Table 1, which are a special case of response surface designs. In (Gurkok et al., 2011) a three-level Box-Behnken design was used in an optimization study. The model developed is a second order model with three independent variables and eight predictor terms. An inspection of the residual plots confirms the residual assumptions. When perturbations are introduced the MAPE is 4.5% and the APE_{90} is 8.3%. Clearly this design is robust with respect to the regression assumptions.

In study (Anunziata & Cussa, 2008) a Box-Behnken was also used to develop a model with 11 predictor terms using 27 data points. The model is given in equation (4).

$$Y = 21.6404 + 1.52833x_1 - 2.52083x_2 - 12.5292x_3 + 8.875x_4 - 2.17x_1x_2 - 0.005x_1x_3 - 9.475x_1x_4 - 0.0125x_2x_3 + 0.825x_2x_4 - 1.525x_3x_4 \quad (4)$$

Inspection of the residual plots does not reveal any violations. However, when applying perturbations to the design we obtained an MAPE of 674%, with a APE_{90} of 1,679%. We believe that the model is hyper-sensitive to measurement error because of low number of data points used in the study.

We observed an inconsistency among Box-Behnken designs with respect to the inclusion of interior points in the design. The design used in (Annuziata and Cussa, 2008) did not include any center points. Whereas, the designs used in (Gurkok et al., 2011), (Baskan & Pala, 2010), and (Dopar et al., 2011) included 6, 5, and 3 center points respectively. The results obtained in (Baskan & Pala, 2010) and (Dopar et al., 2011) are much more reliable when subjected to random perturbations with an MAPE of 33% and 8% respectively.

The last group of regression models examined is those based on a central composite design listed in Table 1. Central composite designs are often used to obtain a quadratic model that will facilitate optimization. As illustrated in Table 3, this group would be ranked second best with respect to the error measures. The model that performed best when subjected to input perturbations was (Mohajeri et al., 2010), which involves 4 input variables and 30 design points; where 16 of the points consist of a 2^4 basic factorial design, 8 points are axial points, and 6 points are center points. The response Y for this model is the weathered crude oil removal percentage and is given in equation (5). Observe that there are 11 predictor terms which implies a run to predictor variable ratio of roughly 2.7. Moreover, there are no predictor terms with three factors.

$$Y = 68.43 - 10.55x_1 + 3.66x_2 + 10.14x_3 + 5.57x_4 + 4.79(x_2)^2 - 14.89(x_3)^2 - 7.08(x_4)^{\wedge 2} \quad (5)$$
$$- 2.28x_1x_2 + 8.63x_1x_3 + 1.81x_3x_4$$

The axial points are identical to the center points except for one factor, which will take on values both below and above the median of the two factorial levels, and typically both outside their range. Interestingly, the number of center points used in the central composite designs of Table 1 is: 12, 3, 6, 6 respectively, and the number of axial points used in these designs is 8, 0, 8 ,8, respectively. The design with 3 center points and 0 axial points was the only "non-robust" central composite design. As in the case of Box-Behnken designs, we see an inconsistency in the number of center and axial points used.

5. Conclusions

While the goal of the study was not intended to be comprehensive in the sense of testing a majority of designs, we present a simulation tool that could be used by investigators to test their designs and developed models for coefficient reliability. The study results in the following conclusions:

- The reliability of a regression model is dependent on the type and parameters of the experimental design. Including numerous design points in the interior of the design space, such as center and axial points, increases the reliability and robustness of the model.
- Scientists should evaluate the robustness of the regression model assumptions before making inferences. Models with very good R^2 and *p*-values may not be very reliable when measurement errors are taken into consideration leading to false inferences. Testing the regression model using simulation would provide a greater degree of confidence in the scientific inferences made. It may also provide new insight into the sensitivity of the scientific process being studied.
- High order terms should be avoided as much as possible. Primary effect terms and two-way interactions and squared term are usually sufficient and lead to models that are much more stable than models with high order terms.

Acknowledgement

The research was funded by National Science Foundation (CBET # 0966741).

References

Abdullah, N. C. N. (2010). Simplex-centroid mixture formulation for optimised composting of kitchen waste. *Bioresource Technol, 101*, 8205-8210.

Anunziata, O. A, & Cussa, J. (2008). Applying response surface design to the optimization of methane activation with ethane over Zn-H_ZSM-11 zeolite. *Chemical Eng. J., 138*, 510-516.

Baskan, M. B., & Pala, A. (2010). A statistical experiment design approach for arsenic removal by coagulation process using aluminum sulfate. *Desalination, 254*, 42-48.

Bhunia, P., & Ghangrekar, M. M. (2007). Statistical modeling and optimization of biomass granulation and COD removal in UASB reactors treating low strength wastewaters. *Bioresource Technology, 99*, 4229-4238.

Chen, L. C, Huang, C. H, Hsiao, M. C, & Tsai, F. R. (2010). Mixture design optimization of the composition of S, C, S_nO_2-codoped TiO_2 for degradation of phenol under visible light. *Chemical Eng. J., 165*, 482-489.

Chen, Y, Lin, C, Jones, G, Fu, S., & Zhan, H. (2009). Enhancing biodegradation of wastewater by microbial consortia with fractional factorial design. *J. Hazardous Mat., 171*, 948-953.

Cornell, J. A. (1981). Experiments with mixtures: designs, models and the analysis of mixture data. New York: Wiley.

Djoudi, W., Aissani, B. F., & Bourouina, B. S. (2007). Optimization of copper cementation process by iron using central composite design experiments. *Chemical Eng, J., 133*, 1-6.

Dobrev, G. T., Pishtiyski, I. G., Stanchev, V. S., & Mircheva, R. (2007) Optimization of nutrient medium containing agricultural wastes for xylanase production by *Aspergillus niger BO3* using optimal composite experimental design. *Bioresource Technology, 98*, 2671-2678.

Dopar, M., Kusic, H., & Koprivanac, N. (2011). Treatment of simulated wastewater by photo-Fenton process. Part 1: The optimization of process parameters using design of experiments (DOE). *Chemical Eng. J., 173*, 267-279

Gleser, L. J. (1992). The importance of assessing measurement reliability in multivariate regression. *J. American*

Statistical Association, 87(419), 696–707.

Gurkok, S., Cekmecelioglu, D., & Ogel, Z. B. (2011). Optimization of culture conditions for *Aspergillus sojae* expressing an *Aspergillus fumigates* β-galactosidase. *Bioresource Technol, 102*, 4925-4929.

Imandi, S. B., Bandaru, V. V. R., Somalanka, S. R., Bandaru, S. R., & Garapati, H. R. (2008). Application of statistical experimental designs for the optimization of medium constituents for the production of citric acid from pineapple waste. *Bioresource Technol, 99*, 4445-4450.

Kahane, L. H. (2008). *Regression Basics* (2nd ed.). Sage.

Landaburu, A. J., Pongracz, E., Peramaki, P., & Keiski, R. L. (2010). Micellar-enhanced ultrafiltration for the removal of cadmium and zinc: Use of response surface methodology to improve understanding of process performance and optimization. *J. Hazardous Mat., 180*, 524-534.

Lima, E. C, Royer, B., Vaghetti, J. C. P., Brasil, J. L., & Simon, N. M., et al. (2007). Adsorption of Cu(II) on Araucaria angustifolia wastes: Determination of the optimal conditions by statistic design of experiments. *J. Hazardous Mat., 140*, 211-220.

Mabilia, R., Scipioni, C., Veglio, F., & Sciano, M. C. T. (2010). Fractional factorial experiments using a test atmosphere to assess the accuracy and precision of a new passive sampler for the determination of formaldehyde in the atmosphere. *Atmospheric Environment, 44*, 3942-3951.

Mohajeri, L., Aziz, H. A., Isa, M. H., & Zahed, M. A. (2010). A statistical experiment design approach for optimizing biodegradation of weathered crude oil in coastal sediments. *Bioresource Technol, 101*, 893-900.

Montgomery, D. C. (2009). *Design and Analysis of Experiments* (7th ed.). New York: Wiley

Pagano, M., & Anoke, S. (2013). Mommy's Baby, Daddy's Maybe: A Closer Look at Regression to the Mean. *Chance Magazine, 27*(1).

Prasad, R. K., & Srivasta, S. N. (2009). Electrochemical degradation of distillery spent wash using catalytic anode: Factorial Design of Experiments. *Chemical Eng. J., 146*, 22-29.

Rispoli, F., Angelov, A., Badia, D., Kumar, A., & Seal, S., et al. (2010). Understanding the toxicity of aggregated zero valent copper nanoparticles again*st Escherichia coli. J. Hazardous Mat., 185*, 212-216.

Saadat, S., & Karimi, J. A. (2011). Optimization of Pb(II) adsorption onto modified walnut shells iusing factorial design and simplex methodologies. *Chemical Eng. J., 173*, 743-749.

Santafe, M. A., Gozalvez, Z. J. M., Lora, G. L., & Garcia, D. J. C. (2005). Mixture design applied to describe the influence of ionic composition on the removal of nitrate ions using nanofiltration. *Desalination, 185*, 289-296.

Srinivasan, A., & Viraraghavan, T. (2010). Oil removal from water by fungal biomass: A factorial design analysis. *J. Hazardous Mat., 175*, 695-702.

Taylor, J. R. (1982). An introduction to error analysis. University Science Books.

Wu, C. F. J., & Hamada, M. (2000). *Experiments: Planning, analysis and parameter design optimization.* New York: Wiley.

20

The CO_2 Gasification Kinetic Study of WEEE Plastic Char Derived from Medium Temperature pyrolysis

Hu Wu[1], Qi An[1], Yafei Shen[1], Noboru Harada[1], Lihao Chen[1] & Kunio Yoshikawa[1]

[1]Department of Environmental Science and Technology, Tokyo Institute of Technology, Yokohama, Japan

Correspondence: Hu Wu, Department of Environmental Science and Technology, Tokyo Institute of Technology, Yokohama, Japan. E-mail: wuhu2010wuhu@gmail.com

Abstract

The isothermal CO_2 gasification characteristics of three chars derived from medium temperature pyrolysis of waste electrical and electronic equipment (WEEE) plastics were studied by using a thermogravimetric analyzer (TGA) within the temperature range of 850−1050°C. Phenolic board (PB), brominated high impact polystyrene (HIPS) and acrylonitrile butadiene styrene (ABS) plastics are widely used for the electric and electronic equipment and were employed as model WEEE plastics for the char sample production in this study. The effects of their physicochemical properties and gasification temperature on the WEEE plastic chars conversion rate, the reactivity indexes and the gasification rate were investigated in detail. The random pore model (RPM), the extended random pore model (eRPM) and the shifted extended random pore model (s-eRPM) have been employed to fit the CO_2 gasification rate curve of WEEE plastic chars, respectively. The kinetic parameters and the correlation coefficients (R^2) were evaluated by RPM, eRPM and s-eRPM, respectively. It was found that the CO_2 gasification reactivity of PB char was the highest, followed by that of HIPS char and the gasification reactivity of ABS char was the lowest, which have a close relationship with their pore and carbon crystal structure properties. In addition, it was found that RMP could fit the gasification reaction rate of HIPS char well whose maximum reaction rate appeared at the char conversion of approximate 0.4. Nevertheless, as for PB char and ABS char, their maximum gasification rate presented at char conversion of around 0.8 and 0.2, respectively. And it was observed that eRMP and s-eRPM could predict their gasification rate of PB char and ABS char very well with higher R^2, respectively.

Keywords: WEEE plastic char, CO_2 gasification, kinetics, random pore model, activation energy

1. Introduction

Waste electrical and electronic equipment (WEEE) are currently considered to be one of the fastest growing solid waste streams in the world. According to the UNEP report (2009), 40 million tons of WEEE were generated and discharged annually and it was expected an alarming growth per year in the future (United Nations Environment Programme, 2009). For example, in the European Union (EU), more than 9 million tons of WEEE were generated in 2005, and it was expected to grow to more than 12 million tons by 2020 (United Nations Environment Programme, 2009). Additionally, it was well known that there are lots of valuable metals and plastics contained in WEEE, which are worthy recyclable feedstock and could be converted into valuable resources and fuels. On the other hand, WEEE contained certain dangerous and hazardous substances, such as toxic metals and brominated flame retardants, which will pose considerable environmental pollution and health risks if treated inadequately (Yang, Sun, Xiang, Hu & Su, 2013; Ongondo, Williams & Cherrett, 2011). Therefore, how to environmentally soundly and cost-effectively reuse, recycle and recover WEEE has drawn plenty of attention through the world.

Pyrolysis has been proposed as a viable processing route for recycling WEEE, which not only can convert WEEE plastics into fuels and chemical feedstock but also easily separate metals and plastic fractions of WEEE (Yang et al, 2013). However, because of the presence of brominated flame retardants in the WEEE plastics, there are large amount of organic and inorganic brominated compounds existing in the pyrolysis oil, thereby reducing the quality and hindering the reuse of pyrolysis oil. Therefore, lots of debromination methods including the addition of effective catalysts and additives in the pyrolysis process of WEEE plastics have been investigated to reduce the bromine content in the oil products (Yang et al, 2013). Bhaskar et al. synthesized the composite Fe-C and Ca-C additives, which could effectively remove the brominated compounds from the pyrolysis oil products (Bhaskar et al, 2002; Bhaskar et al, 2003). Terakado and Hirasawa (2011 & 2013) reported that metal oxides used in the pyrolysis of printed circuit boards containing brominated flame retardants could effectively suppress the formation of HBr and brominated organic compounds. Besides, there were many researches on zeolite catalysts used for the debromination in the pyrolysis of brominated WEEE plastics (Hall & Williams, 2008; Bozi & Blazsó, 2009; Hall, Miskolczi, Onwudili & Williams, 2008). The above research results indicated that the pyrolysis process could effectively convert the WEEE plastic into valuable, storable and transportable oil products used for fuel and chemical feedstock. In addition, they showed that metal recovery from the pyrolysis residues was relatively easy due to their friable and loose structure property (Zhang, Yoshikawa, Nakagome & Kamo, 2013).

However, after the low-medium temperature pyrolysis process, about 20 wt.% of WEEE plastic char remained (Hall & Williams, 2008), which is worth recyclable resource in order to increase the recycle efficiency of WEEE plastics. On the other hand, the existence of the char would be a potential problem for the recycling of further metal screening and refinement. The gasification behaviors of coal char and biomass char have already been widely studied (Zhang, Ashizawa, Kajitani & Miura, 2008; Zhang, Hara, Kajitani & Ashizawa, 2010; Yuan, Chen, Li & Wang, 2011; Jing et al, 2013). For instance, Zhang et al. (2008) have investigated the steam gasification of 14 kinds of biomass chars and found that the extended random pore model could predict the gasification rates of all 14 kinds of biomass chars very well. Jing et al. (2013) studied the CO_2 gasification behaviors of various coal chars and concluded that the CO_2 gasification behaviors had a close relationship with their physicochemical Properties, especially with the CO_2 chemisorbed volumes of chars. With regard to biomass and coal chars, the intrinsic alkali (K and Na), alkaline (Ca and Mg) and transition (Fe) metal contents also play important roles on the char gasification rates (Zhang, et al. 2008; Yuan et al., 2011; Lahijania, Zainala, & Mohamedb, 2012). In addition, the Si and Al metal contents generally exert severe inhibition effect on the gasification reaction (Zhang, et al., 2008; Zhang, et al., 2010; Yuan, et al., 2011). Consequently, for different biomass and coals char containing various amount of alkali and alkaline metal contents, their gasification performance are totally different. However, there are few alkali and alkaline metals exist in the WEEE plastic chars and few attempts have been made to investigate the CO_2 gasification reactivity of WEEE plastic chars derived from the medium-temperature pyrolysis (Gil, Fermoso, Pevida, Pis & Rubiera 2010). Additionally, the CO_2 gasification reactivity of chars is lower than that of O_2 or H_2O gasification, and consequently, it is regarded as the rate-determining step in the gasification process (Jing et al, 2013). The CO_2 gasification of WEEE plastic char might prevent the oxidation of metals existing in the chars. To this end, it is of great significance to understand the CO_2 gasification behavior of WEEE plastic char. The gasification reactivity and kinetic parameters are essential for the efficient and reliable design of the gasification system. Thermogravimetric analysis (TGA) is an ideal instrument for investigating the gasification characteristics of carbonaceous materials because of its high accuracy, simplicity of design and ease of operation (Yuan et al, 2011). By means of TGA, different models can be applied to evaluate the kinetic parameters and the activation energy.

In this study, the isothermal CO_2 gasification behaviors of three WEEE plastics (PB, HIPS and ABS) chars derived from the pyrolysis at 600°C were investigated by using the thermogravimetric analysis (TGA) at the temperatures of 850-1050 °C. Effects of the plastic char types, the reaction temperature on the char conversion, the reactivity index and the gasification rate were investigated in detail. Moreover, RPM, eRPM and s-eRPM were employed to fit the gasification rate of three WEEE plastic char at different temperatures in order to determine their gasification correlation coefficients, kinetic parameters and activation energies, respectively.

2. Experimental

2.1 Preparation of Samples

Three char samples were collected from our previous pyrolysis experiments of three typical WEEE plastics, i.e. phenolic board (PB), brominated high impact polystyrene (HIPS) and acrylonitrile butadiene styrene (ABS), respectively. In brief, 30g of each WEEE plastic was pyrolyzed in a quartz fixed-bed reactor, which was heated to 600 °C at a heating rate of 100 °C/min. The flow rate of the carrier gas (N_2) was 50 mL/min. After the reactor temperature reached 600 °C, it was held at this temperature for 2 hours and was then cooled quickly. Subsequently, three chars were collected, ground and sieved to obtain the powder char sample with a size between 150 and 250 μm. Three char samples were denoted by PB char, HIPS char and ABS char, respectively. The proximate and ultimate analysis were conducted according to ASTM D7582 and ASTM D3176 by DTG-50 TGA and JM 10 elemental analyzer, respectively. And the ash compositions of char samples were analyzed by the XRF. The analysis results are shown in Table 1.

Table 1. The proximate analysis, ultimate analysis and ash analysis of char samples

Sample	PB	HIPS	ABS
Proximate analysis (wt.%)			
Moisture	0.4	1.4	1.4
Volatile matter	21.5	19.1	15.5
Fixed carbon	76.8	53.1	73.8
Ash	1.3	26.5	9.3
Ultimate analysis (wt.%)			
C	90.7	94.6	95.3
H	2.2	2.8	1.4
N	0.3	0.3	1.7
Br	0.0	2.2	1.6
O[a]	6.8	0.0	0.0
Ash content (wt.% in ash)			
Sb	0.0	2.8	60.1
Ti	0.0	73.1	0.5
Fe	18.1	4.2	10.6
Cr	17.4	0.9	3.4
Ca	0.0	2.6	4.6
Cu	20.7	0.0	0.0
Br	0.0	13.4	12.0

[a] By difference

2.2 Analytical Methods

The CHN element analysis was conducted by using a Micro Corder JM 10 Elemental Analyzer. The bromine contents of the char was determined by using an air combustor coupled with Dionex ICS-1100 ion chromatography fitted with a **Shodex ICS1-904E** column according to JIS K 7392. The ash compositions of the char were determined by an energy dispersive X-ray fluorescence spectrometer (XRF) under vacuum mode for precise measurement. A powder X-ray diffraction (XRD) analysis was carried out for the verification of the crystallinity of three chars. XRD measurements were performed using a Rigaku Ultimal V diffractometer with the CuKα radiation ($\lambda = 1.540$) at 40kV and 40 mA. The XRD patterns were accumulated in the range of 5–50° every 0.02° (2θ) with the counting time of 1 s per step. The surface structure property was analyzed by the SEM. Surface area and textural properties of the used catalysts were determined by N_2 physical adsorption at 77 K, applying the Brunauer–Emmett–Teller (BET) method, using a Micromeritics Tristar 3020 equipment.

2.3 Experimental procedure and reactivity measurements

The isothermal CO_2 gasification experiments were conducted using the Shimadzu DTG-50 thermogravimetric analyzer. 10 mg of char sample was filled into a cylindrical alumina crucible and then was heated to the targeted temperature at the heating rate of 30 °C/min under N_2 atmosphere with a flow rate of 150 mL/min. After that, the temperature was kept at the targeted temperature for 10 minutes to maintain the thermal and weight equilibrium. Subsequently, the carrier gas was switched to CO_2 with the same flow rate to carry out the isothermal CO_2 gasification until the char sample weight became constant. The time when CO_2 was injected was defined as t = 0. The carbon conversion (X) and the gasification rate (r) of chars in the gasification experiment were calculated by the following equation:

$$X = \frac{W_0 - W_t}{W_0 - W_a} \tag{1}$$

$$\mathrm{r} = \frac{dX}{dt} \tag{2}$$

Where W_0 (mg) is the initial char weight at the gasification time $t = 0$. W_t (mg) is the instantaneous sample weight at the gasification time of t, and W_a (mg) is the left ash weight after finishing gasification. The reactivity index Rs has been widely used to compare the gasification reaction reactivity between different samples. It is given by $Rs = 0.5/t_{0.5}$ with $t_{0.5}$ being the time required for 50% carbon conversion (Lahijani Zainal, Mohamed & Mohammadi, 2013).

2.4 Kinetic Models

The most widely used oxygen-exchange mechanism for the CO_2 gasification of char, proposed by Reif (Reif, 1952), can be summarized as

$$C_f + CO_2 \leftrightarrow C(O) + CO \tag{3}$$

$$C(O) \rightarrow CO + C_f \tag{4}$$

Where C_f denotes a free carbon active site and C(O) denotes a carbon-oxygen complex.

According to above mechanism, various kinetic models have been proposed and investigated to characterize the CO_2 gasification of char, such as the homogeneous model, the shrinking model, the random pore model and the extended random pore model and the shifted extended random pore. In this work, three typical kinetic models, the random pore model (RPM), the extended random pore model (eRPM) and the shifted extend random pore model (s-eRPM) were used to characterize the CO_2 gasification of char in the absence and presence of carbonate catalysts. RPM, proposed by Bhatia and Perlmutter (1980), was developed on the base of overlapping cylindrical pores mechanism, which means that there are arbitrary pore size distributions in the char particle and all pores grow and coalesce as the reaction progresses. The reaction rate is expressed as:

$$\frac{dX}{dt} = K(1 - X)\sqrt{1 - \Psi \ln(1 - X)} \tag{5}$$

$$K = k_0 P_{CO_2}^n \exp\left(-\frac{E_a}{RT}\right) \tag{6}$$

Where X is the char conversion, K is the rate constant for the reaction on the pore surfaces, Ψ is the pore surface parameter, k_0 (min^{-1}) is the pre-exponential factor, Ea (kJ/mol) is the activation energy, and P_{CO2} (bar) is the CO_2 partial pressure. And the parameter Ψ can be calculated by the following equation:

$$\Psi = \frac{4\pi L_0(1-\varepsilon_0)}{S_0^2} \tag{7}$$

Where S_0, L_0 and ε_0 denote the pore surface area, the pore length and the total volume of the solid porosity, respectively.

RPM have been successfully used for modeling the gasification reaction of chars. However, it failed to describe the reactivity profiles of biomass chars or alkali metals catalyzed char, for which the reactivity increases with the increase of the conversion or exists a maximum gasification rate in higher or lower char conversion range (Zhang et al, 2008; Zhang et al, 2010). Therefore, the eRPM was proposed by Zhang et al (2008) by introducing empirical parameters c and p to better fit the experimental data obtained from the alkali metal catalyzed gasification of coal char and biomass char. The eRPM is expressed in the equation (9).

$$\frac{dX}{dt} = K(1 - X)\sqrt{1 - \varphi \ln(1 - X)}\,(1 + (cX)^p) \tag{8}$$

Where c is a dimensionless constant, and p is a dimensionless power law constant. If c = 0, eRPM becomes the same as RPM. However, it was reported that when the maximum gasification rate appears in the lower char conversion interval, the eRPM also cannot perform well (Yuan et al, 2011; Jing et al, 2013). The shifted extended random pore model (s-eRPM) was proposed by changing the conversion term $(1 + (cX)^p)$ of eRPM from multiplication to division in order to fit the gasification rate when maximum rate appears at lower char conversion (< 0.2) (Jing et al, 2013). The s-eRPM could be written by the following equation.

$$\frac{dX}{dt} = \frac{K(1-X)\sqrt{1-\varphi\ln(1-X)}}{1+(cX)^p} \tag{9}$$

In this study, the rate constant K, the structure parameter Ψ and two semi-empirical parameters c and p are independent of the temperature and estimated by the nonlinear least-squares method in the MATLAB software.

3. Results and Discussions

3.1. Characteristics of Char Samples

From the proximate and ultimate analysis shown in Table 1, it is found that the three chars derived from the pyrolysis at 600 °C still contained certain amount of volatile matter and various amount of ash and fixed carbon. For instance, the weight percent of fixed carbon in the PB char, HIPS char and ABS char is 76.8 wt.%, 53.1 wt.% and 73.8 wt.%, respectively. And the ash content of HIPS char is the highest (26.5 wt.%) while the ash content of

PB char is the lowest (1.3 wt.%). The main compositions of HIPS char ash are Ti and Br elements. And the major components of the ABS char was the Sb and Br. The metal element, such as Ti and Sb, come from the synergist additives and the Br element mainly derived from the brominated flame retardant. The synergist additive and brominated flame retardant are often fed into the WEEE plastic in an attempt to reduce its flammability (Yang et al, 2013; Hall et al, 2008). Additionally, there are some Fe element exist in the ash fractions for three chars, which might come from the steel reactor when the WEEE plastics were pyrolyzed.

The surface morphology characteristics of chars could be a very important parameter to evaluate the CO_2 gasification reactivity of char, which were observed by using SEM and are shown in Figure 1. The morphological structure of PB char showed irregular and sharp-edged property. The PB char exhibited porous and textural structure characteristics, which was in accordance with previous researches (Ke, Yang, Liu, Liu & Dong, 2013). As for the HIPS char, it was observed that HIPS char consisted of small relatively uniform particles and each particle are smooth and compressed. There were some porosity structure during each particle. A significant difference in morphology can be seen for the ABS char. The surface of ABS char was smooth and compact. No obvious pore structure was found in the surface of ABS char, which might be attributed to the agglomeration and melting during the pyrolysis of ABS plastic.

Figure 1. SEM micrographs (3,000X) of WEEE plastic chars: (a) PB Char, (b) HIPS Char, (C) ABS Char

Table 2. The pore structure properties of WEEE plastic chars

Char Samples	BET surface area (m^2/g)	Micropore volume (cm^3/g)	Average pore diameter (nm)
PB Char	309.32	0.136	4.47
HIPS Char	105.91	0.073	8.09
ABS Char	30.16	0.04	12.36

The pore structure characteristics of WEEE plastic chars were analyzed by N_2 adsorption-desorption whose results are shown in Table 2. It indicated that the PB char was more porous than the other two chars. For example, PB char has the largest BET surface area and micropore volume (309 m^2/g and 0.136 cm^3/g, respectively), which are much larger than those of ABS char (30.16 m^2/g and 0.04 cm^3/g, respectively). With regard to the surface area, it was considered to play a vital role in determining the gasification reactivity, which was attributed to the fact that the surface area could provide active sites and the char gasification reactivity was proportional to active sites and active surface area. For the pore size, it was reported that only the pores with diameters larger than 1.5 nm can contribute to the CO_2 gasification reaction (Yuan et al, 2011). Furthermore, the pore size distribution would influence the diffusion of reactants and products (Jing et al, 2013). As shown in Table 2, it was found that the average pore diameter of PB char was much smaller than those of HIPS char and ABS char, which indicated that PB char are richer in micro- and mesopores than those of HIPS char and ABS char. Therefore, the gasification reactivity of PB char is expected to be more active than those of HIPS char and ABS char.

The carbon crystalline structure is another essential feature of char, which is closely associated with the char CO_2 gasification behavior. Liu and co-workers reported that the carbon crystalline structure exerted greater effect on the gasification reactivity than the porosity structure for the chars prepared at different pyrolysis conditions (Jing et al, 2013; Huang, et al, 2009). In the studies by Jing et al. (2013), the carbon crystalline structure performed better than the alkali index and BET surface area on the correlation with the gasification reactivity for the different coal chars. According to the XRD patterns in Figure 2, the crystal plane index C(002) peaks were observed at approximately 24° in the all XRD spectrums of WEEE plastic chars. The C(002) diffraction peak represents the degree of parallel and azimuthal orientation of the aromatic lamellae (Huang, et al, 2009). The disordering of carbon crystalline structure results in the broadening of the C(002) diffraction peak: the more disordering of carbon crystalline structure, the wider the diffraction peak of C(002). It was found that the C(002) diffraction peak of ABS char was narrow and sharp while the C(002) peak of PB char was flat and wide. The width and height of HIPS char were between those of PB char and ABS char. It indicated that the carbon crystalline degree in the three chars was in the sequence of ABS char > HIPS char > PB char. Therefore, it was predicted that the gasification reactivity of three was in the order of PB char > HIPS char > ABS char. Additionally, as for HIPS char, the sharp peaks at around 27.6°, 36.5° and 41.9° represent TiO_2, while in the case of ABS char, the peaks at 28.9°, 32.2° and 44.5° are related to the presence of Sb_2O_3. As mentioned above, the TiO_2 and Sb_2O_3 are commonly employed as synergist and added in WEEE plastic to reduce its flammability (Yang et al, 2013; Hall et al, 2008).

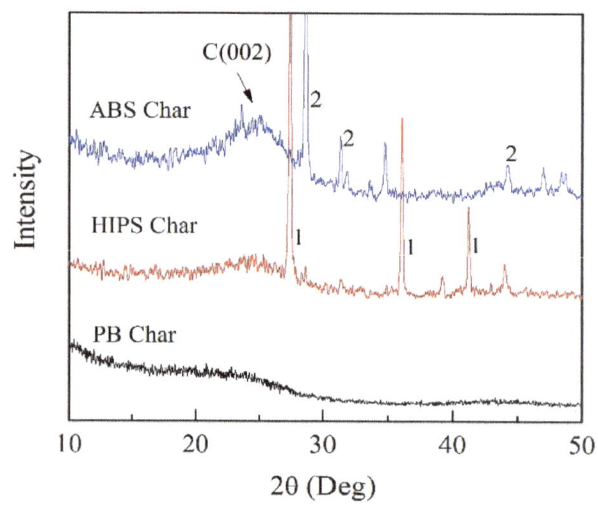

Figure 2. XRD patterns of WEEE plastic chars (1, TiO_2; 2, Sb_2O_3)

According to the analytical results discussed above, the conclusion could be drawn that the characteristics of three WEEE plastic chars are significantly different in the terms of the surface area, pore structure and the carbon crystalline structure. The effect of these properties on the gasification reactivity will be investigated in detail in the following section.

3.2 Effect of the Gasification Temperature on the Char Conversion

It is well known that the gasification temperature is an essential parameter in controlling the gasification reactivity of char. Therefore, the isothermal CO_2 gasification of PB char, HIPS char and ABS char was firstly conducted in the range of 850 – 1050 °C, and the result is shown in Figure 3. It indicated that, for all char samples, with the increase of the gasification temperature, the time for the total carbon conversion decreased. For instance, in the case of HIPS char, the gasification time required for the total carbon conversion was over 200 minutes at 850 °C, while the gasification time decreased to 27 minutes at 1050 °C. The gasification of HIPS at

1050 °C was approximately 7.3 times faster than that at 850 °C. As for PB char, the gasification time at 1050 °C was 11.7 times shorter than that at 850 °C. Meanwhile, in the case of ABS char, the gasification time at 850 °C was 3.8 times longer than that at 1050 °C. It illustrated that the higher the gasification temperature is, the faster the char conversion becomes for all samples. Additionally, it indicated that the gasification behavior of the three samples were significantly different because of their various physicochemical properties. At the gasification temperature of 850 °C, the time for the total char conversion of PB char, HIPS char and ABS char were 175.1, 204.7 and 238.2 minutes, respectively, while when the gasification was 1050 °C, the time for the total char conversion of PB char, HIPS char and ABS char were 15.2, 28.1 and 63.6 minutes, respectively. It was found that under the same gasification temperatures, the gasification reactivity of PB char was found to be the highest, followed by that of HIPS char, and the gasification reactivity of ABS char was the lowest, which might be attributed to its smallest surface area and most compact pore structure property, as shown in Figure 2 and Table 2.

In order to compare the gasification reactivity of three WEEE plastic chars in detail, the reactivity index Rs, proposed by Takarada et al. (1985), have been calculated in the temperature range of 850 °C – 1050 °C. It is given by $Rs = 0.5/t_{0.5}$ with $t_{0.5}$ being the time required for 50% conversion of the fixed carbon. The results are listed in Table 3, which illustrates that the reactivity index is in good agreement with the above mentioned result that the reactivity index of PB char was the highest, followed by that of HIPS and the reactivity index of HIPS was the lowest for all of the temperatures. In addition, as the increase of the gasification temperature, the reactivity index for each char also improved significantly.

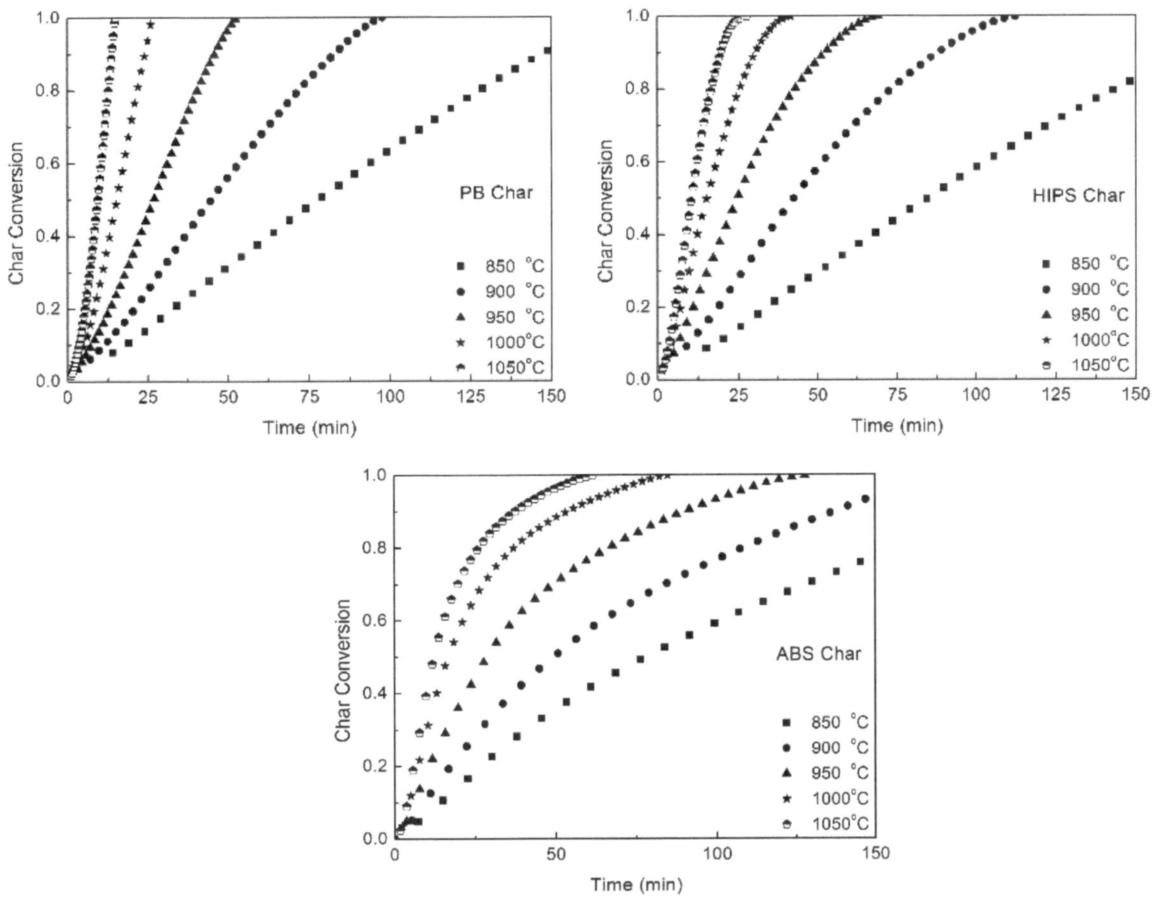

Figure 3. The effect of the gasification temperature on the PB, HIPS and ABS char conversion curves

Table 3. The CO_2 gasification reactivity indexes of PB char, HIPS char and ABS char at different temperatures

Rs	850 °C	900 °C	950 °C	1000 °C	1050 °C
PB Char	0.00640	0.01171	0.02165	0.03373	0.05360
HIPS Char	0.00589	0.01157	0.02045	0.03216	0.04770
ABS Char	0.00569	0.01016	0.01745	0.03014	0.04115

3.3. Kinetic Analyses of CO_2 Gasification of WEEE Plastic Chars

The effect of the temperature on the gasification reaction rate was investigated in the temperature range of 850 °C – 1050 °C, whose result is shown in Figure 4. It is worth mentioning that all of the gasification rates firstly increase to the maximum and then gradually decline with the char conversion. It was reported that this behavior may be attributed to char structure evolution during the reaction process. It is that in the initial stage, the porosity is lower and the original blocked pores are opened and grow, which can improve the surface area available for the reaction. Subsequently, the overlapping of pore structure occurs originating to collapse of the macropore and coalescence of the neighboring pores as gasification proceeds, which would lead to the reduction of the surface area available for the reaction (Ochoa, Cassanello, Bonelli & Cukierman, 2001; Zhang et al, 2008). However, with regard to the different chars, the maximum gasification rate appeared in different carbon conversion ranges. As for the PB char, the maximum gasification rate appeared at the char conversion of approximate 0.8. In the case of HIPS char and ABS char, the maximum gasification rate was obtained when the carbon conversion was about 0.45 and 0.2, respectively. In addition, as expected, the reaction rate results were consistent with that of char conversion, which was sensitive to the gasification temperature and the reactivity was improved as the temperature increased. The experimentally obtained gasification rate results would be used for the kinetic study.

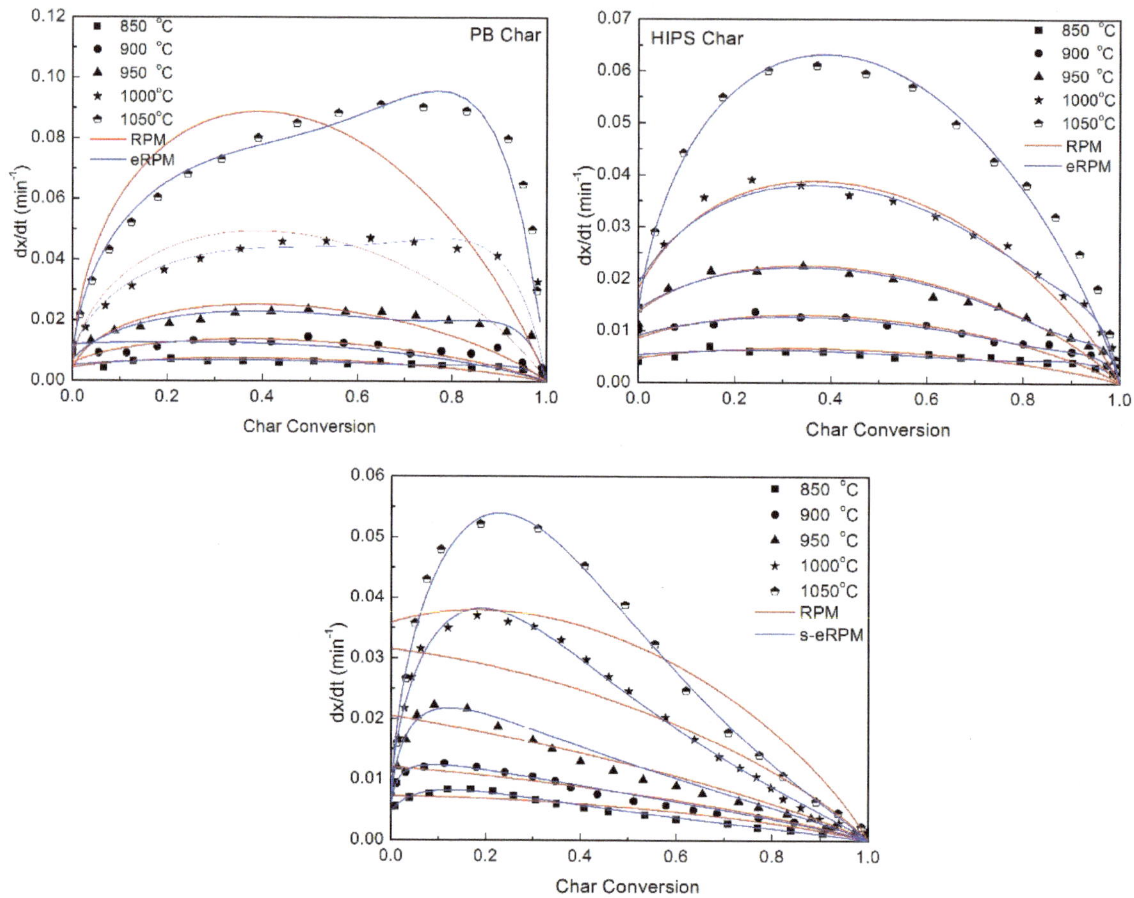

Figure 4. Gasification rates of PB char, HIPS char and ABS char (symbols) and fitting curves of RPM, eRPM and s-eRPM (lines)

Several kinetic models have been applied to simulate the gasification reactions and describe the relationship between the reaction rate and time, such as the homogeneous model, the shrinking core model, the normal distribution model, RPM, eRPM and s-eRPM. It could be obviously found that there was a maximum for the gasification rate of each char as shown in Figure 4. In the above mentioned models, only RPM and eRPM are suitable to simulate this gasification behavior (Zhang et al, 2008; Jing et al, 2013). The equations and the meanings of each parameters of RPM and eRPM have been shown and explained in the session 2.4. The reaction rate of three chars at different temperatures have been simulated by both RPM and eRPM, respectively. The MATLAB software was employed to fit the experimental reaction rate results by the nonlinear least-squares methods in order to estimate the kinetic parameters and the correlation coefficients (R^2). And the curve fitting results and the estimated kinetic parameters results are presented in Figure 4 and Table 4, respectively. From the curve fitting results, as for PB whose maximum gasification rate appeared at the char conversion of about 0.8, eRPM was better than RPM to predict the gasification rate. The maximum gasification rate of HIPS char was obtained at the char conversion of about 0.4. The curve fitting result of RPM and eRPM for the HIPS char gasification rate were almost comparable. Nevertheless, for ABS char, the RPM and eRPM could not simulate the gasification rate very well, because the maximum gasification rate of ABS char appeared at the char conversion of less than 0.2 because of the smallest surface area (Jing et al, 2013). It was reported that when the maximum gasification rate appeared at a lower char conversion, the s-eRPM, as shown in equation 9, was more suitable for the gasification rate curve fitting (Yuan et al, 2011; Jing et al, 2013). It was found that the s-eRPM could fit the gasification rate curve of ABS well with a higher R^2 valve over 0.99. The Jing and co-worker (2013) reported that the better curve fitting results of eRPM and shifted eRPM than that of RPM illustrated that the gasification of coal char is not only affected by the initial pore structure but also the inherent minerals and variation of the pore structure in the gasification process.

Furthermore, Table 4 indicates that the value of Ψ increased as the temperature was raised from 850 °C to 1050 °C, which demonstrates that char porosity development and the progress of reactions occurred inside the pores of chars with the increase of the temperature, which was one of reasons why at a higher temperature, the gasification reactivity was higher.

Table 4. The kinetic parameters and regression coefficients estimated by the RPM and eRPM fitting

Kinetic Parameters									
Sample	T	RPM			e-RPM/s-eRPM[a]				
	℃	K	ψ	R^2	K	ψ	c	p	R^2
PB Char	850	0.0042	14.7	0.931	0.005	7.67	1.2	9.85	0.993
	900	0.0057	29.0	0.972	0.012	1.70	0.6	0.95	0.994
	950	0.0035	286.9	0.951	0.007	62.18	1.2	8.89	0.991
	1000	0.0048	574.9	0.945	0.009	131.73	1.2	5.63	0.992
	1050	0.0035	572.1	0.944	0.009	384.52	1.3	4.45	0.994
HIPS Char	850	0.0046	8.8	0.989	0.005	5.00	1.1	10.47	0.993
	900	0.0084	10.8	0.991	0.009	8.49	1.1	15.71	0.994
	950	0.0137	12.6	0.987	0.014	11.15	1.0	25.50	0.998
	1000	0.0180	23.1	0.990	0.019	19.48	1.0	15.98	0.995
	1050	0.0139	110.2	0.987	0.014	110.20	1.0	117.42	0.996
ABS Char	850	0.0072	1.7	0.871	0.005	48.4	3.5	1.17	0.991
	900	0.0121	1.9	0.866	0.006	196.1	17.1	0.71	0.992
	950	0.0204	2.7	0.894	0.007	213.4	4.7	1.05	0.995
	1000	0.0315	3.4	0.887	0.0093	276.2	2.16	1.71	0.996
	1050	0.0358	4.4	0.853	0.0078	395.0	1.78	2.18	0.997

[a] For PB char and HIPS char, eRPM is used; For ABS char, s-eRPM is employed.

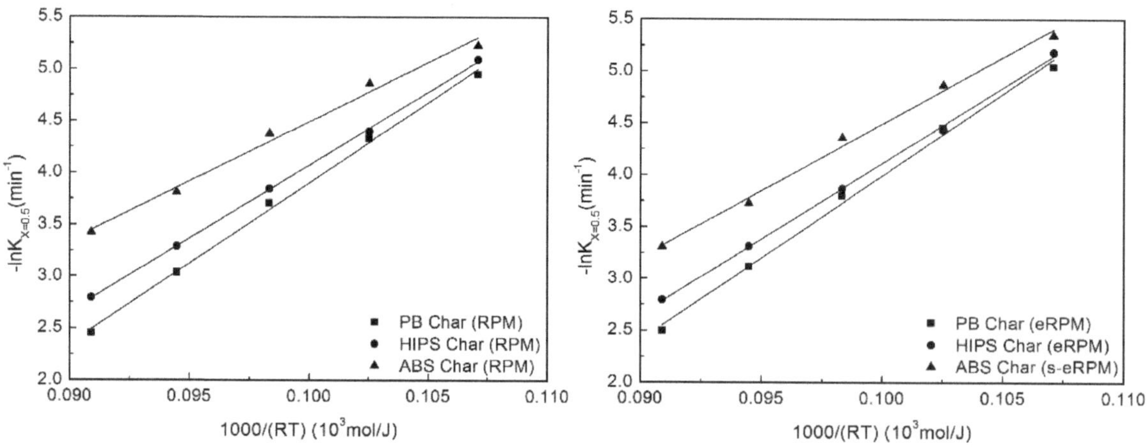

Figure 5. Arrhenius curves of WEEE plastic chars gasification reaction by employing RPM, eRPM and s-eRPM

Table 5. Activation energy E_a and the pre-exponentional factor k_0 calculated by Arrhenius equation

	RPM		eRPM/s-eRPM	
Sample	E_a (kJ/mol)	k_0 (min^{-1})	E_a (kJ/mol)	k_0 (min^{-1})
PB Char	155.11	1.10×10^5	158.44	1.40×10^5
HIPS Char	141.18	2.30×10^4	145.94	3.56×10^4
ABS Char	114.77	1.08×10^3	128.70	4.34×10^3

It have been mentioned that the rate constant K in the above discussed models would become the initial gasification rate dX/dt when char conversion X = 0. Nevertheless, it was reported that the switch of gas results for dX/dt (X = 0) obtained from the TGA is not accurate (Kajitani, Hara & Matsuda, 2002; Yuan et al, 2011; Jing et al, 2013). Consequently, in this study, the gasification rate at char conversion of 0.5 ($r_{0.5}$) was taken as the basis and the $lnr_{0.5}$ versus $1/ (R\cdot T)$ is plotted in Figure 5. According to the RPM, eRPM and s-eRPM, the activation energy E_a and the pre-exponentional factor k_0 could be calculated by the Arrhenius equation, and the results are shown in Table 5. It can be observed that the values of activation energies of three chars obtained from eRPM and s-eRPM are a little higher than that of three chars obtained from RPM. According the best fitting model, the activation energies of PB char, HIPS char and ABS char are 158.44, 145.94 and 128.70 kJ/mol, respectively.

4. Conclusion

The CO_2 gasification of PB char, HIPS char and ABS char were conducted at the temperature range of 850-1050 °C in the TGA. The main conclusions from this work are summarized as follows: (1) As expected, for all three char samples, with the increase of the gasification temperature, the time for the char conversion decreased while the reactivity index and the gasification rate significantly increased. The reactivity indexes of PB char, HIPS char and ABS char at 1050 °C became 8.4, 8.1 and 7.2 times higher than that of them at 850 °C, respectively. (2) At the same gasification temperature, the gasification reactivity of three WEEE plastic chars was in the order of PB char > HIPS char > ABS char, which was attributed to the fact that the surface area of PB char is the largest and lowest carbon crystalline degree. (3) For PB char, its maximum gasification rate appears in the char conversion of approximate 0.8 and the best fitting curve model was eRPM with the highest R^2 (> 0.991). As for HIPS char, RPM and eRPM are comparably suitable to fit the gasification rate curve, and R^2 of RPM and eRPM are > 0.985 and > 0.993, respectively. In the case of ABS char, the maximum gasification rate presents in the lower carbon conversion (< 0.2), and s-eRPM was the most suitable model for predicting the gasification rate with the highest R^2 (> 0.991). (4) The CO_2 gasification activation energies of PB char, HIPS char and ABS char are 158.44, 145.94 and 128.70 kJ/mol, respectively.

References

Bhaskar, T., Matsui, T., Uddin, M. A., Kaneko, J., Muto, A., & Sakata, Y. (2003). Effect of Sb2O3 in brominated heating impact polystyrene (HIPS-Br) on thermal degradation and debromination by iron oxide carbon composite catalyst (Fe-C). *Applied Catalysis B: Environmental, 43*, 229–241. http://dx.doi.10.1016/S0926-3373(02)00306-5

Bhaskar, T., Matsui, T., Kaneko, J., Uddin, M. A., Muto, A., & Sakata, Y. (2002). Novel calcium based sorbent (Ca-C) for the dehalogenation (Br, Cl) process during halogenated mixed plastic (PP/PE/PS/PVC and HIPS-Br) pyrolysis. *Green Chemistry, 4*, 372–375. http://dx.doi.10.1039/b203745a

Bhatia, S., & Perlmutter, D. (1980). A random pore model for fluid-solid reactions: I. Isothermal, kinetic control. *AIChE J, 26*(3), 379–386. http://dx.doi.org/10.1002/aic.690260308

Bozi, J., & Blazsó, M. (2009). Catalytic modification of pyrolysis products of nitrogen-containing polymers over Y zeolites. *Green Chemistry, 11*, 1638–1645. http://dx.doi.10.1039/b913894n

Gil, M. V., Fermoso, J., Pevida, C., Pis, J. J., & Rubiera, F. (2010). Intrinsic char reactivity of plastic waste (PET) during CO_2 gasification. *Fuel Processing Technology, 91*, 1776–1781. http://dx.doi.org/10.1016/j.fuproc.2010.07.019

Hall, W., & Williams, P. T. (2008). Removal of organobromine compounds from the pyrolysis oils of flame retarded plastics using zeolite catalysts. *Journal of Analytical and Applied Pyrolysis, 81*, 139-147. http://dx.doi.10.1016/j.jaap.2007.09.008

Hall, W. J., Miskolczi, N., Onwudili, J., & Williams, P. T. (2008). Thermal Processing of Toxic Flame-Retarded Polymers Using a Waste Fluidized Catalytic Cracker (FCC) Catalyst. Energy & Fuels, 22, 1691–1697. http://dx.doi.10.1021/ef800043g

Huang, Y., Q., Yin, X., L, Wu, C., Z., Wang, C., W., Xie, J., J., Zhou, Z., Q., Ma, L., L. & Li, H., B. (2009). Effects of metal catalysts on CO_2 gasification reactivity of biomass char. *Bitechnology Advances, 27*, 568-572. http://dx.doi.org/10.1016/j.biotechadv.2009.04.013

Jing, X. L., Wang, X. Q., Zhang, Q., Yu, Z. L., Li, C. Y., Huang, J. J., & Fang, Y. (2013). Evaluation of CO_2 Gasification Reactivity of Different Coal Rank Chars by Physicochemical Properties. *Energy & Fuels, 27*, 7287–7293. http://dx.doi.org/10.1021/ef401639v

Jing, X. L., Wang, Z., Q., Yu, Z. L., Zhang, Q., Li, C. Y., & Fang, Y. T. (2013). Experimental and Kinetic Investigations of CO_2 Gasification of Fine Chars Separated from a Pilot-Scale Fluidized-Bed Gasifier. *Energy & Fuels, 27*, 2422-2430. http://dx.doi.org/10.1021/ef4002296

Kajitani, S., Hara, S., & Matsuda, H. (2002). Gasification rate analysis of coal char with a pressurized drop tube furnace. *Fuel, 81*(5), 539–546. http://dx.doi.org/10.1016/S0016-2361(01)00149-1

Kajitani, S., Zhang, Y., Umemoto, S., Ashizawa, M., & Hara. S. (2010). Co-gasification Reactivity of Coal and Woody Biomass in High-Temperature Gasification. *Energy & Fuels, 24*, 145-151. http://dx.doi.org/10.1021/ef900526h

Ke, Y. H., Yang, E. T., Liu, X., Liu, C. L., & Dong, W. S. (2013). Preparation of porous carbons from non-metallic fractions of waste printed circuit boards by chemical and physical activation. *New Carbon Materials, 28* (2), 108-114. http://dx.doi.org/10.1016/S1872-5805(13)60069-4

Kopyscinski, J., Habibi, R., Mims, C. A. & Hill, J. M. (2013). K_2CO_3-Catalyzed CO_2 Gasification of Ash-Free Coal: Kinetic Study. *Energy & Fuels, 27*, 4875–4883. http://dx.doi.org/10.1021/ef400552q

Lahijani, P., Zainal, Z. A., Mohamed, A. R., & Mohammadi, M. (2013). CO_2 gasification reactivity of biomass char: Catalytic influence of alkali,alkaline earth and transition metal salts. *Bioresource Technology, 144*, 288-295. http://dx.doi.org/10.1016/j.biortech.2013.06.059

Lahijania, P., Zainala, Z. A. & Mohamedb, A. R. (2012). Catalytic effect of iron species on CO_2 gasification reactivity of oil palm shell char. *Thermochimica Acta, 546*, 24-31. http://dx.doi.org/10.1016/j.tca.2012.07.023

Micco, G. D., Nasjleti, A., & Bohé, A. E. (2012). Kinetics of the gasification of a Rio Turbio coal under different pyrolysis temperatures. *Fuel, 95*, 537-543. http://dx.doi.org/10.1016/j.fuel.2011.12.057

Ochoal, J., Cassanello, M. C., Bonelli, P. R., & Cukierman, A. L. (2001). CO_2 gasification of Argentinean coal chars: a kinetic characterization. *Fuel Processing Technology, 74*, 161-176. http://dx.doi.org/10.1016/S0378-3820(01)00235-1

Ongondo, F. O., Williams, I. D., & Cherrett, T. J. (2011). How are WEEE doing? A global review of the management of electrical and electronic wastes. *Waste Management, 31*, 714–730. http://dx.doi.10.1016/j.wasman.2010.10.023

Reif, A. E. (1952). The mechnism of the carbon dioxide-carbon reaction. *J. Phys. Chem., 56*, 785. http://dx.org/doi/abs/10.1021/j150498a033

Takarada, T., Tamai, Y., & Tomita, A. (1985). Reactivities of 34 coals under steam gasification. *Fule, 64*(10), 1438-1442. http://dx.doi.org/10.1016/0016-2361(85)90347-3.

Terakado, O., Ohhashi, R., & Hirasawa, M. (2011). Thermal degradation study of tetrabromobisphenol A under the presence metal oxide: Comparison of bromine fixation ability. *Journal of Analytical and Applied Pyrolysis, 91*, 303–309. http://dx.doi.10.1016/j.jaap.2011.03.006

United Nations Environment Programme [UNEP]. (2009). Retrieved from http://www.unep.org/yearbook/2009/

Yang, X. N., Sun, L. S., Xiang, J., Hu, H., & Su, S. (2013). Pyrolysis and dehalogenation of plastics from waste electrical and electronic equipment (WEEE): A review. *Waste Management, 33*, 462–473. http://dx.doi.org/10.1016/j.wasman.2012.07.025

Yuan, S., Chen, X. L., Li, J., & Wang, F. C. (2011). CO_2 Gasification Kinetics of Biomass Char Derived from High-Temperature Rapid Pyrolysis. *Energy & Fuels, 25*, 2314–2321. http://dx.doi.org/10.1021/ef200051z

Zhang, S. Z., Yoshikawa, K., Nakagome, H., & Kamo, T. (2013). Kinetics of the steam gasification of a phenolic circuit board in the presence of carbonates. *Applied Energy. 101*, 815–821. http://dx.doi.org/10.1016/j.apenergy.2012.08.030

Zhang, Y., Ashizawa, M., Kajitani, S., & Miura, K. (2008). Proposal of a semi-empirical kinetic model to reconcile with gasification reactivity profiles of biomass chars. *Fuel, 87*, 475-481. http://dx.doi.org/10.1016/j.fuel.2007.04.026

Zhang, Y., Hara, S., Kajitani, S., & Ashizawa, M. (2010). Modeling of catalytic gasification kinetics of coal char and carbon. *Fuel, 89*, 152-157. http://dx.doi.org/10.1016/j.fuel.2009.06.004

Modelling the Potential Impact of Climate Change on Agricultural Production in the Province of British Columbia

Liliana Perez[1], Trisalyn A. Nelson[2], Mathieu Bourbonnais[2] & Aleck Ostry[3]

[1] Département de géographie, Université de Montréal, Montreal, QC, Canada

[2] Spatial Pattern Analysis and Research Laboratory, Department of Geography, University of Victoria, Victoria, BC, Canada

[3] Department of Geography, University of Victoria, Victoria, BC, Canada

Correspondence: Trisalyn Nelson, Spatial Pattern Analysis and Research Laboratory, Department of Geography, University of Victoria, Victoria, BC, V8W 3R4, Canada. E-mail: trisalyn@uvic.ca

The research is financed by a SSHRC Insight Development Grant

Abstract

The goal of this research was to model the potential impact of climate change on food production, in the Fraser Valley and Peace River regions of British Columbia (BC), using historical crop yield and climate data. We identified eight indicator crops of importance for these regions of BC and utilized historical Census of Agriculture and climate data (temperature and precipitation) to model future potential impacts of climate change on agriculture. We developed three climate change scenarios for these eight indicators crops (extreme, moderate, and business as usual). Under the most extreme climate model scenario the Fraser Valley is expected to experience cooler summers and springs and wetter summers, with incremental increases in oat, blueberry and green bean yields by 2050. These same climate conditions were predicted to decrease the yields for raspberry crops by 2050, while barley and wheat crop yields remain steady. The business as usual scenario, where springs and summers are warmer and summers are wetter in the Fraser Valley, predicted increased barley, oat, wheat, and blueberry crop yields by 2050, while yields of raspberries were predicted to decrease and green bean yields are expected to be steady. Under the more conservative climate change scenario conditions, yields should remain steady for all crops, except green beans where yields will increase by 2050. Future climate conditions for the Peace River area were much different from the Fraser Valley. All three scenarios forecasted warmer and wetter springs and summers with decreased evapotranspiration and moisture deficits. These changes in climate conditions predicted declines in wheat, canola, and barley crop yields by 2050, while incremental increases in oat and dry pea crop yields could be expected by 2050.

Keywords: climate change, food production, Fraser Valley, Peace River, indicator crops, crop yields

1. Introduction

The climate has been changing over the last three decades and is expected continue to do so regardless of any mitigation strategy (Barker, 2007; IPCC, 2007). Temperatures are predicted to increase about 3–5 °C by the middle of the 21st century, while precipitation patterns (amount, seasonality, and intensity) are predicted to shift (Arnell et al., 2004; IPCC, 2007). Agriculture is a climate-dependent activity and hence is highly sensitive to climatic change and climate variability (Lobell & Field, 2007; Rivington et al., 2013). Crop yields are vulnerable to these changes in climate (Rivington et al., 2013; Thornton, Jones, Ericksen, & Challinor, 2011), which can jeopardize food security and economic sustainability that provide the necessary input for sustaining people's livelihoods (FAO, 2012; FAO, WFP & IFAD, 2012). Agriculture is a key driver of national and local economies and largely depends on what can be grown and how efficiently it can be done, taking into consideration variations in climatic trends.

A literature review of the impact of climatic trends on crop yields (Bradshaw, Dolan, & Smit, 2004; Bryant et al., 2000; Joshi, Maharjan, & Luni, 2011; Lobell, Field, Cahill, & Bonfils, 2006) indicated that changes in growing season temperature and precipitation drive annual variation in average yields (Lobell & Field, 2007; Peng et al.,

2004). Crop yields are a function of dynamic, nonlinear interactions between climate, soil water, management, and the physiology of the crop (Semenov & Porter, 1995). For example, changes in the variability of temperature can greatly influence dry matter production as both high and low temperatures decrease the rate of dry matter production and, at extremes, can cause production to cease (Grace, 1988; Ingver, Tamm, Tamm, Kangor, & Koppel,, 2010). Likewise, water deficit occurring immediately before flowering can lead to pollen sterility and will result in a drastic decline in grain yield (Ingver et al., 2010; Nguyen & Sutton, 2009; Thakur, Kumar, Malik, Berger, & Nayyar, 2010). Previous research relating predicted climatic variations to crop response have offered the potential to anticipate changes in crop production early enough to adjust critical decisions (Blench, 2003; Hansen, 2002; Letson et al., 2001).

To study the implications of climate change in British Columbia (BC) agriculture we selected eight indicator crops from three different food groups (grains, fruits, and vegetables). In the grains group we chose barley, spring wheat, oat, canola, and dry peas. Grains have been documented (Joshi et al., 2011; Subedi, Gregory, Summerfield, & Gooding, 1998; Thakur et al., 2010; Willenbockel, 2012) to be very vulnerable to changes in temperature and precipitation; therefore historical data on grain yields can serve as proxy to understand future effects of variable weather conditions. Likewise, fruits such as blueberries and raspberries, as well as vegetables such as green beans are susceptible to water deficit. When water supply is limited during the vegetative and yield formation periods, plant development is usually delayed, causing non-uniform growth and minimizing yield (Glass, Percival, & Proctor, 2005; Hall & Sobey, 2013; Petrova, Matev, & Haitova, 2012). Temporal yield data from fruit and vegetable crops are also used to study the relationship between climate change and agriculture in BC.

1.2 Goal

The goal of this research is to model the potential impact of climate change on BC`s food production, as represented by historical and future crop yield projections. To meet this goal we identify indicator crops of major importance for BC and utilize temporal Census of Agriculture data, available from Statistics Canada, to identify potential impacts of climate change on agriculture. We characterize climate relationships observed over time by relating archived climate information to annual agricultural yields. To reduce redundancy produced by correlated climate predictors, we use Principal Components Analysis (PCA) to derive a new set of uncorrelated climate variables. Quantitative historical values are then used to model potential future yields for 2020 and 2050 under different climate change scenarios.

1.3 Study Area

The province of BC in western Canada spans 944 735 km^2 and has a diversity of landscapes, topography, climatic zones, and geographical features (Holland, 1976). As a result of the province's variety in landforms, agriculture and food industry are diverse. The province is divided into eight agricultural census regions (Figure 1). BC possesses unique agriculture policies directly related to local food production; in particular, the Agricultural Land Reserve (ALR) is a province-wide land preservation policy of nearly five million hectares of protected farmland where farming is encouraged and non-agricultural uses are controlled (Androkovich, Desjardins, Tarzwell, & Tsigaris, 2008; Hanna, 1997). ALR delimitation is based on soil and climate and represents 5% of BC land suitable for farming, with only 1% having the best soil with the highest capability for growing crops.

Table 1. Data summary on indicator crops

Crop	**Region**	**Years**	**Source**
Barley	Peace River Fraser Valley	1991-2009	November 2011 Farm Survey. Statistics Canada, Agriculture Division, Crops Section.
Spring Wheat			
Oat			
Canola	Peace River	1991-2009	November 2011 Farm Survey. Statistics Canada, Agriculture Division, Crops Section.
Dry Peas		1995-2006	
Green Beans	Fraser Valley	1961-2009	Fruit and Vegetable Survey - 3407. Table 001-0013 Area, production and farm value of vegetables, annual.
Blueberries	Fraser Valley	2002-2009	Fruit and Vegetable Survey - 3407. Table 001-0009 Area, production and farm value of fresh and processed fruits, by province, annual.
Raspberries			

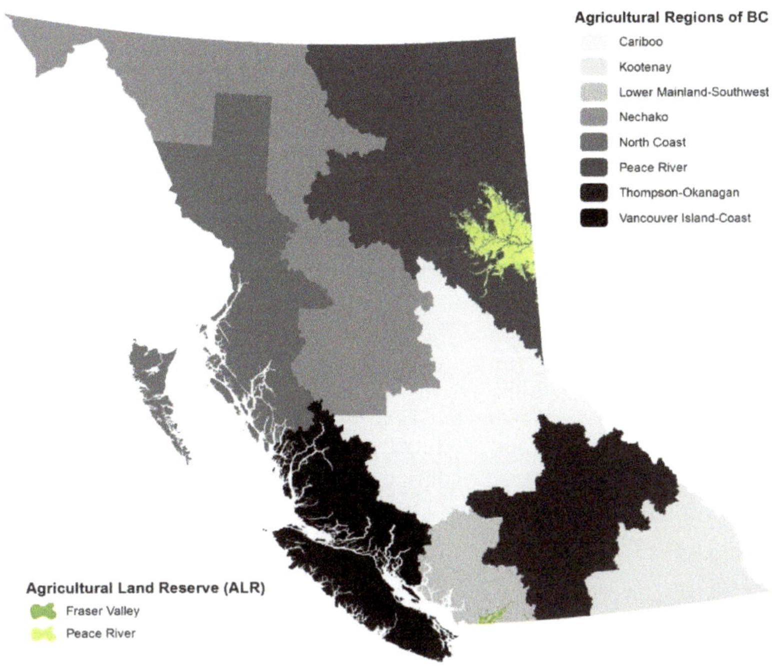

Figure 1. Map of the agricultural regions of British Columbia and the Agricultural Land Reserve of Fraser Valley and Peace River, where the indicator crops are grown.

1.4 Data

The crop yield data used in this paper were provided by Statistics Canada (Table 1). Agricultural Census data provided information on the area of farmland under cultivation for our indicator crops. Data from the farm surveys were disseminated at the provincial scale with no sub-provincial data available. Census and sample survey data were gathered by Statistics Canada from direct questionnaires supplied to farm owners in Canada. Total food production for a given product over a year is calculated by Statistics Canada using extrapolated data from sample surveys. Crop data descriptions are as follows.

1.4.1 Grains

Through Statistics Canada Field Crop Reporting Series, accurate and timely estimates of seeding intentions, seeded and harvested area, production, yield and farm stocks of the principal field crops in Canada were provided at the provincial level (Statistics Canada, 2012a). Field Crop Series data reported farms producing grains in all non-Atlantic provinces; farms were stratified by size and randomly sampled. In BC, grain crops grown in Peace River (Figure 1) were reported separately from the remainder of the province and account for 90% of total grain production excluding forage corn. A large proportion of some grain crops (e.g., oats and barley) are fed to livestock. The estimated proportion of livestock grain was removed from the production data, given that it is not be available for human consumption (Statistics Canada, 2012b).

1.4.2 Fruits and Vegetables

We used data from the annual Spring and Fall Survey of Fruits and Vegetables, conducted by Statistics Canada. Through stratified random samples this survey collected data to provide estimates of the total cultivated area, harvested area, total production, marketed production, and farm gate value of selected fruits and vegetables grown in Canada. The survey excluded farms producing only mushrooms, potatoes, and greenhouse vegetables, as well as farms that were on Indian reserves; likewise, small operations with less than one acre of fruit and less than one acre of vegetables were left out. Annual totals of production and planted area for all fruits, vegetables, and potatoes are released by Statistics Canada each year (Statistics Canada, 2012c; 2012d). Fruits and vegetables in BC are mostly grown in the Lower Mainland-Southwest area (Lower Mainland and Fraser Valley) (Figure 1).

1.4.3 Climate Data:

We used high spatial resolution topographically corrected climate information for BC sourced from the Climate Western North America database version 4.60 (CWNA) (Wang, Hamann, & Spittlehouse, 2010). CWNA utilizes the Parameter-elevation Regressions on Independent Slopes Model (PRISM) approach (Daly et al., 2008; Oregon State University, 2011) and uses up-sampling of climate models employing topography, climate stations, wind patterns, and rain shadow information to provide a continuous coverage of climate data for the province. Four datasets were used. The 49-year data spanning 1961-2009 was used to represent current climate. The three future climate scenarios were based on the Canadian Centre for Climate Modelling and Analysis (CCCma) B1, A1B, and A2 scenarios for the years 2020 and 2050. The future scenario B1 (AR4 – R4) represents the least extreme scenario, A1B (AR4 – R4) represents the business as usual scenario, and A2 (AR4 – R4) represents the most extreme scenario. Model selection was based on using the full range of possible scenarios that were available from respected institutions and were recommended by the Pacific Climate Impacts Consortium (Murdock & Splittlehouse, 2011).

2. Method

2.1 Climate variables

Selected temperature-based variables consisted of Spring and Summer mean temperature, Spring and Summer mean maximum temperature, and Spring and Summer mean minimum temperature. Hybrid climate metrics consisted of Spring and Summer climate moisture deficit and evapotranspiration. Finally, precipitation variables were Spring and Summer mean precipitation.

2.2 Principal Component Analysis

Climate variables such as temperature, precipitation, and evapotranspiration are highly correlated (Trenberth, 2005), leading to problems when applying regression techniques. After testing for correlation between the 12 climatic variables selected for this study, we used Principal Components Analysis (PCA) to derive a new set of uncorrelated climate variables in order to reduce the redundancy created by correlated predictors. PCA is a classical technique used to reduce the dimensionality of a dataset by transforming input data based on a correlation matrix into a set of linear uncorrelated eigenvalues, or principal components (PCs), equal to the number of input variables and accounts for the total variance present in the input data (Jolliffe, 2002). PCs are uncorrelated and ordered such that the kth PC has the kth largest variance among all PCs. The kth PC can be interpreted as the direction that maximizes the variation of the projections of the data points such that it is orthogonal to the first $k-1$ PCs (Abdi & Williams, 2010; Saporta & Niang, 2010). While the first few PCs account for the greatest proportion of the data variance, the last PCs may be as interesting as the first components, since the type or the direction of the shifts are *a priori* unknown (Abdi & Williams, 2010; Jolliffe, 2002; Saporta & Niang, 2010). PCA was conducted using climate variables from the Fraser Valley and Peace River regions. A total of twelve PCs were derived from climate variables in each region.

2.3 Generalized Linear Models

The principal component scores were employed as inputs in generalized linear models (GLMs) with a gamma distribution (Husak, Michaelsen, & Funk, 2007), which were used to predict future crop yields in the Fraser Valley and Peace River regions under different climate scenarios. The gamma distribution was chosen because it is bounded at zero and this takes into account that crops are always present.

First presented by Nelder and Wedderbum (1972), GLMs are a flexible generalization of ordinary linear regression, allowing the linear model to be related to the response variable via a link function and the magnitude of the variance of each measurement to be a function of its predicted value (Gotway & Stroup, 1997; Nelder & Wedderburn, 1972). GLMs permit analyzing data from non-normal distributions and account for non-linear relationships. In this study, the final GLM for each crop was chosen using stepwise model selection based on the second-order Akaike's information criterion (AICc), which includes a bias-correction term to account for a small sample size relative to the number of model parameters (Anderson, Burnham, & Thompson, 2000; Burnham & Anderson, 2002). The model with the lowest AICc score was considered the most parsimonious, thus minimizing estimate bias and optimizing precision (Burnham & Anderson, 2002). Significant PCs retained in final GLMs were assessed to determine which climate variables were influential in the prediction of future crop yields. Forecast of future crop yields for the Fraser Valley and Peace River regions were calculated using the climate scenarios B1, A1B, and A2 and values associated with the 90% confidence interval (CI) are presented as Kg/ha.

3. Results

3.1 Principal Component Analysis

Twelve PCs were drawn from the PCA for each region (Fraser Valley and Peace River). It should be noted that a common rule of thumb for selecting the quantity of PCs is to choose the smallest number of PCs such that a chosen percentage of total variation is exceeded. For the Fraser Valley data, the first 5 PCs cover more than 93%

of the total variation in the data, while for the Peace River dataset the first 4 PCs account for over 92% of the total variation. However, if only four or five components had been chosen, it would have had a detrimental effect on the regression model given that final PCs often represent stable linear relationships between predictor variables that are of interest in climate modelling.

3.2 Generalized Linear Models

The overall variability in crop yields was high and the mean was relatively low, with a positively skewed gamma distribution. Out of the several models generated for each crop and region, a total of eleven models were selected based on their AICc. AICc values and the number of parameter (Ki) per model are reported in Table 2. The reported AICc corresponded to the lower value attained with K parameters, allowing us to determine which model among a set of models was most parsimonious.

Table 2. AICc-selected models for eight different crops in the regions of Fraser Valley and Peace River. *CI* represents the 90% confidence intervals of the predictions in Kg/ha.

Region	Model	AICc	Scenario A2		ScenarioA1B		Scenario B1	
			CI **2020**	*CI* **2050**	*CI* **2020**	*CI* **2050**	*CI* **2020**	*CI* **2050**
Fraser Valley	Barley	308.12	2730.46	3286.22	3283.37	5175.11	2549.75	2894.78
			1946.16	2013.67	1958.96	1897.92	1838.76	1931.5
	Blueberry	154.79	8277.11	11152.66	11340.34	19767.94	8070.06	9606.16
			4972.32	5887.77	6020.75	6817.69	5507.2	5787.11
	Green Beans	662.82	6515.77	6816.78	6940.28	7205.26	6999.72	7828.04
			4991.5	6310.15	6478.13	6632.8	6527.3	6785.83
	Oat	306.44	1958.47	2321.59	2241.29	3543.06	1695.69	1964.69
			1138.05	1240.06	1221.11	1445.43	1011.7	1140.27
	Raspberries	157.78	6705.09	6687.44	7490.87	8353.57	7165.34	7478.56
			4906.72	4080.86	4163.05	1826.83	5145.11	4727.45
	Wheat	303.02	3187.07	3296.26	3213.95	4165.57	3008.59	3086.12
			2242.4	2214.27	2235.79	1943.45	2277.15	2264.57
Peace River	Barley	306.45	4192.23	4872.51	3458.75	3251.29	5828.67	4221.64
			2412.04	2552.77	2183.94	2027.28	2522.36	2452.89
	Canola	277.12	2043.51	2452.29	2430.58	2336.92	2346.43	2313.05
			807.86	460.14	741.19	144.72	371.7	213.47
	Dry Peas	204.44	2416.11	2901.51	2632.67	3294.33	2768.15	2881.91
			1698.82	1698.19	1496.62	974.29	1573.33	1777.48
	Oat	295.41	2902.79	3756.58	3101.85	3499.09	2834.79	3143.49
			1847.08	2064.55	1751.02	1464.07	1763.66	1894.42
	Wheat	291.97	3643.01	4113.9	3315.72	3232.31	4508.41	3653.59
			2364.98	2254.54	2053.81	1721.37	2455.48	2121.15

Table 3. Future trends in climate variables under different emission scenarios

Scenario A2												
Region	**Climate change between 2009 and 2020**											
	Tmax_sp	Tmax_sm	Tmin_sp	Tmin_sm	Tave_sp	Tave_sm	PPT_sp	PPT_sm	Eref_sp	Eref_sm	CMD_sp	CMD_sm
Fraser Valley	↓	↓	↓	↓	↓	↓	↓	↑	↓	↓	↑	↓
Peace River	↑	↓	↑	↑	↑	↑	↑	↑	↑	↓	↓	↓
Scenario A1B												
Region	**Climate change between 2009 and 2020**											

	Tmax_sp	Tmax_sm	Tmin_sp	Tmin_sm	Tave_sp	Tave_sm	PPT_sp	PPT_sm	Eref_sp	Eref_sm	CMD_sp	CMD_sm
Fraser Valley	↑	↑	↑	↑	↑	↑	↓	↑	↑	↓	↑	↑
Peace River	↑	↑	↑	↑	↑	↑	--	↑	↑	↓	↑	↓
Scenario B1												
Region	**Climate change between 2009 and 2020**											
	Tmax_sp	Tmax_sm	Tmin_sp	Tmin_sm	Tave_sp	Tave_sm	PPT_sp	PPT_sm	Eref_sp	Eref_sm	CMD_sp	CMD_sm
Fraser Valley	↑	↓	↑	↑	↑	↓	↓	↑	↑	↓	--	↓
Peace River	↑	↓	↑	↑	↑	↑	↑	↑	↑	↓	↓	↓

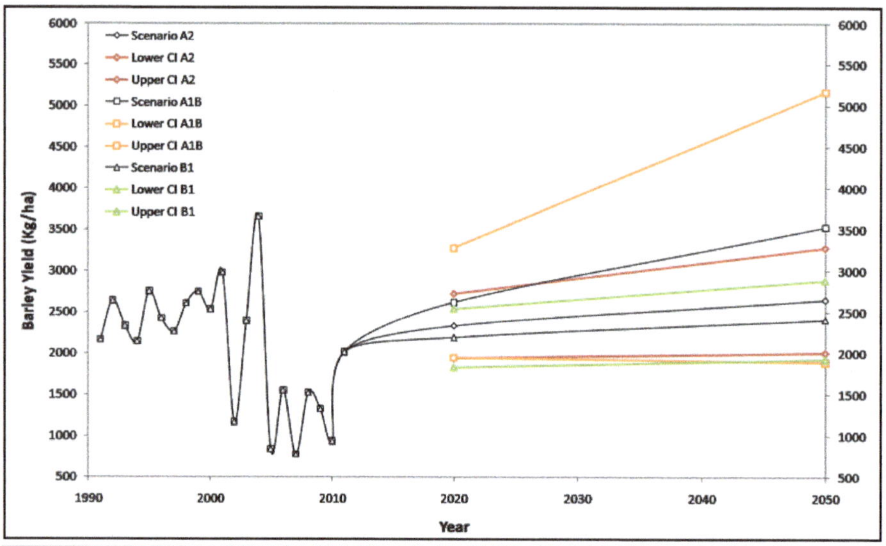

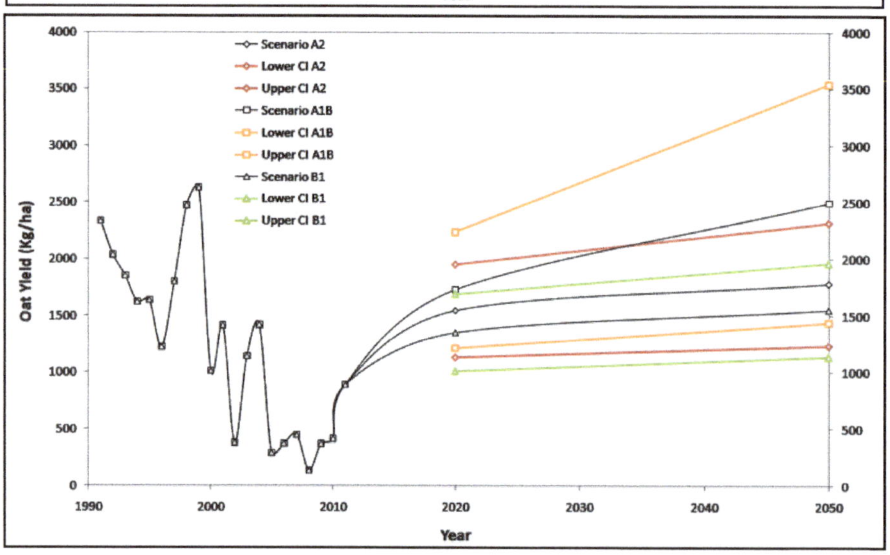

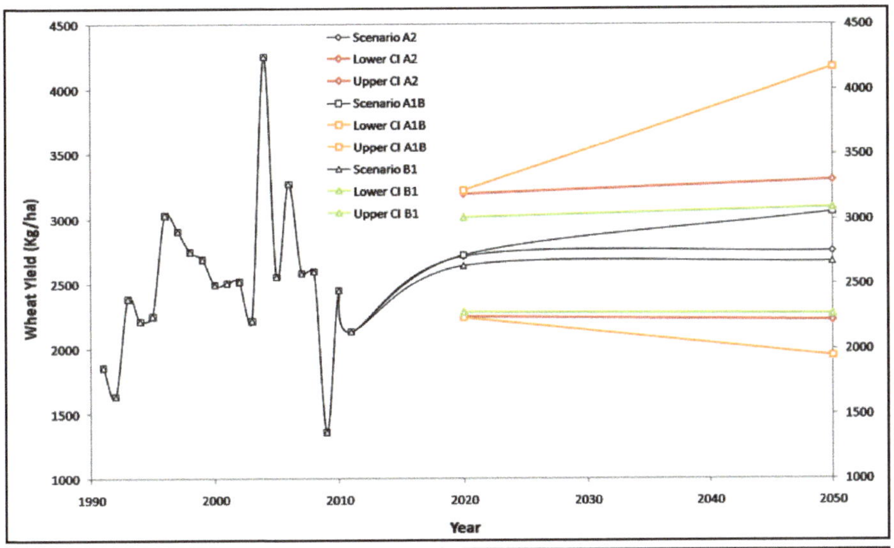
Scenario A2
Lower CI A2
Upper CI A2
Scenario A1B
Lower CI A1B
Upper CI A1B
Scenario B1
Lower CI B1
Upper CI B1
Wheat Yield (Kg/ha)
Year

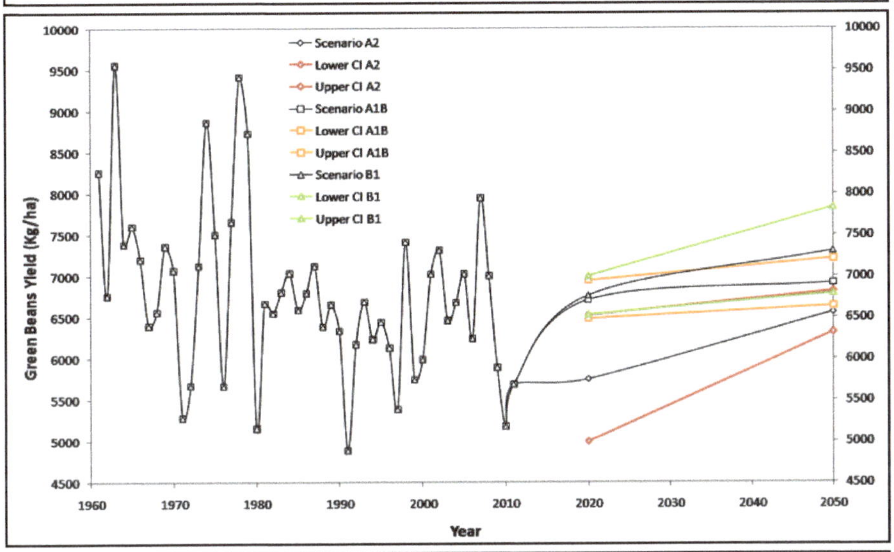
Scenario A2
Lower CI A2
Upper CI A2
Scenario A1B
Lower CI A1B
Upper CI A1B
Scenario B1
Lower CI B1
Upper CI B1
Green Beans Yield (Kg/ha)
Year

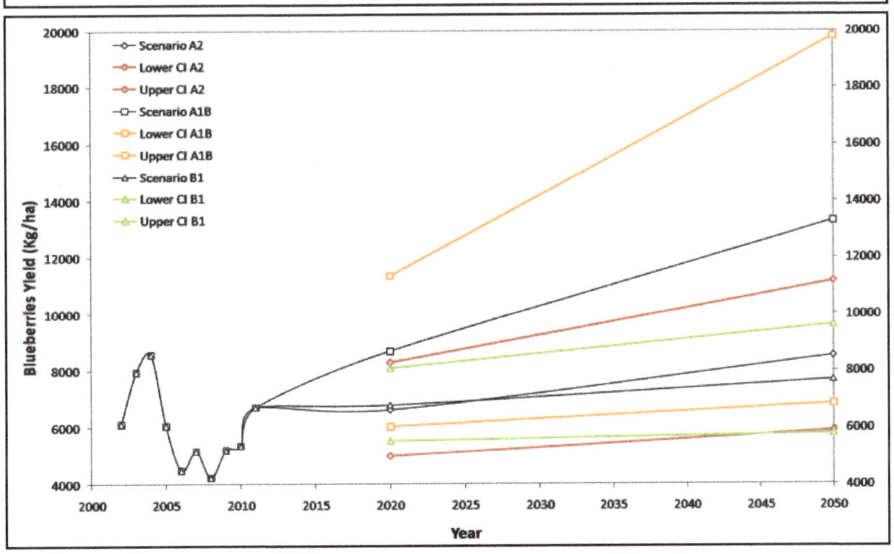
Scenario A2
Lower CI A2
Upper CI A2
Scenario A1B
Lower CI A1B
Upper CI A1B
Scenario B1
Lower CI B1
Upper CI B1
Blueberries Yield (Kg/ha)
Year

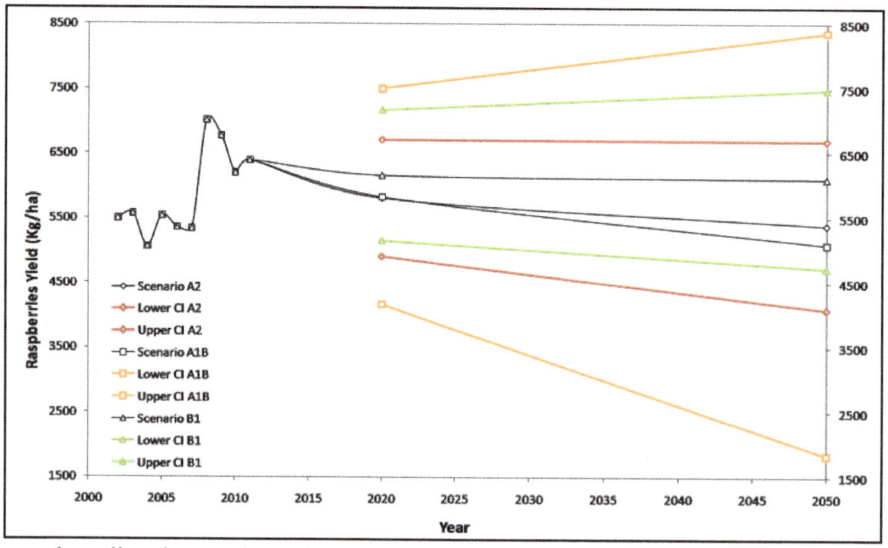

Figure 2. Past and predicted crop yields for the Fraser Valley area. Yield projections for six crops represent scenarios A2, A1B and B1. Colour lines represent upper and lower limits of a 90% confidence interval

Models of climate change showed different trends for the Fraser Valley and Peace River BC under three emission scenarios (Table 3). The most extreme climate model scenario (A2) foresees, for the Fraser Valley, cooler summers and springs and wetter summers, predicting increments in oat, blueberry and green bean yields by 2050 (Figure 2). The same climate conditions are expected to affect raspberry crops by decreasing yields by 2050, while barley and wheat crop yields remain steady through time. The business as usual scenario (A1B), where spring and summers are warmer, and summers are wetter in the Fraser Valley, predicted increased barley, oat, wheat, and blueberry crop yields by 2050, while yields of raspberry crops were predicted to decrease and green bean crop yields were expected to remain stable. The more conservative climate change scenario, the B1 model, forecasts warmer and wetter springs and cooler and dryer summers for the Fraser Valley area. Crop yield predictions under the B1 scenario were all crops have steady yields, except for green bean yields, which would increase by 2050.

Future climate conditions for the Peace River area were different from the Fraser Valley (Table 3). The three scenarios (A2, A1B, B1) forecasted warmer and wetter springs and summers for the three emission scenarios, with decreased evapotranspiration and moisture deficits. These changes in climate conditions predicted declines in wheat, canola, and barley crop yields by 2050, with an increment in oat and dry pea crop yields (Figure 3).

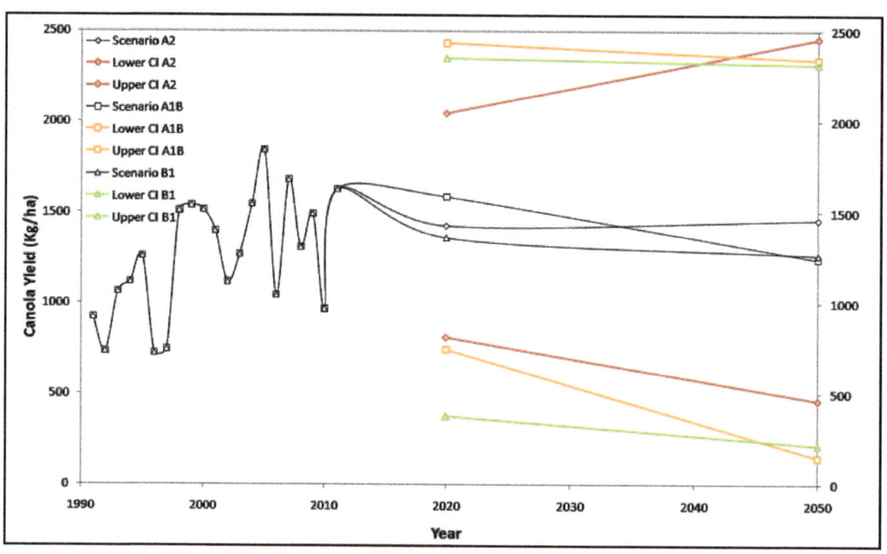

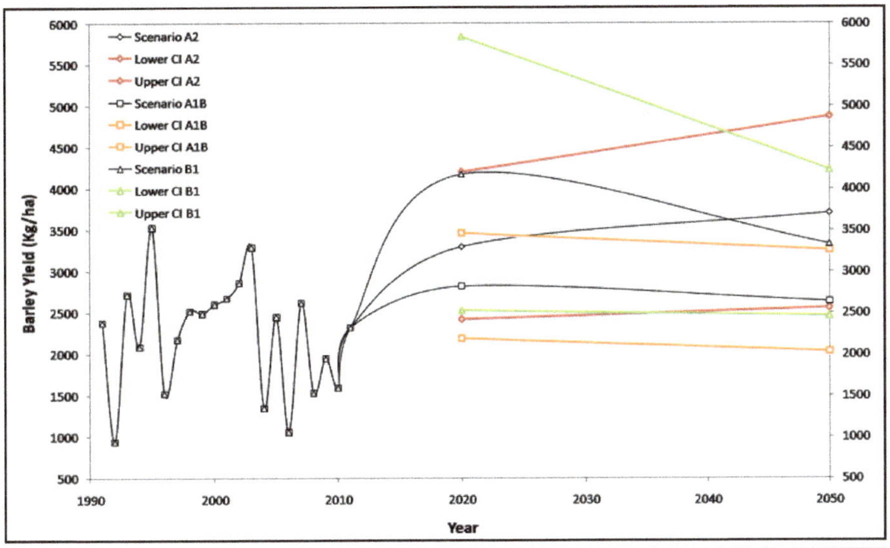
Scenario A2
Lower CI A2
Upper CI A2
Scenario A1B
Lower CI A1B
Upper CI A1B
Scenario B1
Lower CI B1
Upper CI B1
Barley Yield (Kg/ha)
6000
5500
5000
4500
4000
3500
3000
2500
2000
1500
1000
500
1990
2000
2010
2020
2030
2040
2050
Year

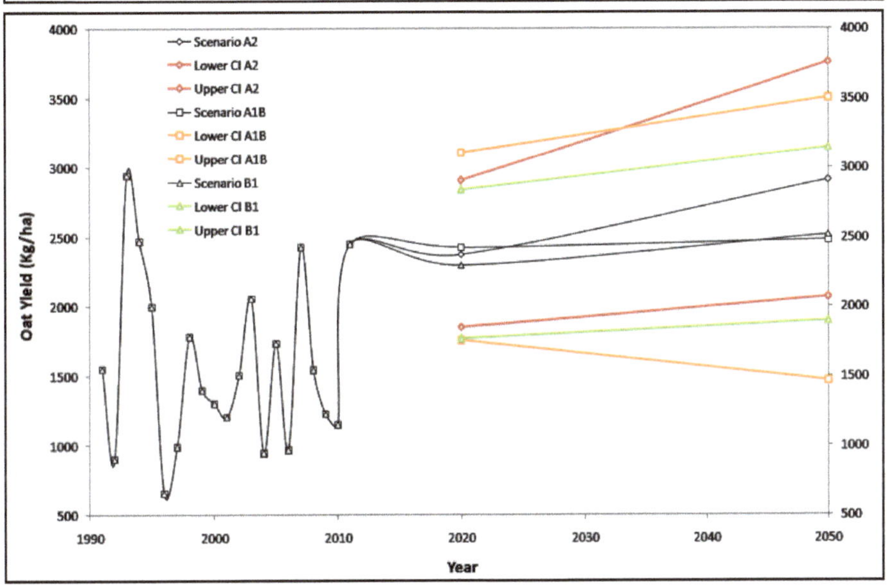
Scenario A2
Lower CI A2
Upper CI A2
Scenario A1B
Lower CI A1B
Upper CI A1B
Scenario B1
Lower CI B1
Upper CI B1
Oat Yield (Kg/ha)
4000
3500
3000
2500
2000
1500
1000
500
1990
2000
2010
2020
2030
2040
2050
Year

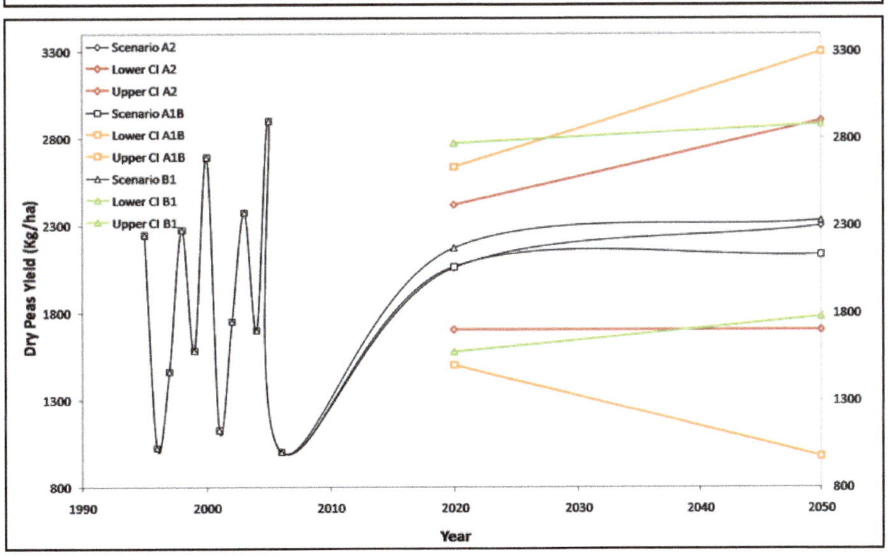
Scenario A2
Lower CI A2
Upper CI A2
Scenario A1B
Lower CI A1B
Upper CI A1B
Scenario B1
Lower CI B1
Upper CI B1
Dry Peas Yield (Kg/ha)
3300
2800
2300
1800
1300
800
1990
2000
2010
2020
2030
2040
2050
Year

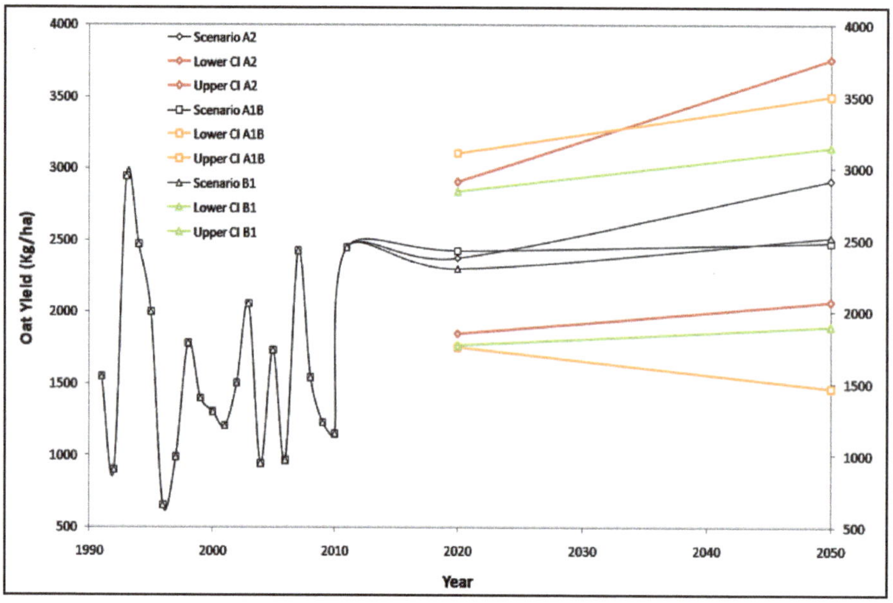

Figure 3. Past and predicted crop yields for the Peace River area. Yield projections for six crops represent scenarios A2, A1B and B1. Colour lines represent upper and lower limits of a 90% confidence interval

4. Discussion

As a result of a high correlation between our climate variable data, we generated twelve uncorrelated PCs to decrease the redundancy product of highly correlated predictors. The development of relationships between the PCs and historical dataset of crop yields allowed us to create a predictive model of yields based on future climate conditions. Despite the fact that it was hard to interpret the directionality of the relationships in our final GLM model coefficients due to the conversion of all climate variables into PCs, overall, the PCs explained a good proportion of the variance in the temperature and precipitation data. Previous studies (e.g., Challinor & Slingo, 2003) incorporating precipitation data in PCs have been found to correlate well with crop yields in other regions.

Underestimation of the uncertainty of a forecast can lead to excessive responses that are inconsistent with a decision maker's risk tolerance, while overestimating uncertainty leads to under confidence and lost opportunity to prepare for adverse conditions or to take advantage of favourable conditions. Specifically for our model, climate variability and crop yield data errors were the major sources of uncertainty in yield forecasting. Likewise, the datasets used in this study showed a lot of variability in year to year crop yields, but less variability with the climate data, creating uncertainty in the long-term forecasting, especially since some of the yield/climate sample sizes were quite small. However, by using GLM as a predictive modeling technique, we were able to forecast yields using less data than what traditional techniques would require. GLMs tend to be more robust than other predictive modeling techniques and are less susceptible to the over-fitting that may occur with small datasets (McCullagh & Nelder, 1989).

Due to confidentiality procedures with the Census of Agriculture, it was not possible to obtain the exact location of farms (Statistics Canada, 2012c). With no spatially identifiable agricultural operations and yields available, the spatial distribution of crop yields (response variable) was not comprehensive. This poses a problem when associating yield data to climate variables in heterogeneous areas. For defining the relationships between climate and crop yields we utilized mean seasonal (spring and summer) values from 1961 to 2009. The use of these datasets in large areas such as the province of BC brings the uncertainty entailed in dynamical downscaling from coarse to fine resolution. This uncertainty in the regional climate response has been documented in numerous contexts (Christensen et al., 2007; Elía et al., 2008). With respect to modelling and statistical downscaling, uncertainties are also associated with imperfect knowledge and/or representation of physical processes, limitations due to the numerical approximation of the model's equations, simplifications and assumptions in the models, and/or approaches and internal model variability (Mearns et al., 2012). Likewise, it is essential to acknowledge that the observed regional climate is sometimes characterized by a high level of uncertainty due to measurement errors and sparseness of stations, especially in remote regions such as northern BC and in regions of complex topography.

Similar to Challinor, Slingo, Wheeler, Craufurd, and Grimes (2003), our models showed that much of the inter-annual variability in crop yields can be explained simply using temperature and precipitation, and that this relationship is demonstrated at a regional scale. For most of the crops, the percent of deviance explained was over 45%. Where crops had lower deviance (e.g., Fraser Valley Green Beans, Wheat, Barley, and Oats) it is possible that they responded more to factors other than climate such as soil conditions and/or available carbon/nitrogen. Previous authors (Moen, Kaiser & Riha, 1994; Reilly et al., 2003) found that increasing within-year temperature variability had the greatest impact on yields if the growing season temperature is outside the optimum for growth. For BC it would be possible to examine whether years of low/high crop yields correlate with low/high temperature and precipitation.

Conclusion

We predicted future crop yields for eight indicator crops under three future climate change scenarios for the Fraser Valley and Peace River regions of BC. In the Fraser Valley, the three climate scenarios predicted different trends in temperature and precipitation, resulting in a range of predicted crop responses. In the most extreme scenario, cooler springs and summers along with wetter summers should result in increased yields of oats, blueberries, and green beans, with decreased raspberry yields. In contrast, the three climate scenarios all predicted warmer and wetter springs and summers for the Peace River region, resulting in predictions of increased yields for oat and dry pea crops, with decreased yields of wheat, canola, and barley crops. Our results have important implications for policy as it is easier to mitigate the potential impacts of climate change on crop yields by instituting policy on farming reform or adaptations at a regional level rather than at the local scale. Using GLMs to model future crop yields allowed us to overcome issues with small sample sizes.

Acknowledgments

This work was funded by a Social Science and Humanities Research Council (SSHRC) Insight grant to Dr. Aleck Ostry and Dr. Trisalyn Nelson.

References

Abdi, H., & Williams, L. J. (2010). Principal component analysis. *Wiley Interdisciplinary Reviews: Computational Statistics 2*, 433–459. http://dx.doi.org/10.1002%2Fwics.101

Anderson, D. R., Burnham, K. P., & Thompson, W. L. (2000). Null Hypothesis Testing: Problems, Prevalence, and an Alternative. *The Journal of Wildlife Management, 64*, 912–923. http://dx.doi.org/10.2307/3803199

Androkovich, R., Desjardins, I., Tarzwell, G., & Tsigaris, P. (2008). Land Preservation in British Columbia : An empirical analysis of the factors underlying public support and willingness to pay. *Journal of Agricultural and Applied Economics, 40*, 999–1013.

Arnell, N. W., Livermore, M. J. L., Kovats, S., Levy, P. E., Nicholls, R., Parry, M. L., & Gaffin, S. R. (2004). Climate and socio-economic scenarios for global-scale climate change impacts assessments: characterising the SRES storylines. *Global Environmental Change, 14*, 3–20. http://dx.doi.org/10.1016/j.gloenvcha.2003.10.004

Barker, T. (2007). Climate Change 2007: An Assessment of the Intergovernmental Panel on Climate Change. *Change,* 446, 12–17.

Blench, R. (2003). Forecasts and farmers: exploring the limitations. In K. L. O'Brien, & C. Vogel (Eds.), *Coping with climate variability: The use of seasonal climate forecasts in southern Africa.* (pp. 59–71). Ashgate Publishing, Ltd.

Bradshaw, B. E. N., Dolan, H., & Smit, B. (2004). Farm-level adaptation to climatic variability and change : crop diversification in the Canadian prairies. *Climatic Change, 67,* 119–141. http://dx.doi.org/10.1007/s10584-004-0710-z

Bryant, C. R., Smit, B., Brklacich, M., Johnston, T. R., Smithers, J., Chiotti, Q., & Singh, B. (2000). Adaptation in Canadian agriculture to climatic variability and change. *Climatic Change, 45*, 181–200. http://dx.doi.org/10.1007/978-94-017-3010-5_10

Burnham, K., & Anderson, D. (2002). *Model selection and multi-model inference: A practical information-theoretic approach.* Springer-Verlag, New York, NY, USA.

Challinor, A. J., Slingo, J. M., Wheeler, T. R., Craufurd, P. Q., & Grimes, I. F. (2003). Toward a combined seasonal weather and crop productivity forecasting system: Determination of the working spatial scale. *Journal of Applied Meteorology, 42,* 175–192.

http://dx.doi.org/10.1175/1520-0450(2003)042<0175:TACSWA>2.0.CO;2

Christensen, J. H., Hewitson, B., Busuioc, A., Chen, A., Gao, X., Held, I., … Whetton, P. (2007). Regional climate projections. In *Climate Change 2007: The Physical Science Basis. Contribution of Working Group I to the Fourth Assessment Report of the Intergovernmental Panel on Climate Change*. (pp. 849–940). Cambridge University Press, Cambridge, UK.

Daly, C., Halbleib, M., Smith, J. I., Gibson, W. P., Doggett, M. K., Taylor, G. H., … Pasteris, P. P. (2008). Physiographically sensitive mapping of climatological temperature and precipitation across the conterminous United States. *International Journal of Climatology*, *28*, 2031–2064. http://dx.doi.org/10.1002/joc.1688

Elía, R., Caya, D., Côté, H., Frigon, A., Biner, S., Giguère, M., … Plummer, D. (2008). Evaluation of uncertainties in the CRCM-simulated North American climate. *Climate Dynamics, 30*, 113–132. http://dx.doi.10.1007/s00382-007-0288-z

FAO. (2012). *The state of food and agriculture 2012. Investing in agriculture for a better future*. FAO, Roma, Italy.

FAO, WFP, IFAD. (2012). *The State of Food Insecurity in the World 2012*. Economic growth is necessary but not sufficient to accelerate reduction of hunger and malnutrition. FAO, Rome, Italy.

Glass, V. M., Percival, D. C., & Proctor, J. T. A. (2005). Tolerance of lowbush blueberries (V*accinium angustifolium* Ait .) to drought stress. I . Soil water and yield component analysis. *Canadian Journal of Plant Science*, *85*, 911–917. http://dx.doi. 10.4141/P03-027

Gotway, C. A., & Stroup, W. W. (1997). A Generalized Linear Model approach to spatial data analysis and prediction. *Journal of Agricultural, Biological, and Environmental Statistics,* 2, 157–178. http://dx.doi.org/10.2307/1400401

Grace, J. (1988). Temperature as a determinant of plant productivity. *Symposia of the Society for Experimental Biology,* *42*, 91–107. Retrieved from http://www.unboundmedicine.com/medline/citation/3270210/Temperature_as_a_determinant_of_plant_productivity_

Hall, H., & Sobey, T. (2013). Climatic requirements. In Funt, R. C., & Hall, H. K. (Eds.), *Raspberries*. Cabi.

Hanna, K. (1997). Regulation and land-use conservation: A case study of the British Columbia Agricultural Land Reserve. *Journal of Soil and Water Conservation*, *52*, 166–170. Retrieved from http://www.jswconline.org/content/52/3/166.extract

Hansen, J. W. (2002). Realizing the potential benefits of climate prediction to agriculture: issues, approaches, challenges. *Agricultural Systems*, *74*, 309–330. http://dx.doi.org/10.1016/S0308-521X(02)00043-4

Holland, S. S. (1976). *Landforms of British Columbia: A physiographic outline*. Retrieved from http://www.empr.gov.bc.ca/Mining/Geoscience/PublicationsCatalogue/BulletinInformation/BulletinsAfter1940/Documents/Bull48-txt.pdf

Husak, G., Michaelsen, J., & Funk, C. (2007). Use of the gamma distribution to represent monthly rainfall in Africa for drought monitoring applications. *International Journal of Climatology*, *27*, 935–944. http://dx.doi.org/10.1002/joc.1441

Ingver, A., Tamm, I., Tamm, Ü., Kangor, T., & Koppel, R. (2010). The characteristics of spring cereals in changing weather in Estonia. *Agronomy Research*, *8*, 553–562. Retrieved from http://agronomy.emu.ee/vol08Spec3/p08s306.pdf

IPCC. (2007). *Climate Change 2007: Mitigation*. Contribution of Working Group III to the Fourth Assessment Report of the Intergovernmental Panel on Climate Change, Climate Change 2007: Mitigation. Contribution of Working Group III to the Fourth Assessment Report of the Intergovernmental Panel on Climate Change. Cambridge University Press, Cambridge.

Jolliffe, I. T. (2002). *Principal Component Analysis*. Springer. Retrieved from http://f3.tiera.ru/2/M_Mathematics/MV_Probability/MVsa_Statistics%20and%20applications/Jolliffe%20I.%20Principal%20Component%20Analysis%20(2ed.,%20Springer,%202002)(518s)_MVsa_.pdf

Joshi, N. P., Maharjan, K. L., & Luni, P. (2011). Effect of climate variables on yield of major food-crops in Nepal: A time-series analysis. *Journal of Contemporary India Studies: Space and Society, Hiroshima University*, 1, 19–26. Retrieved from http://mpra.ub.uni-muenchen.de/35379/1/MPRA_paper_35379.pdf

Letson, D., Llovet, I., Podestá, G., Royce, F., Brescia, V., Lema, D., & Parellada, G. (2001). User perspectives of climate forecasts: crop producers in Pergamino, Argentina. *Climate Research, 19*, 57–67. Retrieved from http://www.int-res.com/articles/cr2002/19/c019p057.pdf

Lobell, D. B., & Field, C. B. (2007). Global scale climate–crop yield relationships and the impacts of recent warming. *Environmental Research Letters, 2*, 014002. http://dx.doi.org/10.1088/1748-9326/2/1/014002

Lobell, D. B., Field, C. B., Cahill, K. N., & Bonfils, C. (2006). Impacts of future climate change on California perennial crop yields: Model projections with climate and crop uncertainties. *Agricultural and Forest Meteorology, 141*, 208–218. http://dx.doi.org/10.1016/j.agrformet.2006.10.006

McCullagh, P., & Nelder, J. A. (1989). *Generalized Linear Models*. Chapman and Hall/CRC.

Mearns, L. O., Arritt, R., Biner, S., Bukovsky, M. S., McGinnis, S., Sain, S., … Snyder, M. (2012). The North American Regional Climate Change Assessment Program: Overview of phase I results. *Bulletin of the American Meteorological Society, 93*, 1337–1362. http://dx.doi.org/10.1175/BAMS-D-11-00223.1

Moen, T. N., Kaiser, H. M., & Riha, S. J. (1994). Regional yield estimation using a crop simulation model: Concepts, methods, and validation. *Agricultural Systems, 46*, 79–92. http://dx.doi.org/10.1016/0308-521X(94)90170-K

Murdock, T. Q., & Splittlehouse, D. L. (2011). *Selecting and using climate change scenarios for British Columbia*. Climate Impacts Consortium, University of Victoria 49.

Nelder, A. J. A., & Wedderburn, R. W. M. (1972). Generalized Linear Models. *Journal of the Royal Statistical Society. Series A, 135*, 370–384. http://dx.doi.org/10.2307/2344614

Nguyen, G. N., & Sutton, B. G. (2009). Water deficit reduced fertility of young microspores resulting in a decline of viable mature pollen and grain set in rice. *Journal of Agronomy and Crop Science, 195*, 11–18. http://dx.doi.org/10.1111/j.1439-037X.2008.00342.x

Oregon State University. (2011). PRISM Climate Group. Retrieved from http://www.prism.oregonstate.edu/

Peng, S., Huang, J., Sheehy, J. E., Laza, R. C., Visperas, R. M., Zhong, X., … Cassman, K. G. (2004). Rice yields decline with higher night temperature from global warming. *Proceedings of the National Academy of Science (PNAS), 101*, 9971–9975. http://dx.doi.org/10.1073/pnas.0403720101

Petrova, R., Matev, A., & Haitova, D. (2012). Influence of the irrigation rate on the vegetative development of green beans. *Agrarni Nauki, 4*, 107–112. Retrieved from http://www.cabdirect.org/abstracts/20133050891.html;jsessionid=AD9E7AD32CFF7CEA1CF7598899DDC54B

Reilly, J., Tubiello, F., McCarl, B., Abler, D., Darwin, R., Fuglie, K., … Rosenzweig, C. (2003). U.S. Agriculture and climate change: New results. *Climatic Change, 57*, 43–67. http://dx.doi.org/10.1023/A:1022103315424

Rivington, M., Matthews, K. B., Buchan, K., Miller, D. G., Bellocchi, G., & Russell, G. (2013). Climate change impacts and adaptation scope for agriculture indicated by agro-meteorological metrics. *Agricultural Systems, 114*, 15–31. http://dx.doi.org/10.1016/j.agsy.2012.08.003

Saporta, G., & Niang, N. (2010). Principal Component Analysis: Application to statistical process control. In:Govaert, G. (Ed.), *Data Analysis*. (pp. 1–23). Wiley. http://dx.doi.org/10.1002/9780470611777.ch1

Semenov, M. A., & Porter, J. R. (1995). Climatic variability and the modelling of crop yields. *Agricultural and Forest Meteorology, 73*, 265–283. http://dx.doi.org/10.1016/0168-1923(94)05078-K

Statistics Canada (2012a). *Field Crop Reporting Series: Definitions, data sources and methods.* Retrieved from www.statcan.gc.ca

Statistics Canada. (2012b). *Field Crop Reporting Series-Catalogue no. 22-002-X.* Retrieved from http://www.statcan.gc.ca/pub/22-002-x/22-002-x2011008-eng.pdf

Statistics Canada. (2012c). *Fruits and Vegetables Survey: Definitions, data sources and methods.* Retrieved from www.statcan.gc.ca

Statistics Canada. (2012d). *Fruit and vegetable production-Catalogue no. 22-003-X.* Retrieved from http://www.statcan.gc.ca/pub/22-003-x/22-003-x2011002-eng.pdf

Subedi, K. D., Gregory, P. J., Summerfield, R. J., & Gooding, M. J. (1998). Cold temperatures and boron deficiency caused grain set failure in spring wheat (*Triticum aestivum* L.). *Field Crops Research, 57*, 277–288. http://dx.doi.org/10.1016/S0378-4290(97)00148-2

Thakur, P., Kumar, S., Malik, J. A., Berger, J. D., & Nayyar, H. (2010). Cold stress effects on reproductive development in grain crops: An overview. *Environmental and Experimental Botany*, *67*, 429–443. http://dx.doi.org/10.1016/j.envexpbot.2009.09.004

Thornton, P. K., Jones, P. G., Ericksen, P. J., & Challinor, A. J. (2011). Agriculture and food systems in sub-Saharan Africa in a 4°C+ world. *Philosophical transactions. Series A, Mathematical, physical, and engineering sciences, 369*, 117–36. http://dx.doi.org/10.1098/rsta.2010.0246

Trenberth, K. E. (2005). Relationships between precipitation and surface temperature. *Geophysical Research Letters, 32*, 4. http://dx.doi.org/10.1029/2005GL022760

Wang, T., Hamann, A., & Splittlehouse, D. (2010). *Climate Western North America.* Retrieved from http://www.genetics.forestry.ubc.ca/cfcg/ClimateWNA/help.htm

Willenbockel, D. (2012). *Extreme weather events and crop price spikes in a changing climate - Illustrative global simulation scenarios.* Retrieved from http://www.iadb.org/intal/intalcdi/PE/2012/10653.pdf

Permissions

All chapters in this book were first published in EER, by Canadian Center of Science and Education; hereby published with permission under the Creative Commons Attribution License or equivalent. Every chapter published in this book has been scrutinized by our experts. Their significance has been extensively debated. The topics covered herein carry significant findings which will fuel the growth of the discipline. They may even be implemented as practical applications or may be referred to as a beginning point for another development.

The contributors of this book come from diverse backgrounds, making this book a truly international effort. This book will bring forth new frontiers with its revolutionizing research information and detailed analysis of the nascent developments around the world.

We would like to thank all the contributing authors for lending their expertise to make the book truly unique. They have played a crucial role in the development of this book. Without their invaluable contributions this book wouldn't have been possible. They have made vital efforts to compile up to date information on the varied aspects of this subject to make this book a valuable addition to the collection of many professionals and students.

This book was conceptualized with the vision of imparting up-to-date information and advanced data in this field. To ensure the same, a matchless editorial board was set up. Every individual on the board went through rigorous rounds of assessment to prove their worth. After which they invested a large part of their time researching and compiling the most relevant data for our readers.

The editorial board has been involved in producing this book since its inception. They have spent rigorous hours researching and exploring the diverse topics which have resulted in the successful publishing of this book. They have passed on their knowledge of decades through this book. To expedite this challenging task, the publisher supported the team at every step. A small team of assistant editors was also appointed to further simplify the editing procedure and attain best results for the readers.

Apart from the editorial board, the designing team has also invested a significant amount of their time in understanding the subject and creating the most relevant covers. They scrutinized every image to scout for the most suitable representation of the subject and create an appropriate cover for the book.

The publishing team has been an ardent support to the editorial, designing and production team. Their endless efforts to recruit the best for this project, has resulted in the accomplishment of this book. They are a veteran in the field of academics and their pool of knowledge is as vast as their experience in printing. Their expertise and guidance has proved useful at every step. Their uncompromising quality standards have made this book an exceptional effort. Their encouragement from time to time has been an inspiration for everyone.

The publisher and the editorial board hope that this book will prove to be a valuable piece of knowledge for researchers, students, practitioners and scholars across the globe.

List of Contributors

F. Morales-Guzmán
Facultad de Ciencias Químicas e Ingeniería, Universidad Autónoma del Estado de Morelos, Av. Universidad 1001. Col. Chamilpa C.P. 62209, Cuernavaca, Morelos, México

R. J. Romero
Centro de Investigaciones en Ingeniería y Ciencias Aplicadas, Universidad Autónoma del Estado de Morelos, Av. Universidad 1001. Col. Chamilpa C.P. 62209, Cuernavaca, Morelos, México

M. L. Domínguez-Patiño
Centro de Investigaciones en Ingeniería y Ciencias Aplicadas, Universidad Autónoma del Estado de Morelos, Av. Universidad 1001. Col. Chamilpa C.P. 62209, Cuernavaca, Morelos, México

A. Rodríguez-Martínez
Centro de Investigaciones en Ingeniería y Ciencias Aplicadas, Universidad Autónoma del Estado de Morelos, Av. Universidad 1001. Col. Chamilpa C.P. 62209, Cuernavaca, Morelos, México

Melgoza-Alemán Rosa María
Facultad de Ciencias Químicas e Ingeniería, Universidad Autónoma del Estado de Morelos, Av. Universidad 1001. Col. Chamilpa C.P. 62209, Cuernavaca, Morelos, México

Clayton Hunter
Department of Earth and Environment, Boston University, Boston, MA, USA

James G. Baldwin
Department of Earth and Environment, Boston University, Boston, MA, USA

Michael L. Mann
Department of Geography, George Washington University, Washington, D.C., USA

Adrian K. James
Environmental Science & Engineering, University of Northern British Columbia, 3333 University Way, Prince George, Canada

Ronald W. Thring
Environmental Science & Engineering, University of Northern British Columbia, 3333 University Way, Prince George, Canada

P. Michael Rutherford
Environmental Science & Engineering, University of Northern British Columbia, 3333 University Way, Prince George, Canada

Steve S. Helle
Environmental Science & Engineering, University of Northern British Columbia, 3333 University Way, Prince George, Canada

Spyridon Achinas
School of Chemical Engineering, National Technical University of Athens, Athens, Greece

Evangelos Kapetanios
Association of Communities and Municipalities of Attica Region, Athens, Greece

Arthur Efeoghene Essaghah
Department of Urban and Regional Planning, Enugu State University of Science & Technology, Agbani, Nigeria

Alabi Michael Oloyede
Department of Geography and Planning, Kogi State University, Anyigba, Nigeria

Maria Kopsakangas-Savolainen
Finnish Environment Institute, Helsinki, Finland
Thule Institute, University of Oulu, Oulu, Finland
Department of Economics and Martti Ahtisaari, Institute of Global Business and Economics, University of Oulu, Oulu, Finland

Rauli Svento
Department of Economics and Martti Ahtisaari, Institute of Global Business and Economics, University of Oulu, Oulu, Finland

Joseph Mawejje
Economic Policy Research Centre, Makerere University Campus, Kampala, Uganda

Ezra Munyambonera
Economic Policy Research Centre, Makerere University Campus, Kampala, Uganda

Lawrence Bategeka
Economic Policy Research Centre, Makerere University Campus, Kampala, Uganda

Mansour Soroush
Norwegian University of Science and Technology- NTNU, Trondheim, Norway

Dag Wessel-Berg
SINTEF, NO-7465, Trondheim, Norway

Ole Torsaeter
Norwegian University of Science and Technology- NTNU, Trondheim, Norway

Jon Kleppe
Norwegian University of Science and Technology- NTNU, Trondheim, Norway

Fabiano Oliveira Chagas
Universidade Federal Fluminense, Rio de Janeiro, Brasil

Osvaldo Luiz Gonçalves Quelhas
Universidade Federal Fluminense, Rio de Janeiro, Brasil

Roberto Guimarães Pereira
Universidade Federal Fluminense, Rio de Janeiro, Brasil

Valdir Lameira
University of Coimbra, Coimbra, Portugal

Ihsan Hamawand
National Centre for Engineering in Agriculture (NCEA), University of Southern Queensland, Toowoomba 4350 QLD, Australia

Behnam Ghasemzadeh
Young Researchers and Elite Club, Tabriz Branch, Islamic Azad University, Tabriz, Iran

Atefeh Fathebaghali
M.A. in Architectural Engineering, Islamic Azad University of Tabriz, Tabriz, Iran

Kasra Rahbaripour
Young Researchers and Elite Club, Tabriz Branch, Islamic Azad University, Tabriz, Iran

Hamed Pouyafar
Department of Art and Architecture, Tabriz Branch, Islamic Azad University, Tabriz, Iran

Felix A. Oginni
Department of Civil Engineering, Osun State University, P.M.B. 9544, Oshogbo, Nigeria

Moses A. Ogunbiyi
Department of Civil Engineering, Osun State University, P.M.B. 9544, Oshogbo, Nigeria

Sodiq O. Balogun
Department of Civil Engineering, Osun State University, P.M.B. 9544, Oshogbo, Nigeria

Takaaki Wajima
Graduate School of Engineering, Chiba University, Chiba, Japan

Yussuf Lukeman
Department of Urban and Regional Planning, University of Lagos, Nigeria

A. I. Bako
Department of Urban and Regional Planning, University of Ilorin, Nigeria

F. K. Omole
Department of Urban and Regional Planning, the Federal University of Technology, Akure, Nigeria

I. I. C. Nwokoro
Department of Urban and Regional Planning, University of Lagos, Nigeria

A. E. Olajuyigbe
Department of Urban and Regional Planning, the Federal University of Technology, Akure, Nigeria

Takuya Ito
Department of Materials and Life Science, Seikei University, Japan

Yusuke Kakuta
Department of Materials and Applied Chemistry, Nihon University, Japan

Katsumi Hirano
Department of Materials and Applied Chemistry, Nihon University, Japan

Toshinori Kojima
Department of Materials and Life Science, Seikei University, Japan

Yaser Souraki
Department of Petroleum Engineering and Applied Geophysics, Norwegian University of Science and Technology - NTNU, Trondheim, Norway

Mohammad Ashrafi
Department of Petroleum Engineering and Applied Geophysics, Norwegian University of Science and Technology - NTNU, Trondheim, Norway

Ole Torsaeter
Department of Petroleum Engineering and Applied Geophysics, Norwegian University of Science and Technology - NTNU, Trondheim, Norway

Zohreh Jalili
Department of Petroleum Engineering and Applied Geophysics, Norwegian University of Science and Technology, N7491, Trondheim, Norway

Vahid Alipour Tabrizy
Department of Reserve Replacements, Åsgard Petroleum Technology, Statoil ASA, Norway

A. Isaac Bamgboye
Department of Agricultural and Environmental Engineering, University of Ibadan, Ibadan, Oyo State, Nigeria

Babajide S. Kosemani
Department of Agricultural and Environmental Engineering, University of Ibadan, Ibadan, Oyo State, Nigeria

Fred J. Rispoli
Department of Mathematics and Computer Sciences, Dowling College, Oakdale, The United States

Vishal Shah
Department of Biology, Dowling College, Oakdale, The United States

Hu Wu
Department of Environmental Science and Technology, Tokyo Institute of Technology, Yokohama, Japan

Qi An
Department of Environmental Science and Technology, Tokyo Institute of Technology, Yokohama, Japan

Yafei Shen
Department of Environmental Science and Technology, Tokyo Institute of Technology, Yokohama, Japan

Noboru Harada
Department of Environmental Science and Technology, Tokyo Institute of Technology, Yokohama, Japan

Lihao Chen
Department of Environmental Science and Technology, Tokyo Institute of Technology, Yokohama, Japan

Kunio Yoshikawa
Department of Environmental Science and Technology, Tokyo Institute of Technology, Yokohama, Japan

Liliana Perez
Département de géographie, Université de Montréal, Montreal, QC, Canada

Trisalyn A. Nelson
Spatial Pattern Analysis and Research Laboratory, Department of Geography, University of Victoria, Victoria, BC, Canada

Mathieu Bourbonnais
Spatial Pattern Analysis and Research Laboratory, Department of Geography, University of Victoria, Victoria, BC, Canada

Aleck Ostry
Department of Geography, University of Victoria, Victoria, BC, Canada